Modern Theories of the Universe
from Herschel to Hubble

MICHAEL J. CROWE
University of Notre Dame

DOVER PUBLICATIONS, INC.
New York

Copyright

Copyright © 1994 by Michael J. Crowe.
All rights reserved under Pan American and International Copyright Conventions.

Bibliographical Note

Modern Theories of the Universe, from Herschel to Hubble is a new work, first published by Dover Publications, Inc., New York, 1994.

Library of Congress Cataloging-in-Publication Data

Crowe, Michael J.
   Modern theories of the universe : from Herschel to Hubble / Michael J. Crowe.
     p.   cm.
   Includes bibliographical references and index.
   ISBN 0-486-27880-8 (pbk.)
   1. Astronomy. 2. Astronomy—History. I. Title.
QB43.2.C76   1994
520—dc20   93-46725
                                                                                                               CIP

Manufactured in the United States of America
Dover Publications, Inc., 31 East 2nd Street, Mineola, N.Y. 11501

To

Michael Thomas Crowe

and

Timothy Dennis Crowe

Preface

Three Points of View

This book can be read more or less simultaneously from three different points of view: the scientific, the historical, and the philosophical.

If read for its **scientific** content, this book provides a solid introduction to the fundamentals of **stellar astronomy,** that is, to the astronomy of the region beyond our solar system. Stellar astronomers deal with the millions of stars visible in telescopes and the arrangements prevailing among those stars. Whereas some books metaphorically take as their subject matter such areas as Lord Byron's universe or the cosmos of the Hebrew prophets, the subject matter treated in this text is, in a quite literal sense, *the* universe. Moreover, a number of the astronomers discussed in these materials expressed views on the history of the universe. Three questions form the special focus for these materials: (1) What is the nature, structure, and size of the Milky Way? (2) What are the nebulous-appearing objects, long called "nebulae," which powerful telescopes reveal are scattered in large numbers throughout the celestial realm? (3) In particular, are these objects "island universes," i.e., groups of vast numbers of stars arranged in a manner comparable to the Milky Way?

Many speculations will be encountered in the various chapters of this volume, but more than speculations emerge. The early chapters lead to and the final chapters present the basic astronomical understanding of the universe as an expanding entity with a history spanning some fifteen billion years and extending to billions of light years in distance. Although details in the present conception of the universe will no doubt change and many unresolved issues remain, the conception of the cosmos that emerges from these chapters has such a wealth of evidence in support of it that a truly major change in it seems improbable.

A special feature of this book is that it contains readings from some of the most creative astronomers of the last three centuries. These selections are in many cases the classic texts that contributed most significantly to the development of stellar astronomy. At some points, relatively elementary mathematics is introduced. Nonetheless, it is a striking feature of these materials that all can be read by persons possessing a sound secondary school training in mathematics and a willingness to use it. This is not to deny that difficult conceptions are involved; these are chiefly of a physical nature. Yet even these should be fully within reach of nearly all readers willing to work through the explanations provided in either the classic selections or the accompanying commentaries. The goal in this regard has been to provide materials that are accessible to persons with minimal background in astronomy.

Second, this book can also be read for its **historical** content. The development of modern stellar astronomy ranks as one of the most impressive achievements of the last three centuries. Moreover, the story of how that development occurred is filled with drama, with unexpected twists and turns, with reversals, and, like a detective story, with the gradual emergence of a particular point of view. The story of stellar astronomy is not, however, presented in its full richness; a number of individuals and events, important in their own right but not essential to the main line of this drama, have not been included. On the other hand, a number of related developments receive treatment, e.g., the evolution of the instruments that contributed to the creation of our modern conception of the cosmos. A central goal of this presentation is to involve the reader as fully as possible in these historical developments, to place the reader on the constantly changing frontier of astronomical research over the last three centuries. One result of this approach is that the astronomical ideas of each period are discussed largely in their own context, rather than being measured against the present view of the universe. In an important sense, this text treats not only stellar astronomy, but also the most important stellar astronomers, the men and women who forged the modern view of the universe.

One advantage of a historical approach is suggested by a passage from the writings of the German historian and philosopher Wilhelm Dilthey (1833–1911). In discussing the uses of history, Dilthey presented an idea that, although phrased in terms of religious history, applies with equal force to the history of scientific thought. He suggested that acquisition of historical experience can offset an unfortunate feature of our lives—that as we grow older, the range of our experiences becomes ever more limited. Nonetheless, Dilthey stated,

> when I run through Luther's letters and writings, the reports of his contemporaries, the records of the religious conferences and councils as well as of his official communications, I live through a religious process of such eruptive power, of such energy, in which it is a matter of life or death, that it lies beyond any possibility of personal experience for a man of our day. But I can relive it.[1]

In short, this book has been designed to offer the reader an opportunity to relive to some extent one of the most dramatic developments in the history of thought. The chance of achieving this result will be enhanced by a commitment to understanding the ideas and arguments formulated by the pioneering astronomers whose writings are contained in this book.

It should be candidly admitted that the reader will encounter some conceptual difficulties that result from the presentation of these ideas in a historical manner. For example, persons conscientiously studying these materials may find themselves repeatedly coming across the term *universe* used in different senses by diverse authors, whereas a traditional astronomical text will, possibly on its first page, provide a definition of that term. No such straightforward definition of *universe* or a number of other important terms is provided in this book, nor can it be, the reason being that the meaning of such terms has changed throughout history. This is itself a point of some historical and philosophical significance.

To put this approach in different terms, we shall be examining both one **product** of science (the theory of the expanding universe), and also the **process** by which that theory

---

[1] Wilhelm Dilthey, *Gesammelte Schriften*, vol. 7, 2nd ed. (Leipzig, 1942), pp. 215-16.

was attained. This approach, in which both the product and process of science are examined, can provide a fresh and more humanized view of science. We shall encounter not only scientific creations, but also scientific creators, not only settled conclusions, but also intense controversies.

Third, this book presents various **philosophical** ideas and issues. The ideas leading to our present conception of the universe as an incredibly vast entity with an astoundingly long period of development presented many challenges to traditional philosophical and theological beliefs. A number of these challenges are discussed in the book and/or illustrated in its Epilogue, which consists of a collection of chronologically arranged quotations from various authors, including many prominent literary, philosophical, religious, and scientific figures, who pondered these issues with special intensity. The Epilogue may be read last or read in conjunction with the various chapters.

The history of stellar astronomy also raises an array of methodological issues. Among these, the one that emerges most vividly in this volume is the nature, significance, and reliability of observation. To help the reader begin to think more deeply about the methodological issues involved in the development of stellar astronomy, an introductory discussion of the methodology of astronomy advocated by Edwin Hubble has been provided. It should be stressed that this presentation is not aimed at indoctrinating the reader into a particular position, but rather at raising a set of issues by providing the convictions concerning them that were reached by one of the most important twentieth-century stellar astronomers.

As a final note, it may be mentioned that it is the author's conviction that these three points of view, the scientific, historical, and philosophical, can be mutually supportive, that one good way of learning science is by studying its historical development and seeing the philosophical ideas involved in it. Moreover, it is very helpful in analyzing methodological issues in science to see them embodied in writings of and debates among practicing scientists.

Acknowledgments

I am indebted first of all to the students who over the last fifteen years have shown an interest in these materials and who, by their perceptive questions, have aided in the numerous revisions they have received. Very warm thanks are due to three scholars, Robert Smith, Mari Williams, and Patrick Wilson, each of whom read the entire manuscript and provided numerous helpful suggestions and corrections. I am also appreciative of the helpful comments made on various parts of these materials by Michael Hoskin and Phillip R. Sloan. A very welcome grant from the Institute for Scholarship in the Liberal Arts of the University of Notre Dame provided funding to permit John P. Bransfield to prepare an index for this volume. To both I am deeply grateful. Warm thanks also to Sherry Reichold for having typed a large portion of the primary materials included in this volume. Finally, special thanks to John W. Grafton of Dover Publications, whose encouragement has helped all phases of the preparation of this volume, and to Alan Weissman, whose careful copy editing has very significantly improved this book.

To the following publishers, institutions, and persons, I am indebted for permission to quote materials: Agnes C. Connolly estate for the passage from Myles Connolly's *Mr. Blue;* Charles Scribner's Sons, an imprint of Macmillan Publishing Company, for passages from Harlow Shapley, *Through Rugged Ways to the Stars* (New York: Charles Scribner's Sons, 1969); Macmillan Publishing Company for passages from Boris Brasol's translation of F. M. Dostoievski's *Diary of a Writer;* Open Court Publishing for the passage from Bertrand Russell's "My Mental Development"; Oxford University Press for the scriptural passages from the Revised Standard Version of the Bible, copyright 1946, 1952, 1971 by the Division of Christian Education of the National Council of the Churches of Christ in the USA; and the Royal Astronomical Society for the paper by Sir Arthur Stanley Eddington that first appeared in 1917 in *Monthly Notices of the Royal Astronomical Society.*

To the following institutions and publishers, I am indebted for permission to use illustrations: Harvard College

Observatory for the photographs of Harlow Shapley, of Henrietta Leavitt, of the Harvard 15" refractor, and of the Small and Large Magellanic Clouds; Lick Observatory for the photograph of NGC 1514; Lund Observatory for its diagram of the Milky Way; *Sky and Telescope* for the picture of Heber Curtis; the Master and Fellows of St. John's College, Cambridge, for the portrait of Sir John Herschel; the Observatories of the Carnegie Institution of Washington for photographs of the 60"- and 100"-aperture Mount Wilson reflectors and for photographs of the Corona Borealis and Hercules regions; Yale University Press for Edwin Hubble's diagram of his classification of nebulae.

Michael J. Crowe
University of Notre Dame

Table of Contents

Chapter One	Methodology and the Telescope	1
Chapter Two	Stellar Astronomy in the Period before William Herschel	15
Chapter Three	Sir William Herschel: Celestial Naturalist	71
Chapter Four	From William Herschel to 1860	146
Chapter Five	The New Astronomy	178
Chapter Six	From the New Astronomy to Henrietta Leavitt	195
Chapter Seven	Background to the Great Debate	233
Chapter Eight	The Great Debate	269
Chapter Nine	The Resolution of the Issues in the Great Debate	328
Epilogue	Humanity and the Universe: Some Quotations	360
Appendix	Laboratory Exercise on Nebulae and on the Milky Way	395
Selected Bibliography of the History of Stellar Astronomy		405
Index		425

Chapter One

Methodology and the Telescope

1. Edwin Hubble and the Methodology of Astronomical Research

Introduction

Although much has been written on scientific method in general, relatively few authors have formulated positions specifically on the methodology of astronomy. One author who attempted this was Edwin Powell Hubble (1887–1953). A number of reasons suggest that Hubble's views merit consideration. First, Hubble is widely recognized as having been one of the most creative astronomers of the twentieth century. In fact, one of Hubble's achievements was to find answers to some of the fundamental astronomical questions we shall be discussing. Second, Hubble's views on the methodology of astronomy were made explicit in a number of his writings; the most important of these for us are:

(1) *The Realm of the Nebulae* (New York: Dover, 1958 reprint of the 1936 original); hereafter *RN*.
(2) "Points of View: Experiment and Experience," *Huntington Library Quarterly*, 3 (April, 1939), 243–50; hereafter "EE".
(3) *The Nature of Science and Other Lectures* (San Marino, Calif.: Huntington Library, 1953); hereafter *NS*.

Hubble's methodological views seem to have been widely accepted; the astronomer Allan Sandage, for example, in his "Preface" to the 1958 reprint of *RN* noted that this book "is important not only in the history of astronomy but also because it is still a source of inspiration in the scientific method in this field." (*RN*, vii) Sandage added that although some minor changes in Hubble's picture of the universe have been necessary, "these are changes in numerical detail and not in fundamental philosophy or direction of attack. Hubble's original approach to observational cosmology remains." (*RN*, vii)

The following materials are an attempt to formulate under

seven categories the main claims made by Hubble in the publications listed above. These claims need not be seen as beyond dispute; in fact, some authorities favor different views. Moreover, the materials included in this volume provide a basis for checking the accuracy of Hubble's claims. Hubble himself stressed the importance of employing a historical approach in examining various broad claims. In his *NS*, while discussing various questions, Hubble states: "When you consider these present-day questions, I earnestly recommend the historical point of view." (*NS*, 47) Although this remark was made in reference to "Social and economic problems . . . which only rarely can be tested by controlled experiments," nonetheless, this historical approach should be valuable in testing broadly philosophical claims. Hubble added the comment that "history is one long record of experiments." Moreover this historical "procedure is so orthodox with sound students that it should be unnecessary to stress its desirability." (*NS*, 47) He admits, however, and laments, that some do not use it.

Edwin Hubble on the Method of Astronomy

Claim (1): The opening sentence of *RN* contains Hubble's first claim: "Science is the one human activity that is truly progressive." (*RN*, 1) To support this claim, he cites a passage from the historian of science George Sarton:

> The saints of today are not necessarily more saintly than those of a thousand years ago; our artists are not necessarily greater than those of early Greece; they are likely to be inferior; and, of course, our men of science are not necessarily more intelligent than those of old; yet one thing is certain, their knowledge is at once more extensive and more accurate. The acquisition and systemization of positive knowledge is the only human activity that is truly cumulative and progressive. (As quoted in *RN*, 2)

In short, Hubble claims that science is the only truly cumulative and progressive area of learning.

Claim (2): Hubble goes on to explain why only science is cumulative and progressive. The reason is that its method entails "strict limitation of subject matter." As he puts it, "Science deals only with judgments concerning which it is possible to obtain universal agreement." ("EE", 244; *RN*, 1) Hubble contrasts the cautious certainty of science with the state of affairs in more subjective areas by stating:

> Now, it is a curious fact that we are most certain about judgments that cannot be demonstrated. Subjective certainty is inversely proportional to objective certainty. Russell put the situation rather bluntly, "Who ever heard a theologian preface his creed, or a politician conclude his speech with an estimate of the probable error of his opinion." ("EE", 249)

In short, Hubble maintains that science deals only with matters concerning which we can secure universal agreement.

Claim (3): But how is this universal agreement to be attained? He answers that it "is secured by means of observations and experiments," for these "tests represent external authorities which all men must acknowledge, by their actions if not by their words, in order to survive." (*RN*, 1) Because of this, science "is barred from the world of values." (*RN*, 1) It can in some ways influence values, but the reverse should not take place. (*RN*, 2)

In short, Hubble contends that "observation and experiment" are the methods whereby science achieves "universal agreement."

Claim (4): Hubble draws a sharp distinction between observation and theory. The observer, he notes, "starts by accumulating an isolated group of data, together with their estimated uncertainties." (*RN*, 2-3) The observer then strives to forge laws from the data, a process that proceeds by "successive approximations." (*RN*, 3) The theoretician enters at this point in order to establish theories based on these laws. The theories must, however, be tested against the data; he notes that "less competent minds are embarrassed by the custom of testing predictions [of theories]." (*RN*, 5)

In short, Hubble asserts that observation and theory are and should be separate realms.

Claim (5): Hubble's distinction between observation and theory leads him to some comparisons between the two activities. In making these comparisons, he qualifies his earlier statement about the progress of science by noting that "observations and the laws which express their relations are permanent contributions to the body of knowledge [whereas] theories change with the spreading background." (RN, 4) Thus for Hubble the enduring portion of science consists in its observations and laws. Theories seem to be ephemeral and in any case are to be sought only after the empirical information is collected. Thus at the end of RN, after noting some astronomical problems for the future, Hubble comments: "The search will continue. Not until the empirical results are exhausted, need we pass on to the dreamy realms of speculation." (RN, 202)

Hubble makes the same point when presenting in RN a sketch of the history of the island universe theory. He notes that early "speculations took many forms and most of them have long since been forgotten." (RN, 23) After noting that Kant in 1755 developed Wright's island universe theory in "a form that endured, essentially unchanged, for the following century and a half" (RN, 23), he asserts that in the resultant debate, "the astronomers took little part . . . : they studied the nebulae." (RN, 25) Finally, the speculations of philosophers, avoided by astronomers, came to be tested against observations. In NS, Hubble contrasts speculators with astronomers, characterizing the latter as:

> . . . explorers [to whom] interpretations are useful only as long as they work. . . . They know that in the past most of the really new fields have been opened by the explorers, using the methods of Galileo and Newton rather than the method of Plato. And they will continue their explorations on the assumption that, in the future as well as the past, new fields will be opened which cannot be predicted from the armchair. (NS, 18)

In short, Hubble argues that whereas astronomical observations and laws are permanent, theories and speculations are ephemeral.

Claim (6): Hubble's emphasis on observation and its permanence leads him to stress the importance of scientific instruments, especially the telescope. In his Preface to RN, he writes: "The conquest of the Realm of the Nebulae is an achievement of great telescopes." (*RN*, x) And in his first chapter he notes that nebulae were long "a favorite subject for speculation," but "now, thanks to great telescopes, we know something of their real size and brightness [and] distances." (*RN*, 20–21) In composing these passages, Hubble was no doubt thinking of the great 100-inch Mount Wilson reflecting telescope, which he used in his research, as well as of the other astronomers who worked on Mount Wilson, for example, Harlow Shapley (for a time) and Adriaan van Maanen.

In short, Hubble stresses that the major source of advancement for modern astronomy has been the great telescopes.

Claim (7): The final point stressed by Hubble in *RN* is the importance of precise terms in astronomy: "The terms always carry the same significance and substitutes are not employed." (*RN*, 7) In the terms used by astronomers, "Variety is sacrificed for precision." (*RN*, 7) One implication of this would seem to be that this practice allows astronomical terms to be permanent rather than ephemeral entities in science.

In short, Hubble contends that astronomy progresses partly because of its use of precise terms, the meaning of which is kept constant.

Conclusion

Such were Hubble's views. Not all authorities hold identical positions. One person holding a somewhat different view is the astronomer Hermann Bondi; see, for example, his "Fact and Inference in Theory and in Observation," *Vistas in Astronomy*, 1 (1955), 155–62. The historian of astronomy Norriss Hetherington has also expressed rather different ideas; see his essay in *Nature*, 306 (Dec. 22–9, 1983), 727–30 or his short book *Science and Objectivity: Episodes in the History of Astronomy* (Ames, Iowa: Iowa State Univ. Press, 1988). Keep Hubble's views in mind, and especially the questions they were designed to answer, as you proceed through the readings that follow.

2. Telescopes: Their Discovery, Design, and Development

Telescopes have played an extremely important role in the development of astronomy. From the time in 1609 when Galileo and Thomas Harriot initiated the use of telescopes in astronomy, these instruments have enriched our knowledge of the universe. Even a telescope providing a magnification of only 30 will in effect transport the astronomer 30 times closer to the object under scrutiny. Consequently, it is very important for an understanding of the development of astronomy to possess a working knowledge of telescope design and function.

Lenses

A telescope is made up of two or more lenses. A lens is a device that bends rays from an object in such a manner as to enhance our ability to see the object. A surface of a lens typically has either a convex or a concave shape. A convex lens curves or bulges out, whereas a concave lens curves in toward its center.

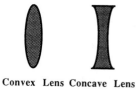

Convex Lens   Concave Lens

The aperture of a lens is its diameter. The focal length of a lens is equal to the distance of the lens from the point, called the focal point, where the lens brings rays to a focus.

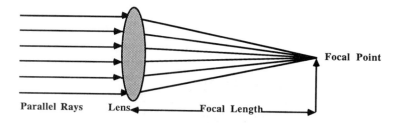

Telescopes contain two or sometimes more lenses. The lens(es) near the eye of the observer is(are) part of the eyepiece of the telescope, whereas the lens at the remote end of the telescope is called the objective lens.

Lenses at times produce distortions in the object viewed.

Two of these, chromatic aberration and spherical aberration, will be of special importance for us.

Chromatic aberration (see diagram) results from the fact that a curved transparent lens bends differently colored rays by differing amounts; any given glass lens will possess different indices of refraction for rays of different colors, producing the result that rays of different color, even when coming from the same point of the object in view, will be bent differently. Thus red rays from a particular region of the object will be brought to a focus at one point, blue rays at another, blurring thereby the image of the object.

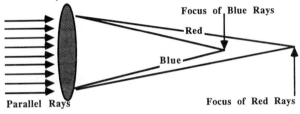

Chromatic Aberration

Spherical aberration results from the fact that most lenses have surfaces that are spherical in form; this is because it is much easier to grind a spherical lens than one having a hyperbolic or parabolic form. Because of spherical aberration, parallel rays striking the lens at different distances from its center are brought to a focus at different points. This results in a blurring of the image of the object seen. This aberration can be reduced by grinding the surface in a parabolic form and also by making telescopes of very long focal length. Long focal length also helps reduce chromatic aberration, but to a lesser extent.

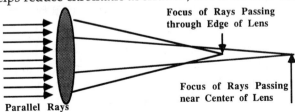

Spherical Aberration

Three Main Types of Telescopes

Although other types of telescopes eventually evolved, nearly all pre-twentieth-century telescopes can be classified in terms of three design types: the Galilean telescope, the Keplerian telescope, and the reflecting telescope. The Galilean telescope, sometimes called an opera glass, consists of a concave eyepiece and a convex objective lens. In 1608, Hans Lipperhey, a Dutch lens maker, applied for a patent for such a telescope. In 1609, Galileo, learning of the Dutch instrument, constructed his own telescope and became the first to use it extensively for astronomical observation. Galilean telescopes give an erect image and are usually shorter than Keplerian refracting telescopes, the reason for this being that in the former type, the eyepiece lies within the focal length of the main lens, whereas the eyepiece lies beyond the focal point in the Keplerian. Galilean telescopes are very limited in size and consequently in usefulness, but provide partial correction for both spherical and chromatic aberration. By the end of the seventeenth century, the best telescopes no longer employed the Galilean optical system.

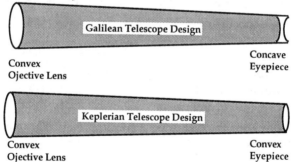

Galilean Telescope Design
Convex Ojective Lens — Concave Eyepiece

Keplerian Telescope Design
Convex Ojective Lens — Convex Eyepiece

The Keplerian or astronomical telescope consists of a convex eyepiece and a convex objective lens. Johannes Kepler in 1611 first designed this type of telescope, which was first constructed by Christopher Scheiner in the 1613–17 period. It produces an inverted image, which is not a problem for astronomers, whereas it is for anyone watching a football game or an opera. It can be made in large size and is the design found in essentially all major post-seventeenth-century refractors.

Such telescopes suffer from spherical and chromatic aberrations, but astronomers gradually learned to correct for these problems.

The primacy accorded refracting telescopes during the seventeenth century began to be challenged late in the eighteenth century by the reflecting telescope, the first of which was constructed by Isaac Newton in 1668, although the idea of making a telescope with an objective lens consisting of a mirror had been proposed earlier. Mirrors, like traditional lenses, bring rays to a focus. In the Newtonian focus reflecting telescope, the rays enter the remote end of the tube, proceed down the tube until they strike a spherically or parabolically curved concave mirror, which reflects the rays back to a second, planar mirror, from which they are reflected to the eyepiece positioned on the side of the tube.

Three Types of Reflecting Telescopes

One reason why Newton developed the reflector is that no chromatic aberration occurs in reflection; thus reflecting telescopes solve one of the most severe problems of telescope design. Moreover, for a mirror, only one surface must be ground, whereas in a refracting telescope, both sides of the objective lens are usually figured. Also reflecting telescopes can be made very large; the largest mirror presently used is nearly eight times larger in diameter than the largest transparent lens. Until after about 1860, the mirrors of nearly all reflecting telescopes were metal, making them both very heavy and more subject to distortions resulting from thermal contraction and expansion than glass lenses. Reflectors come in a number of designs in addition to the form designed by Newton; among these are the Cassegrainian and Herschelian focus types.

In the Cassegrainian design, the second mirror is curved and the eyepiece is placed behind the telescope mirror, in which a small hole is cut so as to allow the rays to pass through to the eyepiece. Cassegrainian telescopes possess the advantage that astronomers observe from the telescope's lower end, rather than its remote end. Reflectors of Herschelian design have the advantage that a second mirror is not needed, thus conserving the amount of light lost in any reflection. This design was created by Sir William Herschel in the 1780s, but is rarely if ever used at present.

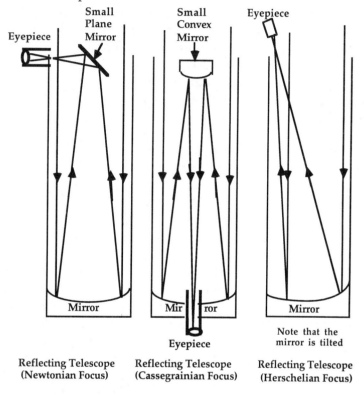

Reflecting Telescope (Newtonian Focus)   Reflecting Telescope (Cassegrainian Focus)   Reflecting Telescope (Herschelian Focus)

Some Factors That Influence the Quality of a Telescope

(1) **Light-Gathering Power and Aperture**: The most important criterion of the quality of a telescope is its power to gather light. This depends on its aperture, i.e., the diameter of its main lens or mirror. The difficulties we experience in trying to see a dim object at night are due to the fact that it supplies insufficient light to register on our retina. We (and cats more obviously) compensate for this by expanding the apertures of our eyes. Similarly, astronomers seek to improve their capability for observing distant objects by building telescopes of large aperture. The mathematics of this is that light-gathering power increases in proportion to the aperture

squared. Let us set the light-gathering power of the eye as 1 and assume that the eye has an aperture of 0.2 inches. Using this data, calculate the light-gathering power for each of the five historic telescopes listed below.

Year	Telescope	Power
1609	Galileo's refracting telescope (a = 3/2 inches)	56
1789	Herschel's 48-inch-aperture reflector	_____
1907	Mount Wilson Observatory 60-inch reflector	_____
1948	Palomar Observatory 200-inch reflector	_____
1976	Crimean Observatory 236-inch reflector	_____

(2) **Focal Length**: Another important criterion of the quality of a telescope is focal length, which is the distance between its objective lens and the image formed by that lens. The focal length of the objective lens is the most important factor influencing magnification in telescopes. The formula for the magnification of a telescope is the focal length of the objective lens divided by the focal length of the eyepiece. Thus if one telescope has twice the focal length of another, it should magnify twice as much, provided that the same eyepiece is used with both telescopes. These facts should make clear that magnification by itself is not a good criterion of the quality of a telescope. The reason is that essentially any magnification can be attained with any particular telescope, simply by employing an eyepiece of sufficiently short focal length. The problem is that telescopes with short focal length objective lenses give far less *useful* magnification that those of longer focal length. Assuming that an eyepiece of focal length 4 inches is applied to each of the telescopes listed below, calculate the magnification that would result.

Telescope	Magnification
Galileo's telescope (focal length = 3 feet)	9
Herschel's largest telescope (f = 40 feet)	_____
Mount Wilson Observatory telescope (f = 25 feet)	_____
Palomar Observatory telescope (f = 55 feet)	_____
Crimean Observatory telescope (f = 100 feet?)	_____

Note on Arc Measure: Angles or arcs are measured in degrees, minutes, seconds, thirds, etc. The moon, for example, is about one-half degree wide; more precisely, it is about 31' (read 31 minutes of arc) wide. We shall shortly encounter quantities measured in seconds of arc; for example, the width of a lunar crater may be 28" (read: 28 seconds of arc). There are 60' in 1°, 60" in 1', etc.

(3) **Resolving Power:** Because of diffraction and other factors, the image of a star or other small object is spread so as to form a small circle. Suppose we have two small star images, or two images from a surface feature on a moon or planet, e.g., a lunar mountain, and suppose that these images are very near each other. If the circles associated with them are too large, the circles will merge so that we see one circle or near circle rather than two separate images. Such distortion is obviously undesirable. The measure of a telescope's ability to resolve fine detail is called its resolution or resolving power and is measured by the limiting arc distance between two objects at which they can still be seen as separate. For example, astronomers test telescopes by seeing how small the angle of separation between two stars can be without preventing the separate stars from being seen as separate. The higher the quality of the telescope, the smaller is its resolving power. For visual telescopes, resolving power is equal to 4.56" divided by the aperture measured in inches. Compare the resolving powers of the five telescopes listed previously.

Galileo's refracting telescope (a = 3/2 inches)	3"
Herschel's 48-inch-aperture reflector	_____
Mount Wilson Observatory 60-inch reflector	_____
Palomar Observatory 200-inch reflector	_____
Crimean Observatory 236-inch reflector	_____

Note: Other factors also influence telescope quality; among these are clarity of atmosphere, quality of mounting, and visual acuity of the observer (or the sensitivity of the photographic plate).

Some Highlights from the Early History of the Telescope

1601   Death of Tycho Brahe, the last major pretelescopic astronomical observer.

1609   Galileo Galilei, hearing about a telescope made in Holland by Hans Lipperhey, constructed his own telescope, which consisted of a concave eyepiece and a convex objective lens. This design is now known as a "Galilean telescope." Galileo demonstrated the usefulness of telescopes by making a number of dramatic discoveries, e.g., mountains on the moon, phases of Venus, moons of Jupiter, and appendages (later recognized as the rings) of Saturn. In the same year, Thomas Harriot also constructed a telescope and used it for astronomical observation.

1611   Johannes Kepler published his *Dioptrice*, in which he gave a theoretical treatment of telescope design; he proposed the construction of a telescope consisting of a convex eyepiece and convex objective lens, i.e., the Keplerian telescope. A few years later, Christopher Scheiner constructed the first telescope of this design.

1667   Colbert and King Louis XIV established the Paris Observatory; a few years later, Jean-Dominique Cassini became its leading astronomical observer. Christiaan Huygens and Olaus Roemer also became associated with the Paris Observatory during this period.

1668   Isaac Newton constructed the first reflecting telescope, designing it in hopes of providing a way around the problem of chromatic aberration, which problem Newton erroneously believed could not be overcome in refracting telescopes.

1673   Johannes Hevelius of Dantzig published his *Machinae coelestis*, in which he described a number of his telescopes, including his 150-foot-focal-length refracting telescope. This was one of many long telescopes Hevelius constructed in hopes of minimizing spherical aberration and maximizing magnification.

1675   England's King Charles II, hoping that it would lead to improvement in navigation, established the Royal Greenwich Observatory near London and appointed John Flamsteed as the King's astronomer. The immediately subsequent Royal Astronomers were Edmond Halley, James Bradley, Nathaniel Bliss, Nevil Maskelyne, John Pond, and George Airy.

1724   Maharaja Jai Singh constructed the famous non-telescopic Jantar Mantar Observatory located near New Delhi in India.

1733   Chester Moor Hall invented the achromatic lens system formed by joining a flint glass lens to a crown glass lens. The dispersion properties of these two types of glass are such that if the lens parts are ground in proper shapes, the refracting characteristics of the lenses offset each other, thereby minimizing chromatic aberration. Shortly after mid-century, John Dollond and others began producing fine quality telescopes with achromatic lens systems.

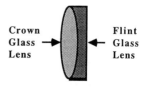

An Achromatic Lens

1789   Sir William Herschel erected his 48-inch-aperture, 40-foot-focal-length reflecting telescope, which until the 1840s remained the world's largest telescope. Herschel's discoveries with reflectors demonstrated the value of this type of telescope as compared to the refractor, thereby creating a rivalry that persisted throughout the nineteenth century.

Chapter Two

Stellar Astronomy in the Period before William Herschel

1. The Stars: Their Apparent Motion and Their Positions

Comprehension of the materials that follow requires an elementary understanding of the apparent motions of the stars and of the methods by which their positions are defined.

When looking at the night sky, one sees hundreds of stars. Over the course of a night, these stars appear to move, even though they retain a fixed position relative to each other. In particular, they move as if they were fixed on the inside of a spherical vault with its axis running through the north polar star, which is known 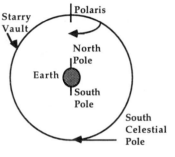 as Polaris and which is the last star in the handle of the Little Dipper. The rotation of the stars on this axis takes approximately 24 hours, with the stars rising in the East and setting in the West. In fact, their apparent daily rotation is due to the rotation of the earth. Nonetheless, their motion is exactly what it would be were the starry vault itself rotating. Their motions can most easily be visualized in terms of the starry vault itself being in motion, which is actually how the heavens appear to us.

For reference purposes, astronomers have defined an imaginary line, the "celestial equator," which is a great circle located 90° down from Polaris. The celestial equator is directly above the earth's equator. Another key reference line among the stars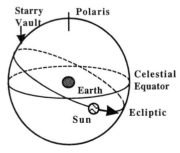

is the "ecliptic," which corresponds to the apparent annual path of the sun. The ecliptic is inclined at about 23.5° to the celestial equator and appears to move with the daily westward motion of the stars.

The next diagram represents the system in motion, including the motion of the sun during the period of one day. The classical planets and the moon all move within about 10° or less of the ecliptic, sometimes moving below it and at other times above.

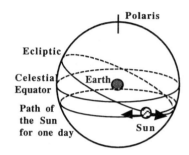

The point on the ecliptic where the sun in the period between winter and spring crosses the celestial equator is defined as the vernal equinox. The point where the sun in the period from summer to autumn crosses the celestial equator is known as the autumnal equinox. The day on which the sun passes through the first of these points is defined as the first day of spring, whereas the day the sun passes through the autumnal equinox is defined as the first day of autumn. Two other reference points are the summer solstice, at which the sun is at the highest point in the sky for inhabitants of the Northern Hemisphere. The fourth point in this group of reference points is the winter solstice. The sun passes through the winter solstice in late December, an event that defines for the Northern Hemisphere the first day of winter.

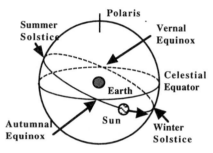

By means of these reference points, astronomers have defined two useful reference systems for indicating the positions of stars, i.e., the Equator System and the Ecliptic System.

Equator System: In the Equator System (see diagram), the position of a celestial object is referred to coordinates based on the celestial equator. The *right ascension* of an object is measured to the east from the Vernal Equinox (VE). The units employed are hours, each 15° corresponding to 1 hour, making

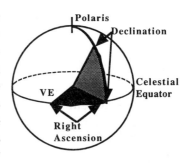

a total of 24 hours. The *declination* of an object is measured in degrees, positive in the direction of Polaris, negative for the southern celestial hemisphere. Consequently, declination values range from +90° to –90°.

Ecliptic System: In the Ecliptic System (diagramed at the right), the position of a celestial object is referred to coordinates based on the ecliptic. The *celestial longitude* of an object is measured to the east along the ecliptic from the Vernal Equinox. The range is from 0° to 360°. The *celestial latitude* of an object is measured up or down from the

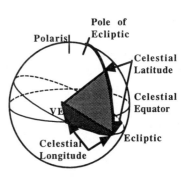

ecliptic, the range being from +90° to –90°.

2. Astronomical Correctional Factors

Important as telescopes are, observations need to be corrected in a number of ways before observations made with them can attain maximum accuracy. This section discusses seven correctional factors that astronomers apply to observations to increase their accuracy. These seven correctional factors are: atmospheric refraction, precession of the equinoxes, proper motion, stellar parallax, aberration of light, nutation, and the personal equation.

Atmospheric Refraction

Refraction is the bending of light rays as they pass through a transparent medium such as glass, water, or air. Because we live at the bottom of a "sea" of air, rays of light from stars or planets are bent as they come through our atmosphere.

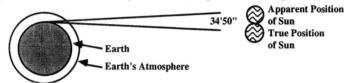

Bending of Light Rays at the Horizon

This effect was known even before the second century A.D., when the astronomer Ptolemy discussed it in his *Optics* but not in his *Almagest*. One of Ptolemy's predecessors, Cleomedes, used atmospheric refraction to explain a lunar eclipse in which both the sun and moon were visible, even though the geometry of lunar eclipses requires that the sun and moon, when viewed from the earth, be separated by 180°.

The magnitude of the effect of atmospheric refraction (see diagram) varies from 0° for rays perpendicular to the atmosphere to 34' 50" for rays at the horizon. At 60° down from the zenith, its magnitude is 1' 40"; at 80°, it is 5' 16"; at 85°, it increases to 9' 45".

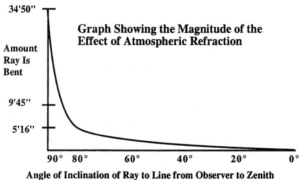

These figures indicate that, when the sun, which is about 31' wide, apparently begins to set, it is already below our horizon. As an example of how this factor is used to correct observations, consider a star measured to be 80° down from the zenith. When this observation is corrected in terms of the figures cited above, the star's position is found to be 80° 5' 16" down from the zenith. Important contributors to the study of atmospheric refraction include Ptolemy, Brahe, Kepler, Descartes (who was the first to publish the law of refraction), J. D. Cassini, Newton, and Bessel.

Precession of the Equinoxes

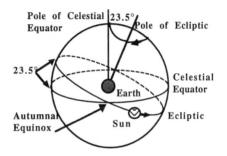

Another correctional factor is precession of the equinoxes. In the second century B.C., the Greek astronomer Hipparchus recognized that the equinoctial points, the points where the celestial equator and ecliptic cross, change slightly relative to the stars; specifically, the celestial equator moves along the ecliptic, making a full circuit in 26,000 years (the Greeks thought 36,000 years). To visualize this, imagine lines from the earth perpendicular to the planes of the celestial equator and ecliptic. These lines extend to the pole of the celestial equator (Polaris) and the pole of the ecliptic. As the next diagram shows, the polar line of the celestial equator turns around the ecliptic's polar line in 26,000 years. Thus the celestial equator slowly changes position during this period.

One result of precession is that Polaris is ceasing to be our pole star. In 13,000 years, a star located 47° (twice 23 1/2°) from Polaris will be our north star. (See diagram.)

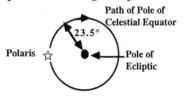

Precessional Movement of Pole of the Celestial Equator

A third way to visualize this phenomenon is to imagine that the earth's axis turns through a circle of radius 23 1/2° in a period of 26,000 years.

This is represented in the diagram at the right, which can also help in understanding the cause of precession. The earth, rather than being truly spherical, bulges at the equator. The moon and sun pull on this equatorial bulge, tending to draw it into the plane of the ecliptic.

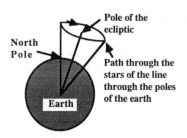

Precessional Movement of the Earth's Axis

This has the effect that the earth, like a spinning top, undergoes a precessional wobble.

The magnitude of the precessional effect is 50" per year. One way in which this effect shows up is that unless a correction for the precessional effect is used, our seasons would make one circuit through the calendar every 26,000 years; that is, 13,000 years from now, winter weather would be occurring when the calendar indicates that it should be summer.

Among the Greeks, Hipparchus studied, and may have been the first to discover, the precession of the equinoxes. Ptolemy erroneously set its value at 1° per century, which led some astronomers, for example, Thabit Ibn Qurra and Copernicus, to believe that the rate of precession varies. Newton was the first astronomer to provide a physical explanation of precession.

Proper Motion

Edmond Halley (1656–1742), who is now most widely known for the comet that bears his name, contributed to many areas of astronomical observation and theory. It was Halley who established that at least some comets are periodic, that after a certain period, such comets again pass by the sun. Having "observed" by inspecting historical records that a bright comet had appeared in 1456, 1531, 1607, and 1682, Halley argued that these four cometary appearances were actually the result of the return of a single bright comet that, he claimed, passes by the sun every 75 to 76 years. This claim led Halley to predict that this comet would return in 1758, which did, in fact, occur, although Halley, having died in 1742, was unable to witness the dramatic fulfillment of his prediction.

Halley, while still an undergraduate at Oxford, had gone to the island of St. Helena in the southern Atlantic to observe the southern-hemisphere stars. Later he assisted Newton in publishing the latter's *Principia* of 1687. He also worked out a method involving the transits of Venus across the sun for measuring the fundamental astronomical unit, the earth-sun distance. In 1720, he succeeded John Flamsteed as England's Astronomer Royal.

The facet of Halley's work that is of present concern is his discovery, announced in a 1718 paper, that at least some of the so-called "fixed stars" are not actually fixed in position relative to each other. Specifically, Halley, drawing on comparisons between ancient and modern positional determinations of stars, asserted that Sirius, Palilicium (also known as Aldebaran), and Arcturus had come to be located about one-half degree more southerly than at Ptolemy's time. The magnitude of this effect, known as the *proper motion* of the stars, varies from star to star, but in most cases is either small or beyond detection. The star with the largest known proper motion is "Barnard's Arrow," which has an annual proper motion 10.3" of arc per year.

Presented next is the paper in which Halley announced this discovery. After reading it, you can test your understanding of it by attempting to answer the following three questions.

(1) Halley discovered proper motion in 1718. Could the discovery have been made earlier; in particular, could it have been made before the invention of the telescope?

(2) What three factors influence the quantity of proper motion that a star will appear to have?

(3) Why are contemporary astronomers interested in ancient records of the positions of stars?

☆ ☆
Edmond Halley, "Considerations on the Change of the Latitudes of Some of the Principal Fixt Stars," *Philosophical Transactions of the Royal Society*, 30 (1718), 736–8.

Having of late had occasion to examine the quantity of the Precession of the Equinoctial Points, I took the pains to compare the Declinations of the fixt Stars delivered by *Ptolomy* in the 3*d* Chapter of the 7*th* Book of his *Almag[est]* as observed by *Timocharis* and *Aristyllus* near 300 Years before *Christ*, and by *Hipparchus* about 170 Years after them, that is about 130 Years before *Christ*, with what we now find: and by the result of very many Calculations, I concluded that the fixt Stars in 1800 Years were advanced somewhat more than 25 degrees in Longitude, or that the Precession is somewhat more than 50"[seconds of arc] *per ann.*[per annum=per year]. But that with so much uncertainty by reason of the imperfect Observations of the Ancients, that I have chosen in my Tables to adhere to the even proportion of five Minutes in six Years, which from other Principles we are assured is very near the Truth. But while I was upon this Enquiry, I was surprized to find the Latitudes of three of the principal Stars in Heaven directly to contradict the supposed greater *Obliquity* of the *Ecliptick*[1], which seems confirmed by the Latitudes of most of the rest: they being set down in the old Catalogue, as if the Plain of the Earth's Orb had changed its Situation, among the fixt Stars, about 20'[20 minutes of arc] since the time of *Hipparchus*. Particularly all the Stars in *Gemini* are put down, those to the *Northward* of the Ecliptick, with so much less Latitude than we find, and those to the *Southward* with so much more *Southerly* Latitude. Yet, the three Stars *Palilicium* or the *Bulls* Eye, *Sirius* and *Arcturus*[2] do contradict this Rule directly: for by it, *Palilicium* being in the days of *Hipparchus* in about 10 *gr.*[degrees] of

---

[1][The obliquity of the ecliptic is the angle between the ecliptic and the celestial equator. This angle, which is currently about 23° 26.5', is gradually decreasing. MJC]

[2][The celestial longitudes and latitudes for these stars are respectively: Palilicium or Aldebaran, 65°, 5° South; Sirius, 100°, 40° South; Arcturus, 200°, 30° North. MJC]

Taurus ought to be about 15 Min. more *Southerly* than at Present, and *Sirius* being then in about 15 [degrees] of *Gemini* ought to be 20 Min. more *Southerly* than now; yet *è contra* *Ptolomy* places the first 20 Min. and the other 22 more *Northerly* in Latitude than we now find them. Nor are these errors of Transcription, but are proved to be right by the declinations of them set down by *Ptolomy*, as observed by *Timocharis, Hipparchus* and himself, which shew that those Latitudes are the same as those Authors intended. As to *Arcturus*, he is too near the Equinoctial Colure,[1] to argue from him concerning the change of the Obliquity of the Eclipstick, but *Ptolomy* gives him 33' more *North* Latitude than he now has; and that greater Latitude is likewise confirmed by the Declinations delivered by the abovesaid Observers. So then all these three Stars are found to be above half a degree more *Southerly* at this time than the Antients reckoned them. When on the contrary at the same time the bright Shoulder of *Orion* has in *Ptolomy* almost a degree more *Southerly* Latitude than at present. What shall we say then? It is scarce credible that the Antients could be deceived in so plain a matter, three Observers confirming each other. Again these Stars being the most conspicuous in Heaven, are in all probability the nearest to the Earth, and if they have any particular Motion of their own, it is most likely to be perceived in them, which in so long a time as 1800 Years may show it self by the alteration of their places, though it be utterly imperceptible in the space of a single Century of Years. Yet as to *Sirius* it may be observed that *Tycho Brahe* makes him 2 Min. more *Northerly* than we now find him, whereas he ought to be above as much more *Southerly* from his Eclipstick, (whose Obliquity he makes 2' 1/2 greater than we esteem it at present) differing in the whole 4 1/2 Min. One half of this difference may perhaps be excused, if refraction were not allowed in this Case by *Tycho;* yet two Minutes, in such a Star as *Sirius*, is somewhat too much for him to be mistaken.

But a further and more evident proof of this change is

---

[1][The equinoctial colure is the great circle that passes through the equinoctial points and the celestial poles. MJC]

drawn from the Observation of the application of the Moon to *Palilicium Anno Christi* 509 *Mart.* [March] 11°[th] when, in the beginning of the Night the Moon was seen to follow that Star very near, and seemed to have Eclipsed it.... Now from the undoubted principles of Astronomy, it was impossible for this to be true at *Athens,* or near it, unless the Latitude of *Palilicium* were much less than we at this time find it. *Vide Ballialdi Astr. Philolaica, pg. 172.*

This Argument seems not unworthy of the *Royal Society's* Consideration, to whom I humbly offer the plain Fact as I find it, and would be glad to have their Opinion.

But whether it were really true, that the Obliquity of the Eclipticks was, in the time of *Hipparchus* and *Ptolemy,* really 22 Min. greater than now, may well be questioned; since *Pappus Alexandrinus,* who lived but about 200 Years after *Ptolemy,* makes it the very same that we do. *Vide Pappi Collect. Lib. VI. Prop. 35.*

☆ ☆

In short, Halley had discovered that at least three "fixed" stars had shifted position by a determinable amount in relation to the positions of other stars. This startling result suggested that many more and possibly all stars are in motion, that they are changing their positions. Moreover, this discovery suggested that astronomers should investigate whether any other proper motions could be detected.

Stellar Parallax and the Aberration of Light

One of the factors involved in the discovery of the aberration of light was the search for stellar parallax. Copernicus was aware that his theory that the earth orbits the sun entails that stars should appear to shift their position. For example, two stars in the plane of the earth's motion should appear to separate as the earth moves toward them, just as the angle between the headlights of an approaching car appears to grow larger. So also, a relatively nearby star above the plane of the earth's orbit should appear against the background of more remote stars to move through a small circle as the earth moves through its orbit (see diagram).

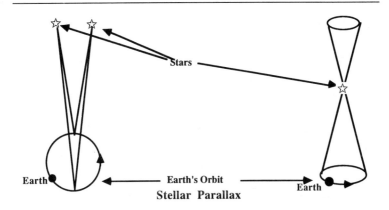

Stellar Parallax

Failing to find such an effect, Copernicus postulated that the stars are so remote that, although this effect called parallax occurs, it is beyond detection. Numerous astronomers of the sixteenth, seventeenth, and eighteenth centuries sought to find this effect; none succeeded. Had it been found, not only would it have proven the Copernican theory, but also would have provided a way to measure the distance to the stars. All pre-1838 efforts to discover parallax failed. It will prove convenient to postpone until a later chapter the discussion of its discovery, focussing at present on a side effect of the search for parallax, i.e., the discovery of the aberration of starlight by James Bradley (1693–1762).

Among the scientists who sought to determine a parallax was Robert Hooke who in 1669 stated that he believed that he had found a parallactic shift of about 15" in gamma Draconis, a star that crosses the heavens near the zenith point for England. James Bradley, working with Samuel Molyneux, began a study of this star in the 1720s, hoping to verify Hooke's claim. Instead, he found that gamma Draconis undergoes a yearly shift of a form quite different from that expected from parallax. In a 1728 paper, Bradley announced his discovery and explained the effect as due to the need for astronomers to shift their telescopes forward very slightly in the direction of the earth's annual motion so as to take account of the time required by the ray of starlight in passing through the tube of the telescope. This is exactly analogous to the fact that when walking in a vertical rainfall, a person will tilt an umbrella into the direction of motion. Otherwise, the person will not be protected by the cylinder of protection provided by the umbrella—that is, the person will get wet. The maximum magnitude of stellar aberration occurs for stars at the pole of the ecliptic; these stars move through a circle of radius 20.5". Although Bradley failed to find parallax, his discovery was the first direct empirical proof of the earth's motion; moreover, it provided a way to measure the speed of light. Another important result announced by Bradley in his 1728 paper was that he was convinced that had a parallactic circle on the order of 1" of arc in diameter been present, he would have detected it. From this he concluded that the nearest star must be at least 400,000 times farther from us than the sun.

Nutation

In 1748, Bradley, who in the interim had become Astronomer Royal, announced a second important discovery. He found a small nutation, i.e., a nodding effect or wobble in the direction of the earth's axis of rotation. This produces a relatively small variation in the larger and longer period precessional wobble. This effect has a period of 18.61 years. The nutational wobble has a radius of 9.2" of arc. Nutation is a gravitational effect of the shift of the nodes of the moon's orbit, which shift also has the period of 18.61 years.

Personal Equation

The next correctional factor is the most curious of all. In 1795, Nevil Maskelyne, who was England's Astronomer Royal at that time, was engaged at Greenwich in making transit measurements, being assisted in this by David Kinnebrook. Transit measurements involve determining very precisely when a star crosses the celestial meridian of a location, i.e., when the star crosses the plane passing through the north and south terrestrial poles and the observer's zenith. Maskelyne and his assistant Kinnebrook made transit measurements by watching the star through a telescope that had a crosshair in the eyepiece running parallel to the meridian. As the star approached the crosshair, they determined the time of crossing by using a very precise clock. Listening for the beats of the clock, which corresponded to seconds of time, they sought to determine exactly when between beats of the clock the star traversed the meridian. Maskelyne, upon checking Kinnebrook's results, found that his measurements differed from his own by about 5/10ths of a second. He informed Kinnebrook of this discrepancy, urging him to reform his technique. Kinnebrook, however, did not improve; in fact, the discrepancy between his and Maskelyne's determinations grew to 8/10ths of a second. As Maskelyne put it in a report on this incident, "therefore, though with great reluctance . . . I parted with him."

The incident passed unnoticed for some years until the astronomer Friedrich Wilhelm Bessel, upon reading in 1816 an account in a history of Greenwich Observatory of the events that led to Maskelyne's decision to fire Kinnebrook, decided to investigate the matter with care. Comparing his own observations with those of other astronomers, he found that his results differed from theirs; in fact, he asserted that each astronomer literally perceives such observations differently. It was hoped for a time that this effect, the so-called personal equation, would be a fixed characteristic of each observer, thereby making it possible to define Maskelyne, for example, as the norm and to correct all of Kinnebrook's observations by a definite amount. Alas it was learned that a person's personal equation may itself vary; in fact, it depends on such factors as the speed, brightness, and direction of motion of the object as well as the observer's expectations. The problem was partly resolved when photographic techniques became available.

Bessel had in effect discovered that, strictly speaking, there are as many heavens as there are astronomers. Not surprisingly, psychologists took great interest in this phenomenon, which led to the result that its subsequent history belongs both to the history of astronomy and to the history of psychology.

Conclusion

What all this implies is that the astronomer derives only raw data from telescopic observations. These results must then be reduced, that is, corrected in terms of the various correctional factors that influence where a celestial object is observed to be located. The laborious process of *reduction of observation* involves such theoretical and empirical parameters as have been discussed. These events also suggest the increasing degree of precision that astronomers were attaining. Whereas the Greeks had detected an effect that attains the magnitude of about 35' of arc, by the eighteenth century, Bradley had discovered the aberration of starlight, which at maximum produces a variance of 20.5" of arc. This represents an improvement by a factor of about 100. By the 1830s, Bessel was attaining results correct to about 1/3", which is 60 times better than eighteenth-century determinations. Using this rough calculation, we see that an improvement in precision of about 6,000 had occurred from antiquity to the 1830s.

3. The Distance of the Stars

How distant are the stars? This question will emerge as crucial for the materials that we shall be discussing. As we shall see, claims concerning the nature of the stars were heavily dependent on beliefs as to their distances. For Claudius Ptolemy the stars were all located on a single sphere, which he positioned immediately outside the orbit of Saturn. Although Ptolemy did not attempt to determine planetary distances in his *Almagest*, he did provide a detailed discussion of this topic in his *Planetary Hypotheses*. In that work, he estimated the mean distance of the sun from the earth as about 1,210 earth radii (hereafter e.r.), which, converted to miles, is about 5 million miles (a value far below the current figure of 92,960,000 miles). He went on to estimate the distances of each of the planets from the earth, basing his calculations on his belief that the planetary orbits with their epicycles are immediately adjacent to each other. In other words, the most remote distance of one planet coincides with the nearest distance of the planet next farthest out. Believing that Saturn is the most remote planet and setting its most extreme distance as 19,865 earth radii, he positioned the stars at 20,000 e.r. Having estimated the arc width for first magnitude stars as 1/20th of that of the sun (a far too large estimate), he calculated that the actual diameter of a star must be about

4 1/2 times greater than the diameter of the earth, but only 1/20th the diameter of the sun. If we convert his figure to modern distance measures, we get that he positioned the stars at about 80 million miles. Viewed from a modern perspective in which the sun's distance from the earth is set at 92,960,000 miles, this distance falls very far short of the mark. Nonetheless, such figures must have struck the ancients as extremely large.

Although Copernicus did not give a quantitative estimate of the distance of the stars, he is certainly a crucial figure in the history of attempts to determine the distance of the stars. In his *De revolutionibus* (1543), he did estimate the mean distance of the sun as 1,142 e.r., which is slightly less than Ptolemy's value. Moreover, for reasons that need not concern us, the geometry of his system entails the result that Saturn, the planet with the largest orbital radius, is substantially nearer the earth than Ptolemy had thought; in fact, its distance from the earth in the Copernican system was almost half that assigned by Ptolemy, which entailed that Copernicus's planetary system was significantly smaller than that of Ptolemy. On the other hand, the claim of Copernicus that the earth orbits the sun entails the conclusion that there must be a stellar parallax, that is, that the stars should shift their apparent position as a result of the motion of the earth. Because no observer at that time had been able to determine a stellar parallax, Copernicus was forced to the conclusion that the stars must be so far distant that their parallax cannot be detected.

The famous observational astronomer Tycho Brahe was very concerned with these questions. He estimated the earth-sun distance at 1,150 e.r., set Saturn's greatest distance at 12,300 e.r., and positioned the stars at 14,000 e.r., arriving thereby at a planetary system that, like that of Copernicus, was substantially smaller than Ptolemy's. Believing that he could detect a stellar parallax if it were as large as 1' of arc, he concluded that his inability to determine a stellar parallax entailed that if Copernicus's heliocentric system were correct, the stars would have to be at least 700 times farther from the sun than Saturn. Taking the radius of Saturn's orbit to be about

9.5 times the radius of the earth's orbit and converting Brahe's figure for the minimum distance of the fixed stars to the so-called astronomical unit (hereafter a.u.), which is the distance of the earth from the sun, we see that he was claiming that the Copernican system entails that the fixed stars must be at least (700 • 9.5 =) 6,650 a.u. distant. Using modern values for the earth-sun distance and for the radius of the earth, we get that 1 a.u. equals about 23,000 e.r. Using this equivalence to convert Brahe's figure for the distance of the stars in the Copernican system to earth radii, we get that 6,650 a.u. equals about 150,000,000 e.r. Given that the volume of a sphere is proportional to its radius cubed, Brahe's determination of the diameter of the Copernican universe entails that it would have to be about 350,000,000 (700^3) times larger in volume than the universe of the geocentric system. The situation was in one way even more striking. Brahe had greatly overestimated the diameter of the brightest stars at about 2' of arc. This entailed that were the Copernican system true, such stars would have diameters hundreds of times larger than the sun's.

Such large distances presented great problems for the Copernicans, who were naturally very anxious to determine a stellar parallax. During the 1670s, the situation was somewhat clarified when attempts were made to determine a more accurate value for the distance of the sun from the earth by measuring a parallax for Mars. The strategy employed was to observe Mars at its closest approach to the earth, at which time Mars is substantially nearer the earth than the sun and hence provides a larger parallax than the solar parallax. By observing Mars from two different locations on earth, they obtained a parallactic triangle and hence the distance of Mars.

Diagram of Method of Measuring the Parallax of Mars (not to scale)

Having found this distance and using Kepler's third law, which gives the relative radii of the Martian and terrestrial orbits, astronomers could calculate the radius of the earth's orbit. They concluded that 1 a.u. = 40 to 90 million miles, the latter figure being fairly close to the modern value (93 million miles). In short, whereas the values assigned by Ptolemy, Copernicus, and Brahe for the earth-sun distance were all about 20 times too low, values available by late seventeenth century were gradually approximating the modern value.

The first person to attract substantial attention for an attempt to measure stellar parallax was Robert Hooke, who over a period of about three months in 1669 used a very long telescope to make four observations of the position of the star gamma Draconis, which, because it approaches very near the zenith for London, presented minimal problems in regard to atmospheric refraction. From these observations, Hooke claimed to have determined a parallax of about 15". Nonetheless, his observations were so few that many doubted his result; moreover, Jean Picard, who a few years later made extensive observations of a star in Lyra, reported that he could not detect any parallactic shift in that star. When in the 1720s, James Bradley made his careful investigation of gamma Draconis that led to his discovery of the aberration of starlight, he not only discredited Hooke's claim, but also (as noted previously) concluded that the stars must be at least 400,000 a.u. distant, basing this conclusion on his conviction that if the parallactic shift were as large as 1", he could have detected it. To get a sense of Bradley's value for the minimum distance of the stars as compared to earlier values, let us convert it to earth radii. This gives Bradley's estimate of the minimum distance of the stars as (400,000 • 23,000=) about 9,200,000,000 e.r., which is (9,200,000,000/20,000=) 460,000 times larger than the distance assigned to the stars by Ptolemy.

Because no one seemed able to measure a stellar parallax and hence attain by this method the distance of the stars, astronomers sought other methods of estimating the distance of the stars. One of these methods was presented in a book entitled *Cosmotheoros*, written by Christiaan Huygens and published 1698. Huygens's method was to place a screen in front of the sun and to cut a small hole in the screen of sufficient size that the brightness of the hole would equal the brightness of the brightest star, Sirius. He found that when he had in this way reduced the brightness of the sun by 27,664 times, the brightness of the hole was equal to that of Sirius. From this he inferred that were the sun positioned at 27,664 a.u. or, using Huygens's value for earth-sun distance, at 660,000,000 e.r., it would shine with the brightness of Sirius. This method of estimating distances was clearly open to serious objections.

Another somewhat comparable method of determining stellar distances had been proposed in 1668 by James Gregory. His method was to find a planet positioned so that its brightness was equal to that of Sirius. He then estimated the portion of the sun's light being reflected to us from that planet. By this method, Gregory cautiously calculated a distance of 83,190 a.u., which if converted to earth radii equals about 1,900,000,000 e.r. Isaac Newton adopted this method to calculate the distance of Sirius as compared to Saturn's distance from us. His calculation, which appeared in 1730 in his *System of the World*, put Sirius 100,000 times more distant from us than Saturn. Using the figure that Saturn is about 9.5 a.u. from us, we get that Sirius, on Newton's estimate, must be about 1,000,000 a.u. distant. As we shall see, the actual distance of the nearest star is about 270,000 a.u., the distance of Sirius being about twice this amount.

Other estimates of the distance of the stars were made during the eighteenth and early nineteenth centuries, the first accurate parallactic determination having been attained in 1838 by F. W. Bessel, whose paper will be considered later.

Given next is a table of comparative estimates of the distance of the stars. It should be noted that it significantly oversimplifies what is stated more precisely above.

Astronomer	Date	Estimated Distance	Comments
Ptolemy	2nd c. A.D.	20,000 e.r.	
Brahe	Late 16th c.	14,000 e.r.	
Copernicans	Late 16th c.	150,000,000 e.r.	Brahe's estimate
Gregory	1668	1,900,000,000 e.r.	Distance of Sirius
Huygens	1698	660,000,000 e.r.	Distance of Sirius
Bradley	1728	9,200,000,000 e.r.	Minimum distance

It is important to realize that Copernicus's system and the distances entailed by it pointed toward a number of dramatic conceptual conclusions and changes, including the following:

(1) The spread of the Copernican system eventually led to the destruction of the starry vault, that is, to the abolition of the notion that the stars are all positioned at a single distance from the earth and on a spherical vault of definite diameter. In opposition to this idea, it came to be realized that the stars are scattered throughout space, some being far nearer to us than others. The first author who represented this idea in a diagram was the English Copernican Thomas Digges, who did so in his 1576 *A Perfit Description of the Caelestiall Orbes.*

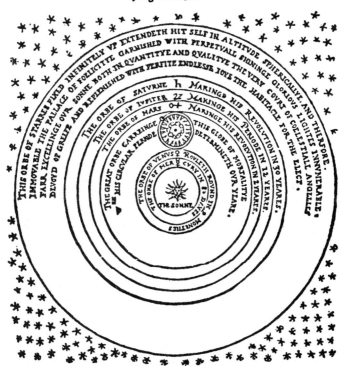

Thomas Digges's Diagram of the Copernican System (1576) Showing the Stars Scattered throughout Space

(2) Astronomers also began to realize that the stars must be very distant and, given their brightness, that they must be very large. In fact, they came to the dramatic conclusion that the

stars are suns; as the poet Edward Young put it in the 1740s, "One sun by day, by night, ten thousand shine. . . ." And if the stars be suns, then it seemed possible, if not probable, that they are encircled by planets. Within fifty years of Copernicus's death, Giordano Bruno drew this inference, even claiming that planets of other suns are inhabited. Although it was not this or his other astronomical claims that led to his execution by the Roman Catholic Inquisition in 1601, he has sometimes been seen (inappropriately) as a martyr for belief in extraterrestrial intelligent life. Even such an outstanding eighteenth-century scientist as Leonhard Euler seems to have accepted the idea, for which there was no direct observational evidence, that planets orbit other suns. Euler's view is suggested by the frontispiece for his *Theoria motuum planetarum et cometarum* (1744).

**Frontispiece of Euler's *Theoria Motuum*
Showing Planets Orbiting Stars**

(3) Another conclusion that gradually became clear was that the heavens as seen by astronomically uninformed individuals or by astronomers in a non-analytical mood differ very substantially from the heavens as analyzed by astronomers. When one looks, for example, at Sirius, one assumes that this process is comparable to looking at, say, a terrestrial light or a sparkling diamond. If, however, we put together Bradley's determination that the stars must be very distant with his measurement of the speed of light, which he found to be very large but finite, we see that when we look at Sirius, we are seeing it not as it now appears, but rather as it was some years ago (using modern values, about nine years ago). And as one looks even deeper into space, one sees objects as they appeared at ever more remote periods of the past. One can consequently easily imagine that a distant star, now shining in our heavens, has in fact long since disappeared or ceased to shine. Or that from a space now appearing dark, the light of a brilliant nova speeds toward us. When one looks at the sky, one is consequently looking back into the past by centuries, millennia, or more.

4. Nebulae and the Milky Way
Some Early Observations and Speculations

Description of Nebulae and the Milky Way

Scattered throughout the heavens are various cloudy or nebulous patches, which astronomers of the eighteenth and nineteenth centuries called nebulae. Although a few of these objects are as wide as a degree or more, the great majority appear far smaller. Moreover, they are very dim; only a few are visible to the naked eye. One can derive some idea of their appearance from drawings, made in the second half of the eighteenth century by Charles Messier, of the two most prominent nebulae visible in the northern celestial hemisphere, the Orion and Andromeda nebulae.

 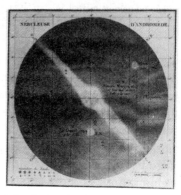

**Messier's Drawings of the Orion Nebula
and of the Andromeda Nebula**

The Milky Way is a vast band of stars that encircles the entire sky. The density of stars in the Milky Way, although varying from location to location, is generally far greater than in other regions of the sky. The edges of the broad band are jagged and uneven, and a few sections of the central region are nearly starless. The band of the Milky Way is not orientated in any obvious relation to the zodiac, through which the planets move. The Milky Way is strikingly evident on a clear, dark night, but it is difficult to see from well-lighted cities.

From Aristotle to Galileo

Scientists showed relatively little interest in the Milky Way throughout most of history. It was, however, prominent in some primitive cosmologies and mythologies and drew the attention of various Greek, Roman, and medieval authors. Aristotle, for example, discussed it as a sublunar phenomenon in his *Meteorology*, Bk. I, Ch. 8, and Claudius Ptolemy treated its appearance in his *Almagest*, Bk. VIII, Ch. 2. Tradition has it that Democritus was the first to suggest that the Milky Way is "the luminescence due to the coalition of many small stars which shine together because of their closeness to one another."[1]

Although Copernicus said little about the Milky Way, the Copernican system, at least gradually, helped astronomers to see the stars and the "Via Lactea" in a new way. This is evident, for example, in the writings of Galileo, who in his *Starry Messenger* (1610) reported on what he saw when he turned his newly invented telescope to the stars and Milky Way. After noting that stars differ in appearance from planets, Galileo stated that in addition to the naked-eye stars, the telescope reveals "a host of other stars ... so numerous as almost to surpass belief. One may, in fact, see more of them than all the stars included among the first six magnitudes [the naked-eye stars]."[2] To illustrate his point, Galileo reported that upon examining the constellation Orion, "... I was overwhelmed by the vast quantities of stars.... There are more than five hundred new stars distributed among the old ones within the limits of one or two degrees of arc. Hence to the three stars in the Belt of Orion and the six in the Sword which were previously known, I have added eighty adjacent stars...."[3] Concerning the Milky Way, Galileo stated that with the telescope,

> ... this has been scrutinized so directly and with such ocular certainty that all the disputes have been resolved, and we are at last freed from wordy debates.... [It] is in fact nothing but a congeries of innumerable stars grouped together in clusters. Upon whatever part of it the telescope is directed, a vast crowd of stars is immediately presented to view. Many of them are rather large and quite bright, while the number of smaller ones is quite beyond calculation.[4]

---

[1] As quoted from Pseudo-Plutarch in Stanley L. Jaki, *The Milky Way: An Elusive Road for Science* (New York: Science History Publications, 1972), p. 10.

[2] Galileo Galilei, *The Starry Messenger* in *Discoveries and Opinions of Galileo*, translated and with an introduction and notes by Stillman Drake (New York: Doubleday Anchor, 1957), p. 47.

[3] Galileo, *Starry Messenger*, p. 47.

[4] Galileo, *Starry Messenger*, p. 49.

Galileo stressed that cloudy patches appear in the Milky Way and also that

> ... several patches of similar aspect shine with faint light here and there throughout the aether, and if the telescope is turned upon any of these it confronts us with a tight mass of stars. And what is even more remarkable, the stars which have been called 'nebulous' by every astronomer up to this time turn out to be groups of very small stars arranged in a wonderful manner. Although each star separately escapes our sight on account of its smallness or the immense distance from us, the mingling of their rays gives rise to that gleam which was formerly believed to be some denser part of the aether that was capable of reflecting rays from stars or from the sun.[1]

Galileo's views caused some controversies; for example, Simon Marius claimed to have resolved the Milky Way before Galileo; in fact, Marius even asserted that he had seen "many distinct stars" in Andromeda. In Galileo's *Assayer* (1618), Galileo again discussed nebulae and the Milky Way:

> We know positively that a nebula is nothing but an aggregate of many minute stars which are invisible to us; yet the field which they occupy does not remain invisible. It shows itself in the aspect of a little whitened patch, coming from the conjunction of the separate radiances that crown each of these little starlets. But since these irradiations exist only in our eyes, it must be that each image of each starlet exists really and distinctly in our eyes. From this, one may deduce another doctrine—that the nebulae and even the Milky Way do not exist in the sky but are a pure sensation of our eyes, in the sense that, if our vision were so acute as to distinguish those minute stars, there would be neither nebulae nor a Milky Way in the sky.[2]

As we proceed to consider additional ideas concerning nebulae and the Milky Way, it will prove helpful to keep the following questions in mind:

---

[1] Galileo, *Starry Messenger*, pp. 49–50.
[2] Galileo Galilei, *The Assayer* in *The Controversy on the Comets of 1618*, trans. by Stillman Drake and C. D. O'Malley (Philadelphia: Univ. of Pennsylvania Press, 1960), p. 201.

(1) What is the Milky Way? Is it resolvable into individual stars? Why does it appear as it does?
(2) What are nebulae? Do different types of nebulae exist? Are all resolvable into stars?

Nebulae: From Halley to Messier

We met Edmond Halley in the discussion of proper motion. In the following short paper, which Halley published in 1715, we encounter him again, in this case presenting a list of six nebulae, most of which will repeatedly reappear in our subsequent readings. As you read Halley's paper on nebulae, pay close attention to his views on the nature of these mysterious objects.

☆☆☆☆☆☆☆☆☆☆☆☆☆☆☆☆☆☆☆☆☆☆☆☆☆☆☆

[Edmond Halley]. "An Account of Several Nebulae or Lucid Spots like Clouds, Lately Discovered among the Fixt Stars by Help of the Telescope," *Philosophical Transactions of the Royal Society*, 29 (1715), 390-2.[1]

In our last[2] we gave a short Account of the several New Stars that have appeared in the Heavens, within the last 150 Years, some of which afford very surprizing Phaenomena. But not less wonderful are certain luminous Spots or Patches, which discover themselves only by the Telescope, and appear to the naked Eye like small Fixt Stars; but in reality are nothing else but the Light coming from an extraordinary great Space in the Ether; through which a lucid *Medium* is diffused, that shines with its own proper Lustre. This seems fully to reconcile that Difficulty which some have moved against the Description *Moses* gives of the Creation, alledging that Light could not be created without the Sun. But in the following instances the

---

[1][My comments on this paper are based mainly on those in Kenneth Glyn Jones, *The Search for the Nebulae* (Chalfont St. Giles, England, 1975), pp. 22-4, which also provides additional commentary. MJC]

[2][A few pages earlier in the same volume of the *Philosophical Transactions* Halley had published a paper calling the attention of astronomers to various novae (new stars) and variable stars. MJC]

contrary is manifest; for some of these bright Spots discover no sign of a Star in the middle of them; and the irregular Form of those that have, shows them not to proceed from the Illumination of a Central Body. These are, as the aforesaid New Stars, Six in Number, all which we will describe in the order of time, as they were discovered; giving their Places in the Sphere of Fixt Stars, to enable the Curious, who are furnished with good Telescopes, to take the Satisfaction of contemplating them.

The first[1] and most considerable is that in the Middle of *Orion's* Sword, marked with θ by *Bayer* in his *Uranometria*, as a single Star of the third Magnitude; and is so accounted by *Ptolemy, Tycho Brahe* and *Hevelius*: but is in reality two very contiguous Stars environed with a very large transparent bright Spot, through which they appear with several others. These are curiously described by *Hugenius* [Christiaan Huygens] in his *Systema Saturnium* pag. 8 who there calls this brightness, *Portentum, cui certe simile aliud nusquam apud reliquas Fixas potuit animadvertere* [a wonderful object, which is certainly unique among the fixed stars]: affirming that he found it by chance in the Year 1656. The Middle of this is at present in [Gemini] 19° 00, with South Lat. 28° 3/4.

About the Year 1661 another of this sort was discovered (if I mistake not) by *Bullialdus. in Cingulo Andromedae.*[2] This is neither in *Tycho* nor *Bayer*, having been omitted, as are many others because of its smallness: But it is inserted into the Catalogue of *Hevelius*, who has improperly called it *Nebulosa* instead of *Nebula*; it has no sign of a Star in it, but appears like a pale Cloud, and seems to emit a radiant Beam into the North East, as that in *Orion* does into the South East. It precedes in Right Ascension the Northern in the Girdle, or ν *Bayero*, about a Degree and three Quarters, and has Longitude at this time [Aries] 24° 00'. with Lat. North 33° 1/3.

---

[1][This is the Orion nebula, which is probably the most famous of all nebulae visible to observers in the Northern Hemisphere. It is M42 in Messier's catalogue of nebulae, which catalogue will be described shortly. MJC]

[2][This is the Andromeda nebula, M31, which is Orion's chief rival among the most spectacular nebulae. MJC]

The Third[1] is near the Eclipticbetween the *Head* and *Bow of Sagittary*, not far from the Point of the Winter Solstice. This it seems was found in the Year 1665 by a *German* Gentleman *M. F. Abraham Ihle*, whilst he attended the Motion of *Saturn* then near his *Aphelion*. This is small but very luminous, and emits a Ray like the former. Its Place at this time is [Capricorn] 4° 1/2 with about half a Degree South Lat.

A fourth[2] was found by M. *Edm. Halley* in the Year 1677, when he was making the Catalogue of the Southern Stars. It is in the *Centaur*, that which *Ptolemy* calls ὁ επι της του νώτον εκφυύσεως [the one who emerges from the (horse's) back] which He names *in dorso Equino Nebulus* [Nebula on the back of a horse] and is *Bayers* ω; It is in appearance between the fourth and fifth Magnitude, and emits but a small Light for its Breadth and is without a radiant Beam; this never rises in *England*, but at this time its Place is [Scorpio] 5° 3/4 with 35° 1/5 South Lat.

A Fifth[3] was discovered by Mr. *G. Kirch* in the Year 1681 preceding the Right Foot of *Antinous*: It is of its self but a small obscure Spot, but has a Star that shines through it, which makes it the more luminous. The Longitude of this is at present [Capricorn] 9°. *circiter* with 17° 1/6. North Latitude.

The Sixth[4] and last was accidentally hit upon by M. *Edm. Halley* in the Constellation of *Hercules*, in the Year 1714. It is nearly in a Right Line with ζ and η of *Bayer*, somewhat nearer to ζ than η: and by comparing its Situation among the Stars, its Place is sufficiently near in [Scorpio] 26° 1/2 with 57°.00. North Lat. This is but a little Patch, but it shews it self to the naked Eye, when the Sky is serene and the Moon absent.

There are undoubtedly more of these which have not yet come to our Knowledge, and some perhaps bigger but though all

---

[1][This object is M22. It is now referred to as a **globular cluster**, i.e., a tightly packed group of stars that displays a spherical symmetry. MJC]

[2][Omega Centauri, a Southern Hemisphere globular cluster, discovered by Halley himself while observing from St. Helena. MJC]

[3][M 11, a **galactic or open cluster**, i.e., a cluster of a dozen or more stars displaying no symmetry. MJC]

[4][The Hercules Cluster, M13, a globular cluster. MJC]

these Spots are in Appearence but little, and most of them but of few Minutes in Diameter; yet since they are among the Fixt Stars, that is, since they have no Annual Parallax, they cannot fail to occupy Spaces immensely great, and perhaps not less than our whole Solar System. In all these so vast Spaces it should seem that there is a perpetual uninterrupted Day, which may furnish Matter of Speculation, as well to the curious Naturalist as to the Astronomer.

☆☆☆☆☆☆☆☆☆☆☆☆☆☆☆☆☆☆☆☆☆☆☆☆☆☆☆☆

A number of astronomers in the decades after Halley's paper published lists of nebulae. Among these astronomers was William Derham (1657–1735), whose list of 21 nebulae appeared in 1733 in the *Philosophical Transactions*. Only 7 of the 21 objects reported by Derham are actually nebulae. The French scientist Pierre-Louis Moreau de Maupertuis became interested in Derham's list; in fact, he republished it in France in one of his books. Maupertuis had noted that Derham described 5 of the nebulae as ellipsoidal (actually only 1 of these 5 appears as elliptical). In the supposed ellipsoidal shape of some nebulae, Maupertuis found support for a theory he favored, that the variations in brightness of variable stars can be explained by the assumption that they are ellipsoidal in form, their varying brightness being a result of the star being alternately seen end-on and full-face.

Another list of nebulae, running to 30 items, was compiled in 1746 by a Swiss astronomer, Jean-Phillippe Loys de Chéseaux (1718–1751), who had himself discovered a number of the objects in his compilation. The most thorough of the early observers of nebulae was Charles Messier (1730–1813), who is remembered today primarily for the list he published around 1780 of 103 nebulae, 42 of which he was the first to report. The numbers assigned by Messier to his 103 nebulae have become a common method of designating the nebulae; Andromeda, for example, is frequently referred to as "M31." Ironically, Messier's passion was for comets rather than nebulae, his interest in nebulae having had its origin in the fact that they are easily mistaken for comets.

Thomas Wright

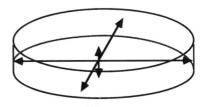

In 1750, Thomas Wright (1711–1786) published a book entitled *An Original Theory or New Hypothesis of the Universe*. Numerous authors over the last two centuries have seen this book as presenting for the first time the current theory that the Milky Way is a pancake-shaped cluster composed of a vast number of stars. To get an idea of this theory, imagine someone located in the interior of a large pancake, the molecules of which correspond to the stars. Also imagine the person looking in any direction parallel to the plane of the pancake. Then the person, if looking within the plane of the pancake, would see large numbers of pancake molecules; however, were the person to look up or down, i.e., in a direction perpendicular to the plane of the pancake, the person would see far fewer molecules. In short, the resulting appearance would be that the person would see a large circular band of molecules. Correspondingly, the appearance of the Milky Way can be explained on the assumption that it is a pancake-shaped structure, somewhere in the midst of which our solar system is located.

In his *Original Theory or New Hypothesis of the Universe.*, Wright employed the next diagram to explain and illustrate the *optical* fact that the appearance of the Milky Way can be explained by assuming that it consists of an extended planar array of stars. At this stage, Wright was discussing optical principles, rather than enunciating a theory of the structure of the Milky Way. Nonetheless, dozens of later authors mistakenly assumed that Wright was presenting the disk or pancake theory of the Milky Way. He was not, however, doing this; in fact, Wright went on to propose two physical theories of the Milky Way, both interesting but alas incorrect.

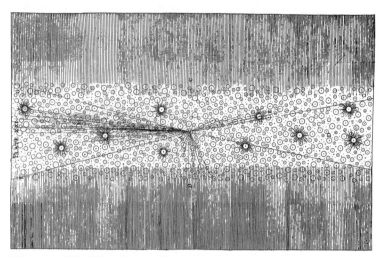

Diagram Used by Wright in Explaining the Optical Principles Involved in the Appearance of the Milky Way

The first theory is that we are positioned within a hollow spherical shell of stars. Thus if we look in a direction tangent to the shell, we see many stars, whereas if we look in a direction perpendicular to the section of its surface in which we are located, we see relatively few stars. His second theory is that the earth is located in a circular, flat ring of stars, shaped like Saturn ring. As before, if we look in a direction within the plane of this ring, we see numerous stars, but the number of stars drops drastically in directions perpendicular to the ring. As was learned from a manuscript written by Wright later in his life but discovered only in the 1960s, he eventually adopted a third (even stranger) theory of the Milky Way! In particular, he claimed that the stars are volcanoes on a solid firmament, adding that the Milky Way "is looked upon as no other than a vast chain of burning mountains forming a flood of fire surrounding the whole starry regions and no how different from other luminous spaces [nebulae], but in [the] number of stars that compose them, or where there are none, in the vast floods of celestial lava that form it."

In short, although Wright failed to attain the correct physical theory of the Milky Way, he deserves credit for having been the first to publish a correct presentation of the optical principles involved in its appearance. Moreover, his book may include (the point is disputed) the idea that nebulae are other Milky Ways or universes.

Immanuel Kant and Johann Lambert

Immanuel Kant

The first person to publish the disk theory of the Milky Way was Immanuel Kant (1724–1804), who included it in his *Allgemeine Naturgeschichte und Theorie des Himmels*, which he published anonymously in 1755 and which has been translated into English as *Universal Natural History and Theory of the Heavens*. A selection from that translation follows. Ironically, Kant believed that he had derived the disk theory from Wright, whose book he knew only from a review of it in a German periodical.

Like many major philosophers, e.g., Plato, Aristotle, Descartes, Leibniz, Peirce, and Whitehead, Kant had a strong interest in science. His special fondness for astronomy is evident in the inscription on his tombstone: "Two things fill my mind with ever new and increasing wonder and awe, the more often and persistently I reflect upon them: *the starry heaven above me and the moral law within me.*" This sentence was taken from Kant's *Critique of Practical Reason,* where the immediately subsequent lines explain what that statement meant to Kant.

> I should not seek and conjecture these two as entities hidden in obscurity or in the boundlessness beyond my sight. I see them in front of me and unite them immediately with the consciousness of my own existence. The first originates in the place that I occupy in the external sense world and extends the nexus in which I stand to an immense size with world

upon worlds and systems of systems, and extends the nexus even further into the limitless time of the periodic movement of these systems, of their beginning and duration. The second originates in my invisible self, my personality, and presents me in a world that has true infinity, only discernible by the faculty of understanding. I am united to this world (thereby also simultaneously to all those visible worlds) not, as I was in the external-sense worlds, in a merely accidental nexus but rather in a universal and necessary one. The first view of an innumerable multitude of worlds nullifies as it were my importance, as an *animal creature* that must again surrender to the planet (a mere point in the cosmos), the matter from which it arose after the animal for a short time (we know not how) has been endowed with life-force. The second, on the contrary, elevates infinitely my worth, as an *intelligence,* through my personality, in which the moral law reveals to me a life independent of animality and even of the whole sense-world, at least as far as can be gleaned by means of this law from the appropriate destiny of my existence, a destiny not confined to the conditions and limits of this life but rather continuing into infinity.[1]

It is a striking fact that Kant and Wright were not the only authors who in the middle years of the eighteenth century published treatises that presented fundamental notions in stellar astronomy. A third pioneer in that area was the mathematician, scientist, and philosopher Johann Lambert (1728–1777), who in 1761 published his *Cosmologische Briefe (Cosmological Letters).* After it had appeared, Lambert learned that Kant had anticipated him in publishing the disk theory of the Milky Way. Consequently, Lambert wrote Kant about this matter. From that letter we learn that Lambert had formulated the disk theory in 1749. Lambert stated:

What gave occasion to my *Cosmological Letters* ... was this: that in the year 1749, on a certain occasion immediately after supper, and contrary to my custom then, I left the company in which I was at the time, and went into another room. I there wrote down my thoughts on a a quarto page, and in the year 1760, when I wrote the *Cosmological Letters,* I had still

---

[1] As translated for this volume by Joseph T. Ross from Kant, *Kritik der praktischen Vernunft,* 2 Theil. in *Kants gesammelte Schriften,* part 1, vol. 5 (Berlin: Georg Reimer, 1913), pp. 161–2.

nothing further on the subject in hand. In the year 1761, I was told at Nürnberg that some years previously an Englishman had printed similar thoughts in letters to certain persons, but that he had not had much success, and that the translation of his Letters, begun at Nürnberg, had not been completed. I answered that I believed my *Cosmological Letters* would not make a great impression, but that perhaps in the future an Astronomer would discover something in the Heavens which could not be otherwise explained.[1]

If one compares Kant's universe with Lambert's, one perceives a number of interesting points. For example, Kant viewed the universe as infinite whereas Lambert argued for a finite cosmos. It is also noteworthy that the books of Kant and Lambert as well as that by Wright are filled with speculations about extraterrestrial life. Kant devoted the entire third section of his book to that topic, even suggesting how the levels of intelligence among extraterrestrials differ in relation to their distances from the sun and from the center of his cosmos. Lambert was no less bold; he pressed his speculations even to the extreme of postulating inhabitants of comets. It is a curious fact that all three of these pioneering texts of stellar astronomy abound with ideas of extraterrestrial life.

In reading the following selection from Kant's book, it will be helpful to keep the following questions in mind.
(1) To what extent did Kant derive his ideas from Wright?
(2) To what extent did Kant's ideas come from Derham and Maupertuis? In general, how strong was Kant's empirical support for his claim that nebulae are other universes comparable to the Milky Way?
(3) What ancient philosophic school can be seen very prominently in the background of Kant's work?
(4) Is evolution a doctrine confined to biology and geology?
(5) Would Kant have objected on religious grounds to Darwin's theory of evolution?
(6) What proof does Kant offer for the existence of God and what is the forcefulness of that proof?

---

[1] As translated and quoted in W. Hastie (ed.), *Kant's Cosmogony* (Glasgow: James Maclehose and Sons, 1900), p. lxx.

☆☆☆☆☆☆☆☆☆☆☆☆☆☆☆☆☆☆☆☆☆☆☆☆☆☆☆

[Selection from] [Immanuel Kant], *Universal Natural History and Theory of the Heavens*, trans. by W. Hastie in W. Hastie (ed.), *Kant's Cosmogony* (Glasgow: James Maclehose and Sons, 1900), pp. 17–34; 53–65.

PREFACE

I have chosen a subject which is capable of exciting an unfavourable prejudice in a great number of my readers at the very outset, both on account of its own intrinsic difficulty, and also from the way they may regard it from the point of view of religion. To discover the system which binds together the great members of the creation in the whole extent of infinitude, and to derive the formation of the heavenly bodies themselves, and the origin of their movements, from the primitive state of nature by mechanical laws, seems to go far beyond the power of human reason. On the other hand, religion threatens to bring a solemn accusation against the audacity which would presume to ascribe to nature by itself results in which the immediate hand of the Supreme Being is rightly recognized; and it is troubled with concern, by finding in the ingenuity of such views an apology for atheism. I see all these difficulties well, and yet am not discouraged. I feel all the strength of the obstacles which rise before me, and yet I do not despair. I have ventured, on the basis of a slight conjecture, to undertake a dangerous expedition; and already I discern the promontories of new lands. Those who will have the boldness to continue the investigation will occupy them, and may have the satisfaction of designating them by their own names.

I did not enter on the prosecution of this undertaking until I saw myself in security regarding the duties of religion. My zeal was redoubled when at every step I saw the clouds disperse that appeared to conceal monsters behind their darkness; and when they were scattered I saw the glory of the Supreme Being break forth with the brightest splendour. As I now know that these efforts are free from everything that is reprehensible, I shall faithfully adduce all that well-disposed or even weak minds may find repellent in my scheme; and I am ready to submit to the judicial severity of the orthodox Areopagus with a frankness which is the mark of an honest conviction. The advocate of the faith may therefore be first allowed to make his reasons heard, in something like the following terms:

'If the structure of the world with all its order and beauty,' he says, 'is only an effect of matter left to its own universal laws of motion, and if the blind mechanics of the natural forces can evolve so glorious a product out of chaos, and can attain to such perfection of themselves, then the proof of the Divine Author which is drawn from the spectacle of the beauty of the universe wholly loses its force. Nature is thus sufficient for itself; the Divine government is unnecessary; Epicurus lives again in the midst of Christendom, and a profane philosophy tramples under foot the faith which furnishes the clear light needed to illuminate it.'

Even if I found some grounds for this objection, yet the conviction which I have of the infallibility of Divine truth is so potent in me that I would hold everything that contradicted it as sufficiently refuted by that truth, and would reject it. But the very harmony and agreement which I find between my system and religion, raises my confidence in the face of all difficulties to an undisturbed tranquillity.

I recognize the great value of those proofs which are drawn from the beauty and perfect arrangement of the universe to establish the existence of a Supremely Wise Creator; and I hold that whoever does not obstinately resist all conviction must be won by those irrefutable reasons. But I assert that the defenders of religion, by using these proofs in a bad way, perpetuate the conflict with the advocates of Naturalism by presenting them unnecessarily with a weak side of their position.

It is usual to signalize and emphasize in nature the harmonies, the beauty, the ends of things, and the perfect relation of means adapted to them. But while nature is thus elevated on this side, the attempt is made on another to belittle it again. This admirable adaptation, it is said, is foreign to nature; abandoned to its own general laws it would bring forth nothing but disorder. These harmonies show an alien hand which has known how to subdue to a wise plan a matter that is wanting in all order or regularity. But I answer, that if the universal laws of the action of matter are themselves a consequence of the supreme plan of the system, they cannot be supposed to have any other destination than just to serve to fulfil the very plan which the Supreme Wisdom has set before itself. And if this is not so, would we not be tempted to believe that matter and its general laws at least are independent, and that the Supremely Wise Power, which has known how to use it in such a glorious way, is indeed great, but not infinite; is indeed powerful, but not all-sufficient?

The defender of religion is afraid that those harmonies which may be explained from a natural tendency of matter, may prove to be independent of Divine Providence. He confesses distinctly that if natural causes could be discovered for all the order of the universe, and that if these causes could bring forth this order from the most general and essential properties of matter, it would be unnecessary to have recourse to a Higher Government at all. The advocate of Naturalism finds his account in not disputing this assumption. He heaps up examples which prove that the general laws of nature are fruitful in perfectly beautiful consequences, and he brings the orthodox believer into danger by adducing reasons which in the believer's hands might become invincible weapons of his faith. I will give some examples. It has often been adduced as one of the clearest proofs of a benevolent Providence which watches over men, that in the hottest climates the sea-breezes, just at the time when the heated soil most needs their cooling breath, are, as it were, called to sweep over the land and refresh it. For instance, in the island of Jamaica, as soon as the sun has risen so high as to throw the most unbearable heat on the soil—just after nine o'clock in the forenoon—a wind begins to rise from the sea, which blows from all sides over the land, and its strength increases in the same proportion as the height of the sun. About one o'clock in the afternoon, when it is naturally hottest, this wind is most violent, and it diminishes again as the sun gradually goes down, so that towards evening the same stillness prevails as at sunrise. Without this desirable arrangement this island would be uninhabitable. The very same benefit is enjoyed by all the coasts of the lands which lie in the Torrid Zone. To them it is most necessary, because, as they are the lowest regions of the dry land, they also suffer from the greatest heat; whereas the regions which are found at a higher altitude in these countries, and which the sea-wind does not reach, have also less need of it, because their higher situation places them in a cooler region of the air. Is not all this beautiful? Are there not here visible ends which are effected by means prudently applied? But, on the other hand, the advocate of Naturalism cannot but find the natural causes of this phenomenon in the most general properties of the air, without needing to suppose special arrangements made for it. He rightly observes that these sea-breezes must go through such periodic movements although there were no man living on such an island, and indeed from no other property than that of the air, and even without any intention specially directed to that end referred to, as it is indispensably necessary merely for the growth of plants, and is brought about just by the elasticity and weight of the air. The heat of the sun breaks up the equilibrium of the air by rarefying that portion of it which lies over the land, and it thereby causes the cooler sea-breeze to drive it out of its position and to occupy its place.

Of what utility are not the winds generally to the earth, and what use does not the acuteness of men make of them! Nevertheless no other arrangements are necessary to produce them than those general conditions of air and heat which must be found upon the earth even apart from these ends.

'If you admit then,' the freethinker here says, 'that useful arrangements and such as point to ends can be derived from the most general and simple laws of nature, and that we have no need of the special government of a Supreme Wisdom, then you must in this see proofs by which you are caught, on your own confession. All nature, especially unorganized nature, is full of such proofs, which enable us to know that matter, while determining itself by the mechanism of its own forces, possesses a certain rightness in its effects, and that it satisfies without compulsion the rules of harmony. And should any one well-disposed to save the good cause of religion contest this capability in the universal laws of nature, he would put himself into embarrassment and by such a defence give unbelief occasion to triumph.'

But let us see how these reasons which, as used in the hands of opponents, are dreaded as prejudicial, are rather in themselves powerful weapons by which to combat them. Matter determining itself according to its most general laws by its natural procedure, or—if any one will so call it—by a blind mechanism, brings forth appropriate effects which appear to constitute the scheme of a Supreme Wisdom. Air, water, heat, when viewed as left to themselves, produce winds and clouds, rains, rivers that water the land, and all those useful consequences without which nature could not but remain desolate, waste, and unfruitful. But they bring forth these effects not by mere chance or by accident, so that they might just as easily have turned out harmful; on the contrary, we see that they are limited by their natural laws so as to act in no other way than they do. What are we then to think of this harmony? How would it be at all possible that things of such diverse nature should tend in combination with each other to effectuate harmonies and beauties so admirably, and even to subserve the ends of such things as are found in some respects outside of the sphere of dead matter (as in being useful to men and animals), unless they acknowledged a common origin, namely, an Infinite Intelligence, an Understanding in which the essential properties of all things have been relatively designed? If their natures were necessary by themselves and independent, what an astonishing chance; or, rather, what an impossibility would it be that they should so exactly fit in to each other with their natural activities and tendencies, just as if a reflective prudent choice had combined them!

Now then, I confidently apply this idea to my present undertaking. I accept the matter of the whole world at the beginning as in a state of general dispersion, and make of it a complete chaos. I see this matter forming itself in accordance with the established laws of attraction, and modifying its movement by repulsion. I enjoy the pleasure, without having recourse to arbitrary hypotheses, of seeing a well-ordered whole produced under the regulation of the established laws of motion, and this whole looks so like that system of the world which we have before our eyes, that I cannot refuse to identify it with it. This unexpected development of the order of nature in the universe begins to become suspicious to me, when such a complicated order is founded upon so poor and simple a foundation. But at last I draw instruction from the view already indicated, namely, that such a development of nature is not a thing unheard of in it; nay, that its inherent essential striving brings such a result necessarily with it, and that this is the most splendid evidence of its dependence on that pre-existing Being who contains in Himself not only the source of these beings themselves but their primary laws of action. This insight redoubles my confidence in the sketch of the system which I have drawn. My confidence increases with every step I make forward, and my timidity vanishes entirely.

But it will be said that the defence of this system is at the same time the defence of the opinions of Epicurus, which have the greatest resemblance to it. I will not deny all agreement whatever with that philosopher. Many have become atheists by the semblance of reasons which, on more exact reflection, would have convinced them most powerfully of the certainty of the existence of the Supreme Being. The consequences which a perverted understanding draws from unimpeachable principles, are frequently very reprehensible; and such were the conclusions of Epicurus, although his scheme exhibited the acuteness of a great thinker.

I will therefore not deny that the theory of Lucretius, or his predecessors, Epicurus, Leucippus, and Democritus, has much resemblance with mine. I assume, like these philosophers, that the first state of nature consisted in a universal diffusion of the primitive matter of all the bodies in space, or of the atoms of matter, as these philosophers have called them. Epicurus asserted a gravity or weight which forced these elementary particles to sink or fall; and this does not seem to differ much from Newton's Attraction, which I accept. He also gave them a certain deviation from the straight line in their falling movement, although he had absurd fancies regarding the causes and consequences of it. This deviation agrees in some degree with the alteration from the falling in a straight line, which we deduce from the repulsion of the particles. Finally, the vortices which arose from the disturbed motion, is also a theory of Leucippus and Democritus, and it will be also found in our scheme. So many points of affinity with a system which constituted the real theory of all denial of God in antiquity do not, however, draw my system into community with its errors. Something true will always be found even in the most nonsensical opinions that have ever obtained the consent of men. A false principle, or a couple of unconsidered conjunctive propositions, lead men away from the high footpath of truth by unnoticed by-paths into the abyss. Notwithstanding the similarity indicated, there yet remains an essential difference between the ancient cosmogony and that which I present, so that the very opposite consequences are to be drawn from mine.

The teachers of the mechanical production of the structure of the world referred to, derive all the order which may be perceived in it from mere chance which made the atoms to meet in such a happy concourse that they constituted a well-ordered whole. Epicurus had the hardihood to maintain that the atoms diverged from their straight motion without a cause, in order that they might encounter one another. All these theorizers pushed this absurdity so far that they even assigned the origin of all animated creatures to this blind concourse, and actually derived reason from the irrational. In my system, on the contrary, I find matter bound to certain necessary laws. Out of its universal dissolution and dissipation I see a beautiful and orderly whole quite naturally developing itself. This does not take place by accident, or of chance; but it is perceived that natural qualities necessarily bring it about. And are we not thereby moved to ask, why matter must just have had laws which aim at order and conformity? Was it possible that many things, each of which has its own nature independent of the others, should determine each other of themselves just in such a way that a well-ordered whole should arise therefrom; and if they do this, is it not an undeniable proof of the community of their origin at first, which must have been a universal Supreme Intelligence, in which the natures of things were devised for common combined purposes?

Matter, which is the primitive constituent of all things, is therefore bound to certain laws, and when it is freely abandoned to these laws it must necessarily bring forth beautiful combinations. It has no freedom to deviate from this perfect plan. Since it is thus subject to a supremely wise purpose, it must necessarily have been put into such harmonious relationships by a First Cause ruling over it; and *there is a God, just because nature even in chaos cannot proceed otherwise than regularly and according to order.*

I have such a good opinion of the honest judgment of those who will do my Essay the honour of examining it, that I am certain that the reasons now adduced, if they do not remove all anxiety about the injurious consequences of my system, will at least put the purity of my intention beyond doubt. If there are, nevertheless, ill-disposed zealots who regard it as a duty worthy of their sacred calling to attach prejudicial interpretations to innocent opinions, I am persuaded that their judgment will have the opposite effect of their intention with reasonable men. Moreover, I am not to be deprived of the right which Descartes has always enjoyed with just judges since he ventured to explain the formation of the heavenly bodies by merely natural laws. I will therefore quote the Authors of the *Universal History*, when they say: "However, we cannot but think the essay of that philosopher who endeavoured to account for the formation of the world in a certain time from a rude matter, by the sole continuation of a motion once impressed, and reduced to a few simple and general laws; or of others, *who have since attempted the same, with more applause, from the original properties of matter, with which it was endued at its creation*, is so far from being criminal or injurious to God, as some have imagined, that it is rather giving a more sublime idea of His infinite wisdom."[1]

I have tried to remove the objections which seemed to threaten my positions from the side of religion. There are

---

[1] Part I. §88. [This quotation is taken from the Introduction to the large English History of the World entitled, *An Universal History from the Earliest Account of Time to the Present*: compiled from Original Authors; and illustrated with Maps, Cuts, Notes, Chronological and other Tables. London, 7 vols. 4°. 1736-44. Vol. I., p. 36c. The Introduction quoted from deals with the Cosmogony or Creation of the World, and it gives an account of all the principal theories ancient and modern, as then understood. The quotation as given in the text is from the original work, but the italics run after Kant's rendering. 'That philosopher' is Descartes; the 'others' are Burnet and Whiston.—WH]

others not less forcible that arise in regard to the subject itself. Thus it will be said, that although it is true that God has put a secret art into the forces of nature so as to enable it to fashion itself out of chaos into a perfect world system, yet will the intelligence of man, so weak as it is in dealing with the commonest objects, be capable of investigating the hidden properties contained in so great an object? Such an undertaking would be equivalent to saying: *'Give me matter only, and I will construct a world out of it.'* Does not the weakness of your insight, which comes to grief on the pettiest objects daily presented to your senses and in your neighbourhood, teach you that it is vain to try to discover the immeasurable, and what took place in nature before there was yet a world? I annihilate this objection by clearly showing that of all the inquiries which can be raised in connection with the study of nature, this is the one in which we can most easily and most certainly reach the ultimate. Just as among all the problems of natural science none can be solved with more correctness and certainty than that of the true constitution of the universe as a whole, the laws of its movements, and the inner mechanism of the revolutions of all the planets—that department of science in which the Newtonian philosophy can furnish such views as are nowhere else to be met with; so I assert, that among all the objects of nature whose first cause is investigated, the origin of the system of the world and the generation of the heavenly bodies, together with the causes of their motions, is that which we may hope to see first thoroughly understood. The reason of this is easy to see. The heavenly bodies are round masses, and therefore have the simplest formation which a body whose origin is sought can possibly have. Their movements are likewise uncomplicated; they are nothing but a free continuation of an impulse once impressed, which, by being combined with the attraction of the body at its centre, becomes circular. Moreover, the space in which they move is empty; the intervals which separate them from each other are exceptionally large, and everything is thus disposed not only for undisturbed motion but also for distinct observation of it in the clearest way. It seems to me that we can here say with intelligent certainty and without audacity: *'Give me matter, and I will construct a world out of it!'* i.e. give me matter and I will show you how a world shall arise out of it. For if we have matter existing endowed with an essential force of attraction, it is not difficult to determine those causes which may have contributed to the arrangement of the system of the world as a whole. We know what is required that a body shall take a spherical figure; and we understand what is required in order that spheres, as orbs moving freely, may assume a circular movement around the centre to which they are drawn. The position of their orbits, in relation to each other, agreement in the direction of their motions, the eccentricity of their paths, may all be referred to the simplest mechanical causes; and we may confidently hope to discover them because they can be reduced to the easiest and clearest principles. But can we boast of the same progress even regarding the lowest plant or an insect? Are we in a position to say: *'Give me matter, and I will show you how a caterpillar can be produced'*? Are we not arrested here at the first step, from ignorance of the real inner conditions of the object and the complication of the manifold constituents existing in it? It should not therefore cause astonishment if I presume to say that the formation of all the heavenly bodies, the cause of their movements, and, in short, the origin of the whole present constitution of the universe, will become intelligible before the production of a single herb or a caterpillar by mechanical causes, will become distinctly and completely understood.

These are the grounds on which I base my confidence that the physical part of universal science may hope in the future to reach the same perfection as that to which Newton has raised the mathematical half of it. Next to the laws by which the universe subsists in the constitution in which it is found, there are perhaps no other laws in the whole sphere of natural science that are capable of such exact mathematical determination as those in accordance with which the system arose. And undoubtedly the hand of a skilful mathematician may here reap no unfruitful fields.

Having thus taken occasion to recommend the subject here considered to a favourable reception, I may now be allowed to explain briefly the manner in which I have treated it. The

First Part deals with a new system of the universe generally. Mr Wright of Durham, whose treatise I have come to know from the Hamburg publication entitled the *Freie Urteile*, of 1751, first suggested ideas that led me to regard the fixed stars not as a mere swarm scattered without visible order, but as a system which has the greatest resemblance with that of the planets; so that just as the planets in their system are found very nearly in a common plane, the fixed stars are also related in their positions, as nearly as possible, to a certain plane which must be conceived as drawn through the whole heavens, and by their being very closely massed in it they present that streak of light which is called the Milky Way. I have become persuaded that because this zone, illuminated by innumerable suns, has very exactly the form of a great circle, our sun must be situated very near this great plane. In exploring the causes of this arrangement, I have found the view to be very probable that the so-called fixed stars may really be slowly moving, wandering stars of a higher order. In confirmation of what will be found regarding this thought in its proper place, I will here quote only a passage from a paper of Mr Bradley on the motion of the fixed stars:

> If a judgment may be formed with regard to this matter, from the result of the comparison of our best modern observations with such as were formerly made with any tolerable degree of exactness, there appears to have been a real change in the position of some of the fixed stars, with respect to each other, and such as seems independent of any motion in our own system, and can only be referred to some motion in the stars themselves. Arcturus affords a strong proof of this: for if its present declination be compared with its place as determined either by Tycho or Flamsteed, the difference will be found to be much greater than what can be suspected to arise from the uncertainty of their observations. It is reasonable to expect that other instances of the like kind must also occur among the great number of the visible stars, because their relative positions may be altered by various means. For if our own solar system be conceived to change its place with respect to absolute space, this might, in process of time, occasion an apparent change in the angular distances of the fixed stars; and in such a case, the places of the nearest stars being more affected than of those that are very remote, their relative positions might seem to alter,

though the stars themselves were really immovable. And on the other hand, if our own system be at rest and any of the stars really in motion, this might likewise vary their apparent positions; and the more so, the nearer they are to us, or the swifter their motions are, or the more proper the direction of the motion is to be rendered perceptible by us. Since then the relative places of the stars may be changed from such a variety of causes, considering that amazing distance at which it is certain that some of them are placed, it may require the observations of many ages to determine the laws of the apparent changes even of a single star. Much more difficult therefore must it be to settle the laws relating to all the most remarkable stars.[1]

I cannot exactly define the boundaries which lie between Mr Wright's system and my own; nor can I point out in what details I have merely imitated his sketch or have carried it out further. Nevertheless, I found afterwards valid reasons for considerably expanding it on one side. I considered the species of nebulous stars, of which De Maupertuis makes mention in his treatise *On the Figure of the Fixed Stars*,[2] which present the form of more or less open ellipses; and I easily persuaded myself that these stars can be nothing else than a mass of many fixed stars. The regular constant roundness of these figures, taught me that an inconceivably numerous host of stars must be here arranged together and grouped around a common centre, because their free positions towards each other would otherwise have presented irregular forms and not exact figures. I also perceived that they must be limited mainly to one plane in the system in which they are found united, because they do not exhibit circular but elliptical figures. And I further saw that, on account of their feeble light, they are removed to an inconceivable distance from us. What I have inferred from these analogies, is presented in the following treatise for the examination of the unprejudiced reader.

---

[1] [This passage is quoted from James Bradley's "A Letter to the Right Honorable George Earl of Macclesfield Concerning an Apparent Motion Observed in Some of the Fixed Stars," *Philosophical Transactions of the Royal Society*, 45 (1745), 39-40. MJC]

[2] [Omitted here is a long footnote in which Kant surveys writings on nebulae by Huygens, Halley, Derham, Maupertuis, and others. MJC]

> In the final four paragraphs of Kant's "Preface" he describes some topics he will treat in later sections of his book, including the origin of the solar system (presented in the second part of his book) and (third part) the idea of extraterrestrial intelligent life. He also comments on the degree of credibility he attributes to the admittedly speculative ideas presented in the various sections of his book. In particular, he states: "Speaking generally, the greatest mathematical precision and mathematical infallibility can never be required from a treatise of this kind. If the system is founded on analogies and harmonies, which are in accordance with the rules of credibility and correct reasoning, it has satisfied all the requirements of its subject."

FIRST PART.
OF THE SYSTEMATIC CONSTITUTION AMONG THE FIXED STARS

The scientific theory of the Universal Constitution of the World has obtained no remarkable addition since the time of Huygens. At the present time nothing more is known than what was already known then, namely, that six planets with ten satellites, all performing the circle of their revolution almost in one plane, and the eternal comets which sweep out on all sides, constitute a system whose centre is the sun, towards which they all fall, around which they perform their movements, and by which they are all illuminated, heated, and vivified; finally, that the fixed stars are so many suns, centres of similar systems, in which everything may be arranged just as grandly and with as much order as in our system; and that the infinite space swarms with worlds, whose number and excellency have a relation to the immensity of their Creator.

The systematic arrangement which was found in the combination of the planets which move around their sun, seemed in the view of astronomers of that time to disappear in the multitude of the fixed stars; and it appeared as if the regulated relation which is found in the smaller solar system, did not rule among the members of the universe as a whole. The fixed stars exhibited no law by which their positions were bounded in relation to each other; and they were looked upon as filling all the heavens and the heaven of heavens without order and without intention. Since the curiosity of man set these limits to itself, he has done nothing further than from these facts to infer, and to admire, the greatness of Him who has revealed Himself in works so inconceivably great.

It was reserved for an Englishman, Mr Wright of Durham, to make a happy step with a remark which does not seem to have been used by himself for any very important purpose, and the useful application of which he has not sufficiently observed. He regarded the Fixed Stars not as a mere swarm scattered without order and without design, but found a systematic constitution in the whole universe and a universal relation of these stars to the ground-plan of the regions of space which they occupy. We would attempt to improve the thought which he thus indicated, and to give to it that modification by which it may become fruitful in important consequences whose complete verification is reserved for future times.

Whoever turns his eye to the starry heavens on a clear night, will perceive that streak or band of light which on account of the multitude of stars that are accumulated there more than elsewhere, and by their getting perceptibly lost in the great distance, presents a uniform light which has been designated by the name *Milky Way*. It is astonishing that the observers of the heavens have not long since been moved by the character of this perceptibly distinctive zone in the heavens, to deduce from it special determinations regarding the position and distribution of the fixed stars. For it is seen to occupy the direction of a great circle, and to pass in uninterrupted connection round the whole heavens: two conditions which imply such a precise destination and present marks so perceptibly different from the indefiniteness of chance, that attentive astronomers ought to have been thereby led, as a matter of course, to seek carefully for the explanation of such a phenomenon.

As the stars are not placed on the apparent hollow sphere of the heavens, and as some are more distant than others from our point of view and are lost in the depths of the heavens, it follows from this, that at the distances at which they are situated away from us, one behind the other, they are not indifferently scattered on all sides, but must have a predominant relation to a certain plane which passes through our point of view and to which they are arranged so as to be found as near it as possible. This relation is such an undoubted phenomenon that even the other stars which are not included in the whitish streak, are yet seen to be more accumulated and closer the nearer their places are to the circle of the Milky Way; so that of the two thousand stars which are perceived by the naked eye, the greatest part of them are found in a not very broad zone whose centre is occupied by the Milky Way.

If we now imagine a plane drawn through the starry heavens and produced indefinitely, and suppose that all the fixed stars and systems have a general relation in their places to this plane so as to be found nearer to it than to other regions, then the eye which is situated in this plane when it looks out to the field of the stars, will perceive on the spherical concavity of the firmament the densest accumulation of stars in the direction of such a plane under the form of a zone illuminated by varied light. This streak of light will advance as a luminous band in the direction of a great circle, because the position of the spectator is in the plane itself. This zone will swarm with stars which, on account of the indistinguishable minuteness of their clear points that cannot be severally discerned and their apparent denseness, will present a uniformly whitish glimmer,—in a word, a Milky Way. The rest of the heavenly host whose relation to the plane described gradually diminishes, or which are situated nearer the position of the spectator, are more scattered, although they are seen to be massed relatively to this same plane. Finally, it follows from all this that our solar world, seeing that this system of the fixed stars is seen from it in the direction of a great circle, is situated in the same great plane and constitutes a system along with the other stars.

In order to penetrate better into the nature of the universal connection which rules in the universe, we will try to discover the cause that has made the positions of the fixed stars come to be in relation to a common plane.

The sun is not limited in the range of its attractive force to the narrow domain of the planetary system. According to all appearance the force of attraction extends *ad infinitum*. The comets, which pass very far beyond the orbit of Saturn, are compelled by the attraction of the sun to return again, and to move in orbits. Although, therefore, it is more in accordance with the nature of a force which seems to be incorporated in the essence of matter, to be unlimited, and it is also actually recognized to be so by those who accept Newton's principles, yet we would only have it granted that this attraction of the sun extends approximatively to the nearest fixed star; also that the fixed stars, as being so many suns, exercise an action around them in a similar range; and, consequently, that the whole host of them are striving to approach each other through their mutual attraction. Thus all the systems of the universe are found so constituted by their mutual approach, which is incessant and is hindered by nothing, that they will fall together sooner or later into one mass, unless this ruin is prevented by the action of the centrifugal forces, as in the globes of our planetary system. These forces, by deflecting the heavenly bodies from falling in a straight line, bring about, when combined with the forces of attraction, their perpetual revolutions, and thereby the structure of the creation is secured from destruction and is adapted for an endless duration.

Thus all the suns of the firmament have movements of revolution, either round one universal centre or round many centres. But we may here apply the analogy which is observed in the revolutions of our Solar System, namely, that the same cause that has communicated to the planets the centrifugal force, in virtue of which they perform their revolutions, has also so directed their orbits that they are all related to one plane. And so also the cause, whatever it may be, which has given the power of revolving to the suns of the upper world, as so many wandering stars of a higher order of worlds, has likewise brought their orbits as much as possible into one plane, and has striven to limit the deviations from it.

Following out this idea, the System of the Fixed Stars may be viewed as represented in some measure by the System of the Planets, if the latter be regarded as indefinitely enlarged. If instead of the six planets with their ten companions, we suppose there to be as many thousands of them; if the twenty-eight or thirty comets which have been observed be multiplied a hundred or a thousand times, and if we suppose these same bodies to be self-luminous, then the eye of the spectator looking at them from the earth would see before it just the very appearance of what is presented to it by the fixed stars of the Milky Way. For these supposed planets, by their proximity to the common plane of their relation, being situated with our earth in the same plane, would present to us a zone densely illuminated by innumerable stars, and its direction would be that of a great circle of the celestial sphere. This streak of light would be seen everywhere thickly sown with stars, although, according to the hypothesis, they are wandering stars and consequently not fixed to one place; for by their displacement there would always be found stars enough on one side, although others had changed from that position.

The breadth of this luminous zone, which represents a sort of zodiac, will be determined by the different degrees of the deviation of the said wandering stars from the plane of their relation, and by the inclination of their orbits to this plane; and as most of them are near this plane, their number will appear more scattered according to the degree of their distance from it. But the comets which occupy all the regions of space without distinction, will cover the field of the heavens on both sides.

The form of the starry heavens is therefore due to no other cause than such a systematic constitution on the great scale as our planetary world has on the small—all the suns constituting a system whose universal relative plane is the Milky Way. Those suns which are least closely related to this plane, will be seen at the side of it; but on that account they are less accumulated, and are much more scattered and fewer in number. They are, so to speak, the comets among the suns.

This new theory attributes to the suns an advancing movement; yet everybody regards them as unmoved, and as having been fixed from the beginning to their places. The designation of them as 'Fixed' Stars, which they have received from that view of them, seems to be established and put beyond doubt by the observation of all the centuries. This difficulty raises an objection which would annihilate the theory here advanced, were it well founded. But in all probability this want of movement is merely apparent. It is either only excessively slow, arising from the great distance from the common centre around which the stars revolve, or it is due to mere imperceptibility, owing to their distance from the place of observation. Let us estimate the probability of this conception by calculating the motion which a fixed star near our sun would have, if we supposed our sun to be the centre of its orbit. If, following Huygens, its distance is assumed to be more than 21,000 times greater than the distance of the sun from the earth, then according to the established law of the periods of revolution—which are in the ratio of the square root of the cube of the distances from the centre—the time which such a star would require to revolve once round the sun would be more than a million and a half years; and this would only produce a displacement of its position by one degree in four thousand years. Now, as there are perhaps very few fixed stars so near the sun as Huygens has conjectured Sirius to be, as the distance of the rest of the heavenly host perhaps immensely exceeds that of Sirius, and would therefore require incomparably longer time for such a periodic revolution; and, moreover, as it is very probable that the movement of the suns of the starry heavens goes round a common centre, whose distance is incomparably great, so that the progression of the stars may therefore be exceedingly slow, it may be inferred with probability from this, that all the time that has passed since men began to make observations on the heavens is perhaps not yet sufficient to make perceptible the alteration which has been produced in their positions. We need not, however, give up the hope yet of discovering even this alteration, in the course of time. Subtle and careful observers, and a comparison of observations taken at a great distance from each other, will be required for it. These observations must be directed especially to the stars of the Milky Way,[1] which is the main plane of all motion. Mr Bradley has observed almost imperceptible displacements of the stars. The ancients have noticed stars in certain places of

---

[1]Likewise to those clusters of stars, of which there are many found together in a small space, as, for example, the Pleiades, which perhaps make up a small system by themselves in the greater system.

the heavens, and we see new ones in others. Who knows but that these are just the former, which have only changed their place? The excellence of our instruments and the perfection of astronomy, give us well-founded hopes for the discovery of such peculiar and remarkable things.[1] The credibility of the fact itself, in accordance with the principles of nature and analogy which well support this hope, is such that they may stimulate the attention of the explorer of nature so as to bring about its realization.

The Milky Way is, so to speak, the zodiac of new stars which alternately show themselves and disappear, almost only in that region of the heavens. If this alteration of their visibility arises from their periodical removal and approach to us, then it appears from the systematic constitution of the stars here indicated, that such a phenomenon must be seen for the most part only in the region of the Milky Way. For as there are stars which revolve in very elongated orbits around other fixed stars like satellites around their planets, analogy with our planetary world, in which only those heavenly bodies that are near the common plane of the movements have companions revolving round them, demands that those stars only which are in the Milky Way will also have suns revolving round them.

I come now to that part of my theory which gives it its greatest charm, by the sublime ideas which it presents of the plan of the creation. The train of thought which has led me to it is short and natural; it consists of the following ideas. If a system of fixed stars which are related in their positions to a common plane, as we have delineated the Milky Way to be, be so far removed from us that the individual stars of which it consists are no longer sensibly distinguishable even by the telescope; if its distance has the same ratio to the distance of the stars of the Milky Way as that of the latter has to the distance of the sun; in short, if such a world of fixed stars is beheld at such an immense distance from the eye of the spectator situated outside of it, then this world will appear

---

[1] De la Hire, in the *Mémoires* of the Paris Academy of the year 1693, remarks that from his own observations, as well as from comparison of them with those of Ricciolus, he has perceived a marked alteration in the positions of the stars of the Pleiades.

under a small angle as a patch of space whose figure will be circular if its plane is presented directly to the eye, and elliptical if it is seen from the side or obliquely. The feebleness of its light, its figure, and the apparent size of its diameter will clearly distinguish such a phenomenon when it is presented, from all the stars that are seen single.

We do not need to look long for this phenomenon among the observations of the astronomers. It has been distinctly perceived by different observers. They have been astonished at its strangeness; and it has given occasion for conjectures, sometimes to strange hypotheses, and at other times to probable conceptions which, however, were just as groundless as the former. It is the 'nebulous' stars which we refer to, or rather a species of them, which M. de Maupertuis thus describes: 'They are,' he says, 'small luminous patches, only a little more brilliant than the dark background of the heavens; they are presented in all quarters; they present the figure of ellipses more or less open; and their light is much feebler than that of any object we can perceive in the heaven.'[1]

The author of the *Astro-Theology* imagined that they were openings in the firmament through which he believed he saw the Empyrean.[2] A philosopher of more enlightened views, M. De Maupertuis, already referred to, in view of their figure and perceptible diameter, holds them to be heavenly bodies of astonishing magnitude which, on account of their great flattening, caused by the rotatory impulse, present elliptical forms when seen obliquely.

Any one will be easily convinced that this latter explanation is likewise untenable. As these nebulous stars must undoubtedly be removed at least as far from us as the other fixed stars, it is not only their magnitude which would be astonishing—seeing that it would necessarily exceed that of the largest stars many thousand times—but it would be strangest of all that, being self-luminous bodies and suns, they should still with this extraordinary magnitude show the

---

[1]*Discours sur la Figure des Astres.* Paris, 1742.
[2][*Astro-Theology, or a Demonstration of the Being and Attributes of God from a Survey of the Heavens*, by W. Derham. London, 1714.]

dullest and feeblest light.

It is far more natural and conceivable to regard them as being not such enormous single stars but systems of many stars, whose distance presents them in such a narrow space that the light which is individually imperceptible from each of them, reaches us, on account of their immense multitude, in a uniform pale glimmer. Their analogy with the stellar system in which we find ourselves, their shape, which is just what it ought to be according to our theory, the feebleness of their light which demands a presupposed infinite distance: all this is in perfect harmony with the view that these elliptical figures are just universes and, so to speak, Milky Ways, like those whose constitution we have just unfolded. And if conjectures, with which analogy and observation perfectly agree in supporting each other, have the same value as formal proofs, then the certainty of these systems must be regarded as established.

The attention of the observers of the heavens, has thus motives enough for occupying itself with this subject. The fixed stars, as we know, are all related to a common plane and thereby form a co-ordinated whole, which is a World of worlds. We see that at immense distances there are more of such star-systems, and that the creation in all the infinite extent of its vastness is everywhere systematic and related in all its members.

It might further be conjectured that these higher universes are not without relation to one another, and that by this mutual relationship they constitute again a still more immense system. In fact, we see that the elliptical figures of these species of nebulous stars, as represented by M. de Maupertuis, have a very near relation to the plane of the Milky Way. Here a wide field is open for discovery, for which observation must give the key. The Nebulous Stars, properly so called, and those about which there is still dispute as to whether they should be so designated, must be examined and tested under the guidance of this theory. When the parts of nature are considered according to their design and a discovered plan, there emerge certain properties in it which are otherwise overlooked and which remain concealed when observation is scattered without guidance over all sorts of objects.

The theory which we have expounded opens up to us a view into the infinite field of creation, and furnishes an idea of the work of God which is in accordance with the infinity of the great Builder of the universe. If the grandeur of a planetary world in which the earth, as a grain of sand, is scarcely perceived, fills the understanding with wonder; with what astonishment are we transported when we behold the infinite multitude of worlds and systems which fill the extension of the Milky Way! But how is this astonishment increased, when we become aware of the fact that all these immense orders of star-worlds again form but one of a number whose termination we do not know, and which perhaps, like the former, is a system inconceivably vast—and yet again but one member in a new combination of numbers! We see the first members of a progressive relationship of worlds and systems, and the first part of this infinite progression enables us already to recognize what must be conjectured of the whole. There is here no end but an abyss of a real immensity, in presence of which all the capability of human conception sinks exhausted, although it is supported by the aid of the science of number. The Wisdom, the Goodness, the Power which have been revealed is infinite; and in the very same proportion are they fruitful and active. The plan of their revelation must therefore, like themselves, be infinite and without bounds.

☆ ☆

Commentary on the Reading from Kant

This selection amply illustrates the boldness of the young Kant's brilliance. Of special note are Kant's bold claims (1) that the Milky Way consists of a vast circular planar array of stars, i.e., the disk theory of the Milky Way, and (2) that the nebulae scattered over the heavens are other Milky Ways, that is, that they consist of vast numbers of stars arranged in an order comparable to that which he attributed to the stars of our Milky Way. The latter idea eventually came to be called the "island universe theory." Surprising aspects of the history of these ideas also emerge. For example, it is evident that Kant believed he had derived the disk theory from Thomas Wright, whose book he knew only from a review of it. In fact, we have seen that Wright had no such theory, although he did claim that the Milky Way can be explained as a perspective effect of our location in an extended planar array of stars. Historically surprising aspects of Kant's second claim are also evident. For example, one source of his evidence for this claim consisted in the discussions of nebulae presented by Huygens, Derham, and Maupertuis. What had struck Kant was the elliptical form attributed to some nebulae. This led him to suggest that nebulae are giant circular disks of stars, which appear elliptical if seen at an angle. The surprising aspect of this is that Derham, in his list of 21 nebulae, described only 5 nebulae as elliptical and (as is now known) only one of these is elliptical. Thus Kant's presentation of the island universe theory rested on very frail observational grounds.[1]

The selection from Kant also contains interesting theological claims. Kant's numerous references to the Epicurean philosophy in the early sections make clear Kant's enthusiasm for that philosophy, especially as expounded by Epicurus's disciple Lucretius. Two central Epicurean doctrines were (1) that the proper mode of explaining natural phenomena, including their development, is by attributing all phenomena to matter in motion, i.e., a form of materialism, and (2) that explanations of events or of aspects of nature in terms of Divine providence are unacceptable. These claims led to the Epicureans being accused of materialism and atheism. The boldness of Kant's approach is evident in the fact that Kant basically endorses the first Epicurean claim, but then moves from it to propose a proof for God's existence. In doing this he challenges earlier authors who attacked Epicurean thought, urging that the goodness and harmony of nature provide evidence of God's providential care. Kant's assertion is that it is misconceived to claim that God directly brings us, for example, cool breezes. Rather what God has done is to subject matter to various laws, which laws, *acting together*, produce the beneficial and beautiful aspects of nature. It is precisely in the fact that the laws work harmoniously with each other that the efforts and solicitude of the Almighty are to be seen.

---

[1]K. G. Jones, "The Observational Basis for Kant's Cosmogony: A Critical Analysis," *Journal for the History of Astronomy*, 2 (1971), 29–34.

Chapter Three

Sir William Herschel: Celestial Naturalist

Sir William Herschel
(Portrait by J. Russell)

Introduction

Sir William Herschel (1738–1822) is usually regarded as the most important astronomer of the period between Newton and the twentieth century. Although he made many contributions to traditional areas of astronomy, he is above all remembered as the foremost pioneer of stellar astronomy.

Nonetheless, the portrayals of Herschel that appear in various histories of astronomy tend to be substantially different from each other, raising questions about the precise nature of his contribution. These questions will be addressed shortly, but are best considered after a discussion of the remarkable manner in which he became an astronomer and a survey of his major contributions to celestial science.

Life

William Herschel was born in 1738 in Hanover, Germany, the son of a musician in the Hanoverian Guard. After elementary schooling, he joined his father's regiment in 1753, serving as an oboist. When his regiment visited England in 1757, Herschel began to learn English. On this journey, he purchased a copy of John Locke's *Essay Concerning Human Understanding*, which indicates the penchant for philosophical discourse that he had acquired from his father. In 1757, Herschel resigned from the military and took up residence in England, where he pursued a career as a music teacher and performer, at first in the north of England. In 1766, he secured a position as organist for the fashionable Octagon Chapel in Bath, where he continued to offer instruction in music and to give performances, sometimes of his own compositions. Herschel's achievements in music were not insignificant; for example, they include composing 24 symphonies, 7 violin concertos, 2 organ concertos, and a variety of other works.

Herschel's interests, however, began to turn to astronomy. In 1766, he made the first entry in his astronomical diary and in 1773 purchased an exposition of astronomy written by James Ferguson. A year earlier, one of his sisters, Caroline Lucretia Herschel (1750–1848), joined him in Bath, serving as his assistant in his musical and eventually in his astronomical

Caroline Herschel

endeavors. Herschel not only studied astronomy but also began building telescopes. In 1776, having already constructed various smaller reflectors, he completed a 12-inch-aperture, 20-foot-focal-length reflecting telescope. A few years later, he attempted to construct a 36-inch-aperture reflector, but the project failed. Using his various telescopes, he began to make what he called "sweeps" of the heavens, observing at first every celestial object down to the fourth magnitude. Then, with characteristic thoroughness, he proceeded to dimmer objects.

His abundant energy and outgoing nature led him to share some of his ideas and observations, as he had earlier shared his music, with his contemporaries. In 1780, he presented a paper to the Bath Philosophical Society "On the Utility of Speculative Inquiries" and, in a more scientific vein, published a paper in the Royal Society's *Transactions* entitled "Astronomical Observations Relating to the Mountains of the Moon." He continued to support himself by his musical endeavors; in fact, at this stage he can only be described as an amateur in astronomy. This is suggested by the passion that then possessed him and that is evident in his lunar observation books for discovering life on the moon.

William Herschel's career was transformed by an event that occurred in 1781 and that resulted in his acquiring an international reputation. On 13 March 1781, while sweeping the heavens, he noticed an object that in time proved to be a previously unknown planet, the planet Uranus. Some years later, the poet John Keats celebrated this discovery by comparing his own excitement at reading a new translation of Homer to Herschel's delight in his discovery by writing:

> Then felt I like some watcher of the skies,
> When a new planet swims into his ken.

Keats's lines seriously misrepresent what Herschel had done. Rather than being an affair of a moment and a chance discovery, Herschel's detection of Uranus took many months, as is suggested by the title, "Account of a Comet," of the paper in which he first announced his discovery. In fact, Herschel for many weeks not only viewed the object as a comet, but also made telescopic measurements of its diameter, which measurements indicated that in a period of about six weeks, the object had doubled in diameter (we now know the observed diameter of Uranus was decreasing slightly during this period). Moreover, his discovery cannot appropriately be described as accidental. On at least twenty occasions, astronomers had recorded the position of Uranus, but without recognizing its planetary nature. The special gift that Herschel possessed for such observation is also indicated by a statement he himself made in a 1782 letter:

> Seeing is in some respects an art that must be learnt. To make a person see with such a power is nearly the same as if I were asked to make him play one of Handel's fugues upon the organ. Many a night have I been practicing to see, and it would be strange if one did not acquire a certain dexterity by such constant practice.[1]

The discovery of Uranus, for which Herschel proposed the name "Georgium Sidus" (George's star) in honor of King George III, led that monarch to offer funding for Herschel's researches, which in turn freed him from his musical responsibilities. As the King's astronomer, Herschel moved from Bath to Slough so as to be near Windsor Castle.

In 1783, Herschel's passion for telescope construction led him to complete an 18.7-inch-aperture, 20-foot-focal-length reflector. In the same year, he published a paper (to be discussed shortly) in which he presented evidence that the sun is moving in the direction of the star lambda Hercules. Major papers (treated subsequently) on the overall structure of the heavens followed in 1784 and 1785. One class of objects that interested Herschel, if few of his contemporaries, consisted of the nebulae, of which about a hundred were known in the mid-1780s. Herschel's interest in these objects, empowered by the large reflecting telescopes he had constructed, produced rich results: in 1786, he published a catalogue of 1000 such objects, most of which he had been the first to discover. His sister Caroline was also active; in the same year, she discovered the

---

[1] Sir William Herschel, *The Scientific Papers of Sir William Herschel*, ed. by J. L. E. Dreyer, vol. 1 (London, 1912), p. xxxiii.

first of the eight comets credited to her. Nor did Herschel neglect the objects of our planetary system: in 1787, he discovered two satellites of Uranus and in 1789 two new Saturnian moons. In 1798, he reported his detection of four additional Uranian satellites, but this report has proven to be spurious.

The most magnificent of Herschel's giant telescopes was completed in 1788 with the aid of extensive funding from the crown. This was Herschel's 48-inch-aperture, 40-foot-focal-length reflector, which remained far and away the largest telescope ever made until the 1840s, when Lord Rosse's Leviathan of Parsonstown surpassed it. It was also in 1788 that Herschel married a wealthy widow, Mary Pitt, their union leading Herschel to enjoy financial security and the birth in 1792 of a son, John Herschel, who eventually distinguished himself not only in astronomy but other areas of science as well.

In 1789, Herschel published a catalogue giving data on a second 1000 nebulae, with a third catalogue of 500 nebulae appearing in 1802. In the previously mentioned papers by Herschel on the construction of the heavens, he had explored the hypothesis that the nebulae are island universes, i.e., extremely remote structures comparable in size and form to our own Milky Way galaxy. Evidence that Herschel uncovered during this period led him in a 1791 paper to back away from this hypothesis and, as will be seen shortly, to advocate another view that involved the notion that many nebulae, rather than consisting of thousand of stars, are composed of a "shining fluid."

In the remaining years of his life, Herschel continued to observe the heavens and also to publish speculations, some of which are now central doctrines in astronomy, and others, such as his belief that the sun is habitable, that have not stood the test of time. As knowledge of his achievements spread, numerous honors came to him. For example, in 1816, he was knighted and in 1821 elected the first president of England's Astronomical Society (now Royal Astronomical Society). Active almost to the last, he died in 1822, his sister Caroline surviving until 1848.

Herschelian Historiography: Some Questions

Although some scholars might question the characterization of William Herschel as the most important astronomer of the eighteenth and nineteenth centuries, few would dispute the magnitude of his contributions or his status as pioneer of stellar astronomy, an area of astronomy that scarcely existed before the 1780s. Yet questions arise concerning the nature of his contributions. For example, it is important to ask: should Herschel be viewed primarily as an observational or as a theoretical astronomer? The remarkable numbers of nebulae, double stars, and even solar system objects discovered by Herschel with his giant telescopes have at times led to the characterization of him as above all an observer of the heavens. And his spectacular achievements in telescope construction reinforce this interpretation. Nonetheless, his writings also contain numerous speculations. Regarding the latter, the astronomer Arthur Stanley Eddington commented that "It cannot be denied that [Herschel] was given to jumping to conclusions in a way which, when it comes off, we describe as profound insight, and when it does not come off, we call wild-cat speculation."[1] An allied issue is the nature of Herschel's methodology. Should he be characterized as primarily an empiricist or did speculation play a central role in his creative work?[2] It is true that unlike such astronomers as his contemporary Laplace who employed sophisticated mathematics in their researches, the autodidact Herschel made relatively little use of higher mathematics. Another important question to keep in mind while reading the selections from Herschel that appear in this chapter is: how did Herschel's views change over the four decades he devoted to astronomical research? Finally, it may be asked: is the characterization of Herschel suggested in the subtitle of this chapter as a "Celestial Naturalist" appropriate? Implied by

---

[1] As quoted in Michael Hoskin, *William Herschel and the Construction of the Heavens* (London: Oldbourne, 1963), p. 15.

[2] I have taken a stance on this issue in M. J. Crowe, *The Extraterrestrial Life Debate 1750–1900: The Idea of a Plurality of Worlds from Kant to Lowell* (Cambridge: Cambridge Univ. Press, 1988), pp. 61–70.

that subtitle is that Herschel may most justifiably be seen as having employed an approach that shared more similarities with that practiced by such contemporary naturalists as the famous geologist James Hutton, who formed various hypotheses about the evolution and stratifications of the earth's surface, than with the scientific methods used by Laplace and other mathematical astronomers of the Newtonian tradition.

Synopsis of Herschel's 1783 Paper "On the Proper Motion of the Sun and Solar System..."

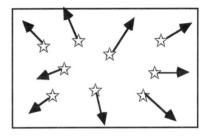

The proper motion of stars had been detected in 1718 by Edmond Halley. In 1783, Herschel published a paper in which from an analysis of about a dozen already known proper motions of stars, he claimed that the proper motions of these stars, rather than being random, reveal a common component, which he attributed to the motion of the sun itself. In particular, he hypothesized that the sun is moving in the direction of the star λ Hercules. To obtain an idea of his approach in this paper, examine the accompanying diagram, designed to portray the proper motions of various stars. Each vector represents the direction and magnitude of the proper motion of the star located at its base. Can you form a well-grounded estimate of the direction of the solar motion in relation to these stars? Would it make more sense to assume that the sun is moving toward or away from this collection of stars? What help do you suppose Herschel's giant telescopes provided him in his research for this paper?

Herschel's 1784 Paper on "... the Construction of the Heavens"

In December 1781, Herschel received a copy of a list of 68 nebulae compiled by Charles Messier and published in 1780 in an almanac *(Connoissance des temps)* for the year 1783. In the 1784 version of the almanac (published in 1781), Messier expanded his list of nebulae to 103 objects, Herschel eventually receiving a copy of this as well. Herschel's first recorded observation of a nebula, the Orion nebula, is dated 1774. When he compared his observation of Orion with previous drawings of the nebula, he concluded that it had changed in form during the intervening period. This was significant because, were an observable change to occur in a nebula, it would indicate that the object cannot consist of a vast number of stars and be located at such a great distance as to make the stars very dim. Rather the object must be located at a relatively moderate distance, have a correspondingly modest size, and not be composed of stars. Herschel observed the Orion nebula at various times during the 1770s, but as of the time when he received Messier's 1780 list, he had recorded observations in his journals of a total of only four nebulae. Despite the fact that he indicates in his paper that immediately after December, 1781, he began observing nebulae, his records show that he observed no nebulae from then until August 1782 and began looking for them in a systematic fashion only after March, 1783.

Herschel's 18.7-inch-Aperture, 20-foot-Focal-Length Reflector

On 23 October 1783, Herschel was able to begin observing with his 18.7-inch-aperture reflector. It was with this instrument that he achieved his first important successes in locating these faint objects. One of the questions that was foremost in his mind when observing the nebulae was whether they can be resolved into individual stars or whether they consist of some sort of unresolvable material. If the former were true, then it would follow—at least if one accepts Herschel's assumption that all stars are of approximately equal brightness—that the nebulae must be very distant. This in turn would support Kant's hypothesis that nebulae are island universes, comparable in structure to our Milky Way.

☆ ☆

William Herschel,
"Account of Some Observations Tending to Investigate the Construction of the Heavens"

[Read June 17, 1784 and published originally in the *Philosophical Transactions of the Royal Society*, 74 (1784), 437–451 and reprinted here from The *Scientific Papers of William Herschel*, ed. by J. L. E. Dreyer, vol. 1 (London, 1912), pp. 157–66.]

In a former paper I mentioned, that a more powerful instrument was preparing for continuing my reviews of the heavens. The telescope I have lately completed, though far inferior in size to the one I had undertaken to construct when that paper was written, is of the Newtonian form, the object speculum being of 20 feet focal length, and its aperture $18\frac{7}{10}$ inches. The apparatus on which it is mounted is contrived so as at present to confine the instrument to a meridional situation, and by its motions to give the right-ascension and declination of a celestial object in a coarse way; which, however, is sufficiently accurate to point out the place of the object, so that it may be found again. It will not be necessary to enter into a more particular description of the apparatus, since the account I have now the honour of communicating to the Royal Society regards rather the performance of the telescope than its construction.

It would, perhaps, have been more eligible to have waited longer, in order to complete the discoveries that seem to lie within the reach of this instrument, and are already, in some respects, pointed out to me by it. By taking more time I should undoubtedly be enabled to speak more confidently of the *interior construction* of the heavens, and its various *nebulous and sidereal strata* (to borrow a term from the natural historian) of which this paper can as yet only give a few outlines, or rather hints. As an apology, however, for this prematurity, it may be said, that the end of all discoveries being communication, we can never be too ready in giving facts and observations, whatever we may be in reasoning upon them.

Hitherto the sidereal heavens have, not inadequately for the purpose designed, been represented by the concave surface of a sphere, in the center of which the eye of an observer might be supposed to be placed. It is true, the various magnitudes of the fixed stars even then plainly suggested to us, and would have better suited the idea of an expanded firmament of three dimensions; but the observations upon which I am now going to enter still farther illustrate and enforce the necessity of considering the heavens in this point of view. In future, therefore, we shall look upon those regions into which we may now penetrate by means of such large telescopes, as a naturalist regards a rich extent of ground or chain of mountains, containing strata variously inclined and directed, as well as consisting of very different materials. A surface of a globe or map, therefore, will but ill delineate the interior parts of the heavens.

It may well be expected, that the great advantage of a large aperture would be most sensibly perceived with all those objects that require much light, such as the very small and immensely distant fixed stars, the very faint nebulae, the close and compressed clusters of stars, and the remote planets.

On applying the telescope to a part of the *via lactea* [Milky Way], I found that it completely resolved the whole whitish appearance into small stars, which my former telescopes had not light enough to effect. The portion of this extensive tract, which it has hitherto been convenient for me to observe, is that immediately about the hand and club of Orion. The glorious multitude of stars of all possible sizes that presented themselves here to my view was truly astonishing; but, as the dazzling brightness of glittering stars may easily mislead us so far as to estimate their number greater than it really is, I endeavoured to ascertain this point by counting many fields, and computing, from a mean of them, what a certain given portion of the milky way might contain. Among many trials of this sort I found, last January the 18th, that six fields, promiscuously taken, contained 110, 60, 70, 90, 70, and 74 stars each. I then tried to pick out the most vacant place that was to be found in that neighbourhood, and counted 63 stars. A mean of the first six gives 79 stars for each field. Hence, by allowing 15 minutes of a great circle for the diameter of my field of view, we gather, that a belt of 15 degrees long and two broad, or the quantity which I have often seen pass through the field of my telescope in one hour's time, could not well contain less than fifty thousand stars, that were large enough to be distinctly numbered. But, besides these, I suspected at least twice as many more, which, for want of light, I could only see now and then by faint glittering and interrupted glimpses.

The excellent collection of nebulae and clusters of stars which has lately been given in the *Connoissance des Temps* for 1783 and 1784, leads me next to a subject which, indeed, must open a new view of the heavens. As soon as the first of these volumes came to my hands, I applied my former 20-feet reflector of 12 inches aperture to them; and saw, with the greatest pleasure, that most of the nebulae, which I had an opportunity of examining in proper situations, yielded to the force of my light and power, and were resolved into stars. For instance, the 2d, 5, 9, 10, 12, 13, 14, 15, 16, 19, 22, 24, 28, 30, 31, 37, 51, 52, 53, 55, 56, 62, 65, 66, 67, 71, 72, 74, 92, all which are said to be nebulae without stars, have either plainly appeared to be nothing but stars, or at least to contain stars, and to shew every other indication of consisting of them entirely.[1] I have examined them with a careful scrutiny of various powers and light, and generally in the meridian. I should mention, that five of the above, *viz.* the 16th, 24, 37, 52, 67, are called clusters of stars containing nebulosity; but my instrument resolving also that portion of them which is called nebulous into stars of a much smaller size, I have placed them into the above number. To these may be added the 1st, 3d, 27, 33, 57, 79, 81, 82, 101,

---

[1][It is now known that Messier objects 31, 51, 65, 66, and 74 could not have been resolved by Herschel. MJC]

which in my 7, 10, and 20-feet reflectors shewed a mottled kind of nebulosity, which I shall call resolvable;[1] so that I expect my present telescope will, perhaps, render the stars visible of which I suppose them to be composed. Here I might point out many precautions necessary to be taken with the very best instruments, in order to succeed in the resolution of the most difficult of them; but reserving this at present too extensive subject for a future opportunity, I proceed to speak of the effects of my last instrument with regard to nebulae.

My present pursuits, as I observed before, requiring this telescope to act as a fixed instrument, I found it not convenient to apply it to any other of the nebulae in the *Connoissance des Temps* but such as came in turn; nor, indeed, was it necessary to take any particular pains to look for them, it being utterly impossible that any one of them should escape my observation when it passed the field of view of my telescope. The few which I have already had an opportunity of examining, shew plainly that those most excellent French astronomers, Mess. MESSIER and MECHAIN, saw only the more luminous part of their nebulae; the feeble shape of the remainder, for want of light, escaping their notice. The difference will appear when we compare my observation of the 98th nebula with that in the *Connoissance des Temps* for 1784, which runs thus: "Nébuleuse sans étoile, d'une lumière extremement foible, au dessus de l'aile boréale de la Vierge, sur le parallèle et près de l'étoile N° 6, cinquième grandeur, de la chevelure de Bérénice, suivant FLAMSTEED. M. MECHAIN la vit le 15 Mars, 1781."[2] My observation of the 30th of December, 1783, is thus: A large, extended, fine nebula. Its situation shews it to be M. MESSIER'S 98th; but from the description it appears, that that gentleman has not seen the whole of it, for its feeble branches extend above a quarter of a degree, of which no notice is taken. Near

---

[1][It is now known that among these nine objects that Herschel called resolvable, only Messier objects 3 and 79 could have been resolved by him. MJC]

[2]["Nebula without star, of an extremely weak light, above the northern wing of Virgo, on the parallel of and near the star N° 6, fifth magnitude, of Coma Berenices, after FLAMSTEED. M. MECHAIN saw it on March 15, 1781." MJC]

the middle of it are a few stars visible, and more suspected.[1] My field of view will not quite take in the whole nebula. See fig. I, Plate I. Again, N° 53, "Nébuleuse sans étoiles, decouverte au-dessous et près de la chevelure de Bérénice, à peu de distance de l'étoile quarante-deuxieme de cette constellation, suivant FLAMSTEED. Cette nébuleuse est ronde et apparente, &c."[2] My observation of the 170th Sweep runs thus: A cluster of very close stars; one of the most beautiful objects I remember to have seen in the heavens. The cluster appears under the form of a solid ball, consisting of small stars, quite compressed into one blaze of light, with a great number of loose ones surrounding it, and distinctly visible in the general mass. See fig. 2.

When I began my present series of observations, I surmised, that several nebulae might yet remain undiscovered,[3] for want of sufficient light to detect them; and was, therefore, in hopes of making a valuable addition to the clusters of stars and nebulae already collected and given us in the work before referred to, which amount to 103. The event has plainly proved that my expectations were well founded: for I have already found 466 new nebulae and clusters of stars, none of which, to my present knowledge, have been seen before by any person; most of them, indeed, are not within the reach of the best common telescopes now in use. In all probability many more are still in reserve; and as I am pursuing this track, I shall make them up into separate catalogues, of about two or three hundred at a time, and have the honour of presenting them in that form to the Royal Society.

---

[1][M98 could not have been resolved by Herschel.]

[2]["Nebula without stars, discovered below and near Coma Berenices, a small distance from the star forty two of this constellation, after FLAMSTEED." MJC]

[3][This must rank as one of the greatest understatements of all time. MJC]

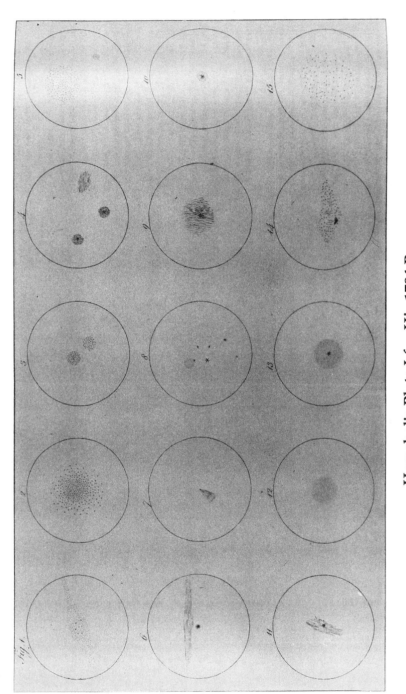

Herschel's Plate I for His 1784 Paper

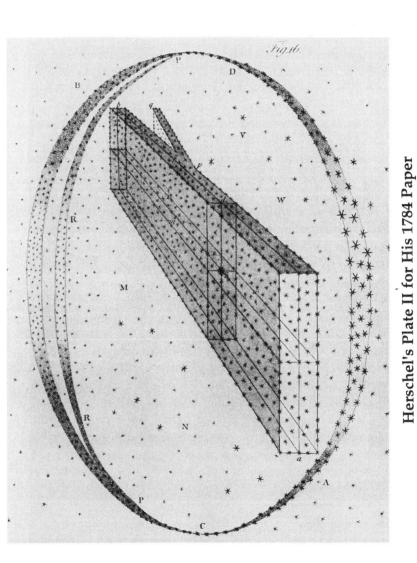

Herschel's Plate II for His 1784 Paper

A very remarkable circumstance attending the nebulae and clusters of stars is, that they are arranged into strata, which seem to run on to a great length; and some of them I have already been able to pursue, so as to guess pretty well at their form and direction. It is probable enough, that they may surround the whole apparent sphere of the heavens, not unlike the milky way, which undoubtedly is nothing but a stratum of fixed stars. And as this latter immense starry bed is not of equal breadth or lustre in every part, nor runs on in one straight direction, but is curved and even divided into two streams along a very considerable portion of it; we may likewise expect the greatest variety in the strata of the clusters of stars and nebulae. One of these nebulous beds is so rich, that, in passing through a section of it, in the time of only 36 minutes, I detected no less than 31 nebulae, all distinctly visible upon a fine blue sky. Their situation and shape, as well as condition, seem to denote the greatest variety imaginable. In another stratum, or perhaps a different branch of the former, I have seen double and treble nebulae, variously arranged; large ones with small, seeming attendants; narrow but much extended, lucid nebulae or bright dashes; some of the shape of a fan, resembling an electric brush, issuing from a lucid point; others of the cometic shape, with a seeming nucleus in the center; or like cloudy stars, surrounded with a nebulous atmosphere; a different sort again contain a nebulosity of the milky kind, like that wonderful, inexplicable phaenomenon about θ Orionis; while others shine with a fainter, mottled kind of light, which denotes their being resolvable into stars. See fig. 3, &c. But it would be too extensive at present to enter more minutely into such circumstances, therefore I proceed with the subject of nebulous and sidereal strata.

It is very probable, that the great stratum, called the milky way, is that in which the sun is placed, though perhaps not in the very center of its thickness. We gather this from the appearance of the Galaxy, which seems to encompass the whole heavens, as it certainly must do if the sun is within the same. For, suppose a number of stars arranged between two parallel planes, indefinitely extended every way, but at a given considerable distance from each other; and, calling this a sidereal stratum, an eye placed somewhere within it will see all the stars in the direction of the planes of the stratum projected into a great circle, which will appear lucid on account of the accumulation of the stars; while the rest of the heavens, at the sides, will only seem to be scattered over with constellations, more or less crowded, according to the distance of the planes or number of stars contained in the thickness or sides of the stratum.

Thus, in fig. 16 (Plate II.) an eye at S within the stratum *ab*, will see the stars in the direction of its length *ab*, or height *cd*, with all those in the intermediate situations, projected into the lucid circle ACBD; while those in the sides *mv*, *nw*, will be seen scattered over the remaining part of the heavens at MVNW.

If the eye were placed somewhere without the stratum, at no very great distance, the appearance of the stars within it would assume the form of one of the less circles of the sphere, which would be more or less contracted to the distance of the eye; and if this distance were exceedingly increased, the whole stratum might at last be drawn together into a lucid spot of any shape, according to the position, length, and height of the stratum.

Let us now suppose, that a branch, or smaller stratum, should run out from the former, in a certain direction, and let it also be contained between two parallel planes extended indefinitely onwards, but so that the eye may be placed in the great stratum somewhere before the separation, and not far from the place where the strata are still united. Then will this second stratum not be projected into a bright circle like the former, but will be seen as a lucid branch proceeding from the first, and returning to it again at a certain distance less than a semi-circle.

Thus, in the same figure, the stars in the small stratum *pq* will be projected into a bright arch at PRRP, which, after its separation from the circle CBD, unites with it again at P.

What has been instanced in parallel planes may easily be applied to strata irregularly bounded, and running in various directions; for their projections will of consequence vary according to the quantities of the variations in the strata and the distance of the eye from the same. And thus any kind of curvatures, as well as various different degrees of brightness, may be produced in the projections.

From appearances then, as I observed before, we may infer, that the sun is most likely placed in one of the great strata of the fixed stars, and very probably not far from the place where some smaller stratum branches out from it. Such a supposition will satisfactorily, and with great simplicity, account for all the phaenomena of the milky way, which, according to this hypothesis, is no other than the appearance of the projection of the stars contained in this stratum and its secondary branch. As a farther inducement to look on the Galaxy in this point of view, let it be considered, that we can no longer doubt of its whitish appearance arising from the mixed lustre of the numberless stars that compose it. Now, should we imagine it to be an irregular ring of stars, in the center nearly of which we must then suppose the sun to be placed, it will appear not a little extraordinary, that the sun, being a fixed star like those which compose this imagined ring, should just be in the center of such a multitude of celestial bodies, without any apparent reason for this singular distinction; whereas, on our supposition, every star in this stratum, not very near the termination of its length or height, will be so placed as also to have its own Galaxy, with only such variations in the form and lustre of it, as may arise from the particular situation of each star.

> In a section omitted at this point, Herschel presents in a preliminary form a method of analysis he called "star-gaging." This method consists of counting the number of stars telescopically visible in each of various directions. After tallying the results, a depth is assigned to the Milky Way in each direction proportional to the number of stars seen in that direction. In his discussion, Herschel stresses that the number of stars visible in any direction will depend on the quality of the telescope used. One of the conclusions he draws from this analysis is that our sun is located within the Milky Way.

If the sun should be placed in the great sidereal stratum of the milky way, and, as we have surmised above, not far from the branching out of a secondary stratum, it will very naturally lead us to guess at the cause of the probable motion of the solar system: for the very bright, great node of the Via Lactis, or union of the two strata about Cepheus and Cassiopeia, and the Scorpion and Sagittarius, points out a conflux of stars manifestly quite sufficient to occasion a tendency towards that node in any star situated at no very great distance; and the secondary branch of the Galaxy not being much less than a semi-circle seems to indicate such a situation of our solar system in the great undivided stratum as the most probable.

What has been said in a former paper on the subject of the solar motion seems also to support this supposed situation of the sun; for the apex there assigned lies nearly in the direction of a motion of the sun towards the node of the strata. Besides, the joining stratum making a pretty large angle at the junction with the primary one, it may easily be admitted, that the motion of a star in the great stratum, especially if situated considerably towards the side farthest from the small stratum, will be turned sufficiently out of the straight direction of the great stratum towards the secondary one. But I find myself insensibly led to say more on this subject than I am as yet authorised to do; I will, therefore, return to those observations which have suggested the idea of celestial strata.

In my late observations on nebulae I soon found, that I generally detected them in certain directions rather than in others; that the spaces preceding them were generally quite deprived of their stars, so as often to afford many fields without a single star in it; that the nebulae generally appeared some time after among stars of a certain considerable size, and but seldom among very small stars; that when I came to one nebula, I generally found several more in the neighbourhood; that afterwards a considerable time passed before I came to another parcel; and these events being often repeated in different altitudes of my instrument, and some of them at a considerable distance from each other, it occurred to me, that the intermediate spaces between the sweeps might also contain nebulae; and finding this to hold good more than once, I ventured to give notice to my assistant at the clock, "to prepare, since I expected in a few minutes to come at a stratum of the nebulae, finding myself already" (as I then figuratively expressed it) "on nebulous ground." In this I succeeded immediately; so that I now can venture to point out several not far distant places, where I shall soon carry my telescope, in expectation of meeting with many nebulae. But how far these circumstances of vacant places preceding and following the nebulous strata, and their being as it were contained in a bed of stars, sparingly scattered between them, may hold good in more distant portions of the heavens, and which I have not yet been able to visit in any regular manner, I ought by no means to hazard a conjecture. The subject is new, and we must attend to observations, and be guided by them, before we form general opinions.

Before I conclude, I may, however, venture to add a few particulars about the direction of some of the capital strata or their branches. The well known nebula of Cancer,[1] visible to the naked eye, is probably one belonging to a certain stratum, in which I suppose it to be so placed as to lie nearest to us. This stratum I shall call that of Cancer. It runs from ε Cancri towards the south over the 67th nebula of the *Connoissance des Temps*, which is a very beautiful and pretty much compressed cluster of stars, easily to be seen by any good telescope, and in which I have observed above 200 stars at once in the field of view of my great reflector, with a power of 157. This cluster appearing so plainly with any good, common telescope, and being so near to the one which may be seen by the naked eye, denotes it to be probably the next in distance to that within the quartile formed by γ, δ, η, θ; from the 67th nebula the stratum of Cancer proceeds towards the head of Hydra; but I have not yet had time to trace it farther than the equator.

Another stratum, which perhaps approaches nearer to the solar system than any of the rest, and whose situation is nearly at rectangles to the great sidereal stratum in which the sun is placed, is that of Coma Berenices, as I shall call it. I suppose the Coma itself to be one of the clusters in it, and that, on account of its nearness, it appears to be so scattered. It has many capital nebulae very near it; and in all probability this stratum runs on a very considerable way. It may, perhaps, even make the circuit of the heavens, though very likely not in one of the great circles of the sphere: for, unless it should chance to

---

[1][This is M44, also known as Praesepe. MJC]

intersect the great sidereal stratum of the milky way beforementioned, in the very place in which the sun is stationed, such an appearance could hardly be produced. However, if the stratum of Coma Berenices should extend so far as (by taking in the assistance of M. MESSIER'S and M. MECHAIN'S excellent observations of scattered nebulae, and some detached former observations of my own) I apprehend it may, the direction of it towards the north lies probably, with some windings, through the great Bear onwards to Cassiopeia; thence through the girdle of Andromeda and the northern Fish, proceeding towards Cetus; while towards the south it passes through the Virgin, probably on to the tail of Hydra and the head of Centaurus. But, notwithstanding I have already fully ascertained the existence and direction of this stratum for more than 30 degrees of a great circle, and found it almost every where equally rich in fine nebulae, it still might be dangerous to proceed in more extensive conjectures, that have as yet no more than a precarious foundation. I shall therefore wait till the observations in which I am at present engaged shall furnish me with proper materials for the disquisition of so new a subject. And though my single endeavours should not succeed in a work that seems to require the joint effort of every astronomer, yet so much we may venture to hope, that, by applying ourselves with all our powers to the improvement of telescopes, which I look upon as yet in their infant state, and turning them with assiduity to the study of the heavens, we shall in time obtain some faint knowledge of, and perhaps be able partly to delineate, *the Interior Construction of the Universe.*

☆☆☆☆☆☆☆☆☆☆☆☆☆☆☆☆☆☆☆☆☆☆☆☆☆☆☆☆☆

Herschel's 1785 Paper: "On the Construction of the Heavens"

In 1785, Herschel presented another major paper on the stellar portion of the heavens. What follows is an abbreviated version of that paper. In reading it, be sure to note the interesting methodological statement in its second paragraph. After distinguishing five forms of nebulae, Herschel presents various ideas concerning the construction of the heavens, setting these out as no more than hypotheses that agreed with what was known about gravitational forces. Using Newtonian ideas about gravitation, although in a very qualitative way, he boldly speculates not only about cosmology (the study of the present form of the universe), but also about cosmogony (the study of the development of the universe). He then presents an array of empirical information, which he claims supports the theoretical views he has developed. One of the high points of the paper is the diagram of the structure of the Milky Way that he derived from his gages. His method in constructing this diagram rested on the assumptions that the stars are more or less homogeneously distributed, and that his telescopes would "fathom" (reach to) the most distant stars in the Milky Way. On this basis, he puts forth a suggestion about the actual size of the Milky Way in relation to the distance of the very bright star Sirius, which, because of its brightness, he assumed to be the nearest star. In the final portion of the paper, he comments on the structure of various nebulous objects that he had observed with his giant telescopes. Among the objects on which he comments are Orion, which is pictured below in a photograph made late in the nineteenth century by Henry Draper, and also what Herschel calls "planetary nebulae," a type of object that was given extended attention in one of his subsequent papers.

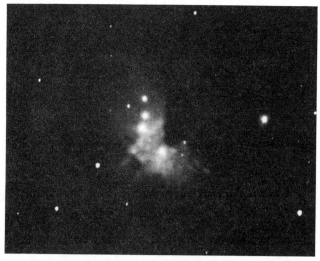

M42: The Great Nebula in Orion
(Photograph Taken in 1882 by Henry Draper)

☆☆☆☆☆☆☆☆☆☆☆☆☆☆☆☆☆☆☆☆☆☆☆☆☆☆☆☆
William Herschel,
"On the Construction of the Heavens"

[Read February 3, 1785 and originally published in *Philosophical Transactions of the Royal Society*, 75 (1785), 213–66 and reprinted here from *The Scientific Papers of William Herschel*, ed. by J. L. E. Dreyer, vol. 1 (London, 1912), pp. 223–59.]

The subject of the Construction of the Heavens, on which I have so lately ventured to deliver my thoughts to this Society, is of so extensive and important a nature, that we cannot exert too much attention in our endeavours to throw all possible light upon it; I shall, therefore, now attempt to pursue the delineations of which a faint outline was begun in my former paper.

By continuing to observe the heavens with my last constructed, and since that time much improved instrument, I am now enabled to bring more confirmation to several parts that were before but weakly supported, and also to offer a few still further extended hints, such as they present themselves to my present view. But first let me mention that, if we would hope to make any progress in an investigation of this delicate nature, we ought to avoid two opposite extremes of which I can hardly say which is the most dangerous. If we indulge a fanciful imagination and build worlds of our own, we must not wonder at our going wide from the path of truth and nature; but these will vanish like the Cartesian vortices, that soon gave way when better theories were offered. On the other hand, if we add observation to observation, without attempting to draw not only certain conclusions, but also conjectural views from them, we offend against the very end for which only observations ought to be made. I will endeavour to keep a proper medium; but if I should deviate from that, I could wish not to fall into the latter error.

That the milky way is a most extensive stratum of stars of various sizes admits no longer of the least doubt; and that our sun is actually one of the heavenly bodies belonging to it is as evident. I have now viewed and gaged this shining zone in almost every direction, and find it composed of stars whose number, by the account of these gages, constantly increases and decreases in proportion to its apparent brightness to the naked eye. But in order to develop the ideas of the universe, that have been suggested by my late observations, it will be best to take the subject from a point of view at a considerable distance both of space and of time.

Theoretical view.

Let us then suppose numberless stars of various sizes, scattered over an indefinite portion of space in such a manner as to be almost equally distributed throughout the whole. The laws of attraction, which no doubt extend to the remotest regions of the fixed stars, will operate in such a manner as most probably to produce the following remarkable effects.

Formation of nebula.

Form I. In the first place, since we have supposed the stars to be of various sizes, it will frequently happen that a star, being considerably larger than its neighbouring ones, will attract them more than they will be attracted by others that are immediately around them; by which means they will be, in time, as it were, condensed about a center; or, in other words, form themselves into a cluster of stars of almost a globular figure, more or less regularly so, according to the size and original distance of the surrounding stars. The perturbations of these mutual attractions must undoubtedly be very intricate, as we may easily comprehend by considering what SIR ISAAC NEWTON says in the first book of his *Principia*, in the 38th and following problems; but in order to apply this great author's reasoning of bodies moving in ellipses to such as are here, for a while, supposed to have no other motion than what their mutual gravity has imparted to them, we must suppose the conjugate axes of these ellipses indefinitely diminished, whereby the ellipses will become straight lines.

Form II. The next case, which will also happen almost as frequently as the former, is where a few stars, though not superior in size to the rest, may chance to be rather nearer each other than the surrounding ones; for here also will be formed a prevailing attraction in the combined center of gravity of them all, which will occasion the neighbouring stars to draw together; not indeed so as to form a regular or globular figure, but however in such a manner as to be condensed towards the common center of gravity of the whole irregular cluster. And this construction admits of the utmost variety of shapes, according to the number and situation of the stars which first gave rise to the condensation of the rest.

Form III. From the composition and repeated conjunction of both the foregoing forms, a third may be derived, when many large stars, or combined small ones, are situated in long extended, regular, or crooked rows, hooks, or branches; for they will also draw the surrounding ones, so as to produce figures of condensed stars coarsely similar to the former which gave rise to these condensations.

Form IV. We may likewise admit of still more extensive combinations; when, at the same time that a cluster of stars is forming in one part of space, there may be another collecting in a different, but perhaps not far distant quarter, which may occasion a mutual approach towards their common center of gravity.

[Form] V. In the last place, as a natural consequence of the former cases, there will be formed great cavities or vacancies by the retreat of the stars towards the various centers which attract them; so that upon the whole there is evidently a field of the greatest variety for the mutual and combined attractions of the heavenly bodies to exert themselves in. I shall, therefore, without extending myself farther upon this subject, proceed to a few considerations, that will naturally occur to every one who may view this subject in the light I have here done.

Objections considered.

At first sight then it will seem as if a system, such as it has been displayed in the foregoing paragraphs, would evidently tend to a general destruction, by the shock of one star's falling upon another. It would here be a sufficient answer to say, that if observation should prove this really to be the system of the universe, there is no doubt but that the great Author of it has amply provided for the preservation of the whole, though it should not appear to us in what manner this is effected. But I shall moreover point out several circumstances that do manifestly tend to a general preservation; as, in the first place, the indefinite extent of the sidereal heavens, which must produce a balance that will effectually secure all the great parts of the whole from approaching to each other. There remains then only to see how the particular stars belonging to separate clusters will be preserved from rushing on to their centers of attraction. And here I must observe, that though I have before, by way of rendering the case more simple, considered the stars as being originally at rest, I intended not to exclude projectile forces; and the admission of them will prove such a barrier against the seeming destructive power of attraction as to secure from it all the stars belonging to a cluster, if not for ever, at least for millions of ages. Besides, we ought perhaps to look upon such clusters, and the destruction of now and then a star, in some thousands of ages, as perhaps the very means by which the whole is preserved and renewed. These clusters may be the *Laboratories* of the universe, if I may so express myself, wherein the most salutary remedies for the decay of the whole are prepared.

Optical appearances.

From this theoretical view of the heavens, which has been taken, as we observed, from a point not less distant in time than in space, we will now retreat to our own retired station, in one of the planets attending a star in its great combination with numberless others; and in order to investigate what will be the appearances from this contracted situation, let us begin with the naked eye. The stars of the first magnitude being in all probability the nearest, will furnish us with a step to begin our scale; setting off, therefore, with the distance of Sirius or Arcturus, for instance, as unity, we will at present suppose, that those of the second magnitude are at double, and those of the third at treble the distance and so forth. It is not necessary critically to examine what quantity of light or magnitude of a star intitles it to be estimated of such or such a proportional distance, as the common coarse estimation will answer our present purpose as well; taking it then for granted, that a star of the seventh magnitude is about seven times as far as one of the first, it follows, that an observer, who is inclosed in a globular cluster of stars, and not far from the center, will never be able, with the naked eye, to see to the end of it: for, since, according to the above estimations, he can only extend his view to about seven times the distance of Sirius, it cannot be expected that his eyes should reach the borders of a cluster which has perhaps not less than fifty stars in depth everywhere around him. The whole universe, therefore, to him will be comprised in a set of constellations, richly ornamented with scattered stars of all sizes. Or if the united brightness of a neighbouring cluster of stars should, in a remarkable clear night, reach his sight, it will put on the appearance of a small, faint, whitish, nebulous cloud, not to be perceived without the greatest attention. To pass by other situations, let him be placed in a much extended stratum, or branching cluster of millions of stars, such as may fall under the III form of nebulae considered in a foregoing paragraph. Here also the heavens will not only be richly scattered over with brilliant constellations, but in shining zone or milky way will be perceived to surround the whole sphere of the heavens, owing to the combined light of those stars which are too small, that is too remote to be seen. Our observer's sight will be so confined, that he will imagine this single collection of stars, of which he does not even perceive the thousandth part, to be the whole contents of the heavens. Allowing him now the use of a common telescope, he begins to suspect that all the milkiness of the bright path which surrounds the sphere may be owing to stars. He perceives a few clusters of them in various parts of the heavens, and finds also that there are a kind of nebulous patches; but still his views are not extended so far as to reach to the end of the stratum in which he is situated, so that he looks upon these patches as belonging to that system which to him seems to comprehend every celestial object. He now increases his power of vision, and, applying himself to a close observation, finds that the milky way is indeed no other than a collection of very small stars. He perceives that those objects which had been called nebulae are evidently nothing but clusters of stars. He finds their number increase upon him, and when he resolves one nebula into stars he discovers ten new ones which he cannot resolve. He then forms the idea of immense strata of fixed stars, of clusters of stars and of nebulae[1]; till, going on with such interesting observations, he now perceives that all these appearances must naturally arise from the confined situation in which we are placed. *Confined* it may justly be called, though in no less a space than what before appeared to be the whole region of the fixed stars; but which now has assumed the shape of a crookedly branching nebula; not, indeed, one of the least, but perhaps very far from being the most considerable of those numberless clusters that enter into the construction of the heavens.

Result of Observations.

I shall now endeavour to shew, that the theoretical view of the system of the universe, which has been exposed in the foregoing part of this paper, is perfectly consistent with facts, and seems to be confirmed and established by a series of observations. It will appear, that many hundreds of nebulae of the first and second forms are actually to be seen in the heavens, and their places will hereafter be pointed out. Many of the third form will be described, and instances of the fourth related. A few of the cavities mentioned in the fifth will be particularised, though many more have already been observed; so that, upon the whole, I believe, it will be found, that the foregoing theoretical view, with all its consequential appearances, as seen by an eye inclosed in one of the nebulae, is no other than a drawing from nature, wherein the features of the original have been closely copied; and I hope the resemblance will not be called a bad one, when it shall be considered how very limited must be the pencil of an inhabitant of so small and retired a portion of an indefinite system in attempting the picture of so unbounded an extent.

---

[1]See a former paper on the Construction of the Heavens.

> A section of Herschel's paper has been deleted at this point. In it, he provides a mathematical analysis relating the number of stars in a given direction to the depth of space in that direction. He also presents extensive empirical information concerning the number of stars in various directions, i.e., his gages, which data he uses in estimating the depth of space in those directions. In counting the number of stars in any direction, he avoids counting stars in clusters lest such data distort his result. The unit of distance he employs in this technique is the distance of Sirius, which is the brightest and, he correspondingly assumes, the nearest star. On this basis, he assigns the Milky Way a width of about 100 times the distance of Sirius, whereas he places the most remote stars at 497 times the distance of Sirius. He expresses his goal in this investigation as being to show "that the stupendous sidereal system we inhabit, this extensive stratum and its secondary branch, consisting of many millions of stars, is, in all probability, a detached Nebula," i.e., that it is separate from other nebula. In this context, he remarks concerning the Andromeda nebula that it is "very evident that the united lustre of millions of stars, such as I suppose the nebula in Andromeda to be, will reach our sight in the shape of a very small, faint nebulosity...." In other words, he is saying that the Andromeda nebula is an island universe comparable to the Milky Way. And he adds: "Among the great number of nebulae which I have now already seen, amounting to more than 900, there are many which in all probability are equally extensive with that which we inhabit; and yet they are all separated from each other by very considerable intervals. Some indeed there are that seem to be double and treble; and though with most of these it may be, that they are at a very great distance from each other, yet we allow that some such conjunctions really are to be found...."

Section of our sidereal system.

By taking out of this table the visual rays which answer to the gages, and applying lines proportional to them around a point, according to their respective right ascensions and north polar distances, we may delineate a solid by means of the ends of these lines, which will give us so many points in its surface; I shall, however, content myself at present with a section only.

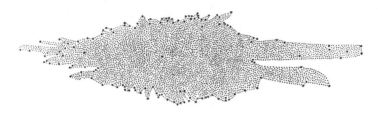

I have taken one which passes through the poles of our system, and is at rectangles to the conjunction of the branches which I have called its length. The name of poles seemed to me not improperly applied to those points which are 90 degrees distant from a circle passing along the milky way, and the north pole is here assumed to be situated in R.A. 186° and P.D. 58°. The section represented [above] is one which makes an angle of 35 degrees with our equator, crossing it in 124 1/2 and 304 1/2 degrees. A celestial globe, adjusted to the latitude of 55° north, and having σ Ceti near the meridian, will have the plane of this section pointed out by the horizon, and the gages which have been used in this delineation are those which in table I. [omitted] are marked by asterisks. When the visual rays answering to them are taken out of the second table, they must be projected on the plane of the horizon of the latitude which has been pointed out; and this may be done accurately enough for the present purpose by a globe adjusted as above directed; for as gages, exactly to the plane of the section, were often wanting. I have used many at some small distance above and below the same, for the sake of obtaining more delineating points; and in the figure the stars at the borders which are larger than the rest are those pointed out by the gages. The intermediate parts are filled up by smaller stars arranged in straight lines between the gaged ones. The delineating points, though pretty numerous, are not so close as we might wish: it is however to be hoped that in some future time this branch of astronomy will become more cultivated, so that we may have gages for every quarter of a degree of the heavens at least, and these often repeated in the most favourable circumstances. And whenever that shall be the case, the delineations may then be repeated with all the accuracy that long experience may enable us to introduce; for, this subject being so new, I look upon what is here given partly as only an example to illustrate the spirit of the method. From this figure however, which I hope is not a very inaccurate one, we may see that our nebula, as we observed before, is of the third form; that is: *A very extensive, branching, compound Congeries of many millions of stars*; which most probably owes its origin to many remarkably large as well as pretty closely scattered small stars, that may have drawn together the rest. Now, to have some idea of the wonderful extent of this system, I must observe that this section of it is drawn upon a scale where the distance of Sirius is no more than the 80th part of an inch;[1] so that probably all the stars, which in the finest nights we are able to distinguish with the naked eye, may be comprehended within a sphere, drawn round the large star near the middle, representing our situation in the nebula, of less than half a quarter of an inch radius.

The Origin of nebulous Strata.

If it were possible to distinguish between the parts of an indefinitely extended whole, the nebula we inhabit might be said to be one that has fewer marks of profound antiquity upon it than the rest. To explain this idea perhaps more clearly we should recollect that the condensation of clusters of stars has been ascribed to a gradual approach; and whoever reflects on the numbers of ages that must have past before some of the clusters, that will be found in my intended catalogue of them, could be so far condensed as we find them at present, will not wonder if I ascribe a certain air of youth and vigour to many very regularly scattered regions of our sidereal stratum. There are moreover many places in it where there is the greatest reason to believe that the stars, if we may judge from appearances, are now drawing towards various secondary centers, and will in time separate into different clusters, so as to occasion many sub-divisions. Hence we may surmise that when a nebulous stratum consists chiefly of nebulae of the first and second form, it probably owes its origin to what may be called the decay of a great compound nebula of the third form; and that the sub-divisions, which happened to it in length of time, occasioned all the small nebulae which sprung from it to lie in a certain range, according as they were detached from the primary one. In like manner our system, after numbers of ages,

---

[1][On the scale adopted in this book, the 170th part of an inch. MJC]

may very possibly become divided so as to give rise to a stratum of two or three hundred nebulae; for it would not be difficult to point out so many beginning or gathering clusters in it.[1] This view of the present subject throws a considerable light upon the appearance of that remarkable collection of many hundreds of nebulae which are to be seen in what I have called the nebulous stratum of Coma Berenices. It appears from the extended and branching figure of our nebula, that there is room for the decomposed small nebulae of a large, reduced, former great one to approach nearer to us in the sides than in other parts. Nay, possibly, there might originally be another very large joining branch, which in time became separated by the condensation of the stars; and this may be the reason of the little remaining breadth of our system in that very place: for the nebulae of the stratum of the Coma are brightest and most crowded just opposite our situation, or in the pole of our system. As soon as this idea was suggested, I tried also the opposite pole, where accordingly I have met with a great number of nebulae, though under a much more scattered form.

An Opening in the heavens.

Some parts of our system indeed seem already to have sustained greater ravages of time than others, if this way of expressing myself may be allowed; for instance, in the body of the Scorpion is an opening, or hole,[2] which is probably owing to this cause. I found it while I was gaging in the parallel from 112 to 114 degrees of north polar distance. As I approached the milky way, the gages had been gradually running up from 9.7 to

---

[1] Mr. MITCHELL has also considered the stars as gathered together into groups (*Phil. Trans.* vol. LVII. p. 240); which idea agrees with the subdivision of our great system here pointed out. He founds an elegant proof of this on the computation of probabilities, and mentions the Pleiades, the Praesepe Cancri, and the nebula (or cluster of stars) in the hilt of Perseus's sword, as instances.

[2] [Caroline Herschel recorded that once when observing a dark region in Scorpio, her brother exclaimed: "Hier ist wahrhaftig ein Loch in Himmel!" ("Here is truly a hole in the heavens!") A good example of such a dark region is the so-called "Coal Sack," a large, very dark region of the Milky Way. MJC]

17.1; when, all of a sudden, they fell down to nothing, a very few pretty large stars excepted, which made them shew 0.5, 0.7, 1.1, 1.4, 1.8; after which they again rose to 4.7, 13.5, 20.3, and soon after to 41.1. This opening is at least 4 degrees broad, but its height I have not yet ascertained. It is remarkable, that the 80th *Nébuleuse sans étoiles* of the *Connoissance des Temps*, which is one of the richest and most compressed clusters of small stars I remember to have seen, is situated just on the western border of it, and would almost authorize a suspicion that the stars, of which it is composed, were collected from that place, and had left the vacancy. What adds not a little to this surmise is, that the same phaenomenon is once more repeated with the fourth cluster of stars of the *Connoissance des Temps*; which is also on the western border of another vacancy, and has moreover a small, miniature cluster, or easily resolvable nebula of about 2 1/2 minutes in diameter, north following it, at no very great distance.

Phaenomena at the Poles of our Nebula.

I ought to observe, that there is a remarkable purity or clearness in the heavens when we look out of our stratum at the sides; that is, towards Leo, Virgo, and Coma Berenices, on one hand, and towards Cetus on the other; whereas the ground of the heavens becomes troubled as we approach towards the length or height of it. It was a good while before I could trace the cause of these phaenomena; but since I have been acquainted with the shape of our system, it is plain that these troubled appearances, when we approach to the sides, are easily to be explained by ascribing them to some of the distant, straggling stars, that yield hardly light enough to be distinguished. And I have, indeed, often experienced this to be actually the cause, by examining these troubled spots for a long while together, when, at last, I generally perceived the stars which occasioned them. But when we look towards the poles of our system, where the visual ray does not graze along the side, the straggling stars of course will be very few in number; and therefore the ground of the heavens will assume that purity which I have always observed to take place in those regions.

Enumeration of very compound Nebulae or Milky-Ways

As we are used to call the appearance of the heavens, where it is surrounded with a bright zone, the Milky-Way, it may not be amiss to point out some other very remarkable Nebulae which cannot well be less, but are probably much larger than our own system; and, being also extended, the inhabitants of the planets that attend the stars which compose them must likewise perceive the same phaenomena. For which reason they may also be called milky-ways by way of distinction.

My opinion of their size is grounded on the following observations. There are many round nebulae, of the first form, of about five or six minutes in diameter, the stars of which I can see very distinctly; and on comparing them with the visual ray calculated from some of my long gages, I suppose, by the appearance of the small stars in those gages, that the centers of these round nebulae may be 600 times the distance of Sirius from us.

In estimating the distance of such clusters I consulted rather the comparatively apparent size of the stars than their mutual distance; for the condensation in these clusters being probably much greater than in our own system, if we were to overlook this circumstance and calculate by their apparent compression, where, in about six minutes diameter, there are perhaps ten or more stars in the line of measures, we should find, that on the supposition of an equal scattering of the stars throughout all nebulae, the distance of the center of such a cluster from us could not be less than 6000 times the distance of Sirius. And, perhaps, in putting it, by the apparent size of the stars, at 600 only, I may have considerably underrated it; but my argument, if that should be the case, will be so much the stronger. Now to proceed.

Some of these round nebulae have others near them, perfectly similar in form, colour, and the distribution of stars, but of only half the diameter: and the stars in them seem to be doubly crowded, and only at about half the distance from each other: they are indeed so small as not to be visible without the utmost attention. I suppose these miniature nebulae to be at double the distance of the first. An instance, equally remarkable and instructive, is a case where, in the neighbourhood of two such nebulae as have been mentioned, I met with a third, similar, resolvable, but much smaller and fainter nebula. The stars of it are no longer to be perceived; but a resemblance of colour with the former two, and its diminished size and light, may well permit us to place it at full twice the distance of the second, or about four or five times that of the first. And yet the nebulosity is not of the milky kind; nor is it so much as difficulty resolvable, or colourless. Now, in a few of the extended nebulae, the light changes gradually so as from the resolvable to approach to the milky kind; which appears to me an indication that the milky light of nebulae is owing to their much greater distance. A nebula, therefore, whose light is perfectly milky, cannot well be supposed to be at less than six or eight thousand times the distance of Sirius; and though the numbers here assumed are not to be taken otherwise than as very coarse estimates, yet an extended nebula, which in an oblique situation, where it is possibly fore-shortened by one-half, two-thirds, or three-fourths of its length, subtends a degree or more in diameter, cannot be otherwise than of a wonderful magnitude, and may well outvie our milky-way in grandeur.

The first I shall mention is a milky Ray of more than a degree in length. It takes *k* (FL. 52) Cygni into its extent, to the north of which it is crookedly bent so as to be convex towards the following side; and the light of it is pretty intense. To the south of *k* it is more diffused, less bright, and loses itself with some extension in two branches, I believe; but for want of light I could not determine this circumstance. The northern half is near two minutes broad, but the southern is not sufficiently defined to ascertain its breadth.

The next is an extremely faint milky Ray, above 3/4 degree long, and 8 or 10' broad; extended from north preceding to south following. It makes an angle of about 30 or 40 degrees with the meridian, and contains three or four places that are brighter than the rest. The stars of the Galaxy are scattered over it in the same manner as over the rest of the heavens. It follows ε Cygni 11.5 minutes in time, and is 2° 19' more south.

The third is a branching Nebulosity of about a degree and a

half in right ascension, and about 48' extent in polar distance. The following part of it is divided into several streams and windings, which, after separating, meet each other again towards the south. It precedes ζ Cygni 16' in time, and is 1° 16' more north. I suppose this to be joined to the preceding one; but having observed them in different sweeps, there was no opportunity of tracing their connection.[1]

The fourth is a faint, extended milky Ray of about 17' in length, and 12' in breadth. It is brightest and broadest in the middle, and the ends lose themselves. It has a small, round, very faint nebula just north of it; and also, in another place a spot, brighter than the rest, almost detached enough to form a different nebula, but probably belonging to the great one. The Ray precedes α Trianguli 18'.8 in time, and is 55' more north. Another observation of the same, in a finer evening, mentions its extending much farther towards the south, and that the breadth of it probably is not less than half a degree; but being shaded away by imperceptible gradations, it is difficult exactly to assign its limits.[2]

The fifth is a Streak of light about 27' long, and in the brightest part 3 or 4' broad. The extent is nearly in the meridian, or a little from south preceding to north following. It follows β Ceti 5'.9 in time, and is 2° 43' more south. The situation is so low, that it would probably appear of a much greater extent in a higher altitude.[3]

The sixth is an extensive milky Nebulosity divided into two parts; the most north being the strongest. Its extent exceeds 15'; the southern part is followed by a parcel of stars which I suppose to be the 8th of the *Connoissance des Temps*.[4]

---

[1][These and the other nebulae discussed by Herschel in this section have been identified by David Dewhirst in the notes he provided for the reprinting of this paper in Michael Hoskin, *William Herschel and the Construction of the Heavens* (London, 1963). Herschel's first, second, and third objects are part of the "Veil" nebula. MJC]

[2][This is Messier 33, although Herschel does not seem to be aware of this fact. MJC]

[3][This is NGC 247, meaning object 247 in the *New Galactic Catalogue*, a later list of nebulous objects. MJC]

[4][This is the "Lagoon" nebula, which is NGC6523–33. MJC]

The seventh is a wonderful, extensive Nebulosity of the milky kind. There are several stars visible in it, but they can have no connection with that nebulosity, and are, doubtless, belonging to our own system scattered before it. It is the 17th of the *Connoissance des Temps*.[1]

In the list of these must also be reckoned the beautiful Nebula of Orion. Its extent is much above one degree; the eastern branch passes between two very small stars, and runs on till it meets a very bright one. Close to the four small stars, which can have no connection with the nebula, is a total blackness; and within the open part, towards the north-east, is a distinct, small, faint nebula, of an extended shape, at a distance from the border of the great one, to which it runs in a parallel direction, resembling the shoals that are seen near the coasts of some islands.[2]

The ninth is that in the girdle of Andromeda,[3] which is undoubtedly the nearest of all the great nebulae; its extent is above a degree and a half in length and, in even one of the narrowest places, not less than 16' in breadth. The brightest part of it approaches to the resolvable nebulosity, and begins to shew a faint red colour; which, from many observations on the colour and magnitude of nebulae, I believe to be an indication that its distance in this coloured part does not exceed 2000 times the distance of Sirius. There is a very considerable, broad, pretty faint, small nebula near it; my Sister discovered it August 27, 1783, with a Newtonian 2-feet sweeper. It shews the same faint colour with the great one, and is, no doubt, in the neighbourhood of it. It is not the 32d of the *Connoissance des Temps*; which is a pretty large round nebula, much condensed in the middle, and south following the great one; but this is about two-thirds of a degree north preceding it, in a line parallel to β

---

[1][This is M17, also known as the "Omega" or "Horseshoe" nebula. MJC]

[2][This is the Orion nebula, M42. To derive an idea of its apparent size, note Herschel's remark that it extends "much above one degree" in conjunction with the fact that the moon and sun are each about one half degree in diameter. MJC]

[3][The Andromeda nebula, M31, surpasses even Orion in width. MJC]

and ν Andromedae.[1]

To these may be added the nebula in Vulpecula:[2] for, though its appearance is not large, it is probably a double stratum of stars of a very great extent, one end whereof is turned towards us. That it is thus situated may be surmised from its containing, in different parts, nearly all the three nebulosities; *viz.* the resolvable, the coloured but irresolvable, and a tincture of the milky kind. Now, what great length must be required to produce these effects may easily be conceived when, in all probability, our whole system, of about 800 stars in diameter, if it were seen at such a distance that one end of it might assume the resolvable nebulosity, would not, at the other end, present us with the irresolvable, much less with the colourless and milky sort of nebulosities.

A Perforated Nebula, or Ring of Stars.

Among the curiosities of the heavens should be placed a nebula, that has a regular, concentric, dark spot in the middle, and is probably a Ring of stars.[3] It is of an oval shape, the shorter axis being to the longer as about 83 to 100; so that, if the stars form a circle, its inclination to a line drawn from the sun to the center of this nebula must be about 56 degrees.

FIG. 5.

The light is of the resolvable kind, and in the northern side three very faint stars may be seen, as also one or two in the southern part. The vertices of the longer axis seem less bright and not so well defined as the rest. There are several small stars very near, but none that seem to belong to it. It is the 57th of the *Connoissance des Temps*. Fig. 5 is a representation of it.

---

[1][The two nebulae that Herschel mentions as near the Andromeda nebula are NGC 205, which, as Herschel states, was discovered by Caroline Herschel, and M33. MJC]

[2][This is the "Dumbbell" nebula, M27. MJC]

[3][This object is the "Ring" nebula, M57, located in the constellation Lyra. Herschel was incorrect in believing that he had succeeded in resolving it. MJC]

Planetary Nebulae

I shall conclude this paper with an account of a few heavenly bodies, that from their singular appearance leave me almost in doubt where to class them.

The first precedes ν Aquarii 5'.4 in time, and is 1' more north.[1] Its place, with regard to a small star Sept. 7, 1782, was, Distance 8' 13" 51'''; but on account of the low situation, and other unfavourable circumstances, the measure cannot be very exact. August 25, 1783, Distance 7' 5" 11''', very exact, and to my satisfaction; the light being thrown in by an opaque-microscopic-illumination. Sept. 20, 1783, Position 41° 24' south preceding the same star; very exact, and by the same kind of illumination. Oct. 17, 1783, Distance 6' 55" 7'''; a second measure 6' 56" 11''', as exact as possible. Oct. 23, 1783, Position 42° 57'; a second measure 42° 57'; a second measure 42° 45'; single lens; power 71; opaque-microscopic-illumination. Nov. 14, 1783, Distance 7' 4" 35'''. Nov. 12, 1784, Distance 7' 22" 35'''; Position 38° 39'. Its diameter is about 10 or 15". I have examined it with the powers of 71, 227, 278, 460, and 932; and it follows the laws of magnifying, so that its body is no illusion of light. It is a little oval, and in the 7-feet reflector pretty well defined, but not sharp on the edges. In the 20-feet, of 18.7 inch aperture, it is much better defined, and has much of a planetary appearance, being all over of an uniform brightness, in which it differs from nebulae: its light seems however to be of the starry nature, which suffers not nearly so much as the planetary disks are known to do, when much magnified.

A description of five additional planetary nebulae has been omitted.

The planetary appearance of the two first is so remarkable, that we can hardly suppose them to be nebulae; their light is so uniform, as well as vivid, the diameters so small and well defined, as to make it almost improbable they should belong to that species of bodies. On the other hand, the effect of different powers seems to be much against their light's being of a planetary nature, since it preserves its brightness nearly in

---

[1][This is NGC 7009. Herschel, as he states, called such objects planetary nebulae because, being circular, they to some extent resemble planets in appearance. MJC]

the same manner as the stars do in similar trials. If we would suppose them to be single stars with large diameters we shall find it difficult to account for their not being brighter; unless we should admit that the intrinsic light of some stars may be very much inferior to that of the generality, which however can hardly be imagined to extend to such a degree. We might suspect them to be comets about their aphelion, if the brightness as well as magnitude of the diameters did not oppose this idea; so that after all, we can hardly find any hypothesis so probable as that of their being Nebulae; but then they must consist of stars that are compressed and accumulated in the highest degree. If it were not perhaps too hazardous to pursue a former surmise of a renewal in what I figuratively called the Laboratories of the universe, the stars forming these extraordinary nebulae, by some decay or waste of nature, being no longer fit for their former purposes, and having their projectile forces, if any such they had, retarded in each others' atmosphere, may rush at last together, and either in succession, or by one general tremendous shock, unite into a new body. Perhaps the extraordinary and sudden blaze of a new star in Cassiopea's chair, in 1572, might possibly be of such a nature.[1] But lest I should be led too far from the path of observation, to which I am resolved to limit myself, I shall only point out a considerable use that may be made of these curious bodies. If a little attention to them should prove that, having no annual parallax, they belong most probably to the class of nebulae, they may then be expected to keep their situation better than any one of the stars belonging to our system, on account of their being probably at a very great distance. Now to have a fixed point somewhere in the heavens, to which the motions of the rest may be referred, is certainly of considerable consequence in Astronomy; and both these bodies are bright and small enough to answer that end.[2]

☆☆☆☆☆☆☆☆☆☆☆☆☆☆☆☆☆☆☆☆☆☆☆☆☆☆☆☆

---

[1][The new star seen in "Cassiopea's [sic] chair" was detected and much discussed by Tycho Brahe. MJC]

[2][Herschel added at this point a footnote in which he enthusiastically reported his detection of two additional planetary nebulae. MJC]

Herschel's 1789 "Catalogue of a Thousand New Nebulae and Clusters of Stars..."

Herschel's 48-Inch-Aperture Reflecting Telescope

Herschel continued to observe and theorize about nebulae during the last half of the 1780s. In 1788, he completed his gigantic 48-inch-aperture, 40-foot-focal-length reflecting telescope (see picture), which further increased his ability to see dim objects. The remarkable character of his researches is reflected in a comment made by the novelist Fanny Burney after she had visited Herschel in 1786; she exclaimed that Herschel "has discovered fifteen hundred universes! How many more he may find who can conjecture?" In 1789, Herschel published a catalogue of a second thousand new nebulae and star clusters that he had discovered. Joined to this catalogue was a discussion of his observations in which he presented a new view of the stellar cosmos. The focus of the paper is a class of objects known as "globular clusters." These objects (see illustration) appear circular in form, but the number of stars composing them drops off as one looks away from the object's center.

**M 13: The Globular Cluster in Hercules
(Photograph Taken in 1895 by Isaac Roberts)**

In both the previous paper and the one that follows, Herschel mentions one of his more important predecessors, Rev. John Michell (1724–1793). In 1767 Michell had published a paper in which he raised the question whether the stars appear to be randomly distributed in the heavens. He asked, for example, whether the number of double stars, stars that from our point of view appear to be separated from each other by a very small angle, is what one would expect, were the stars randomly scattered over the heavens. The conclusion that Michell reached is that substantially more double stars occur than one would expect on the basis of assuming a random distribution. From this Michell concluded that a significant portion of double stars are stars physically adjacent to each other, possibly linked gravitationally, rather than being stars in approximately the same line of sight, but with one far more distant than the other. Gradually astronomers began to make a distinction between two types of double stars: gravitational doubles and optical doubles. The former group consists of double stars that are linked gravitationally, whereas doubles in the latter group consist of those that from our perspective lie in the same line of sight. Michell's statistical argument pointed to the existence of some gravitational doubles, but it was another matter to determine which double stars are, in fact, gravitational doubles. As Herschel notes, Michell extended his analysis to the cluster of stars known as the Pleiades, concluding that the odds are 500,000 to 1 against the idea that these six stars simply happen to lie in nearly the same line of sight. Michell's paper marked the introduction of statistical and probabilistic considerations into astronomy.

In reading Herschel's 1789 paper, it will be productive to keep the following questions in mind:
(1) In his 1784 paper, Herschel had urged that the stars be seen not as positioned on a starry vault, but rather as spread through three dimensions. In that context, what new element enters in this paper?
(2) In this paper, Herschel faces a problem comparable to that faced by someone given access to a greenhouse for one day and the task of determining the life history of the rose. How does Herschel resolve this problem?

☆☆☆☆☆☆☆☆☆☆☆☆☆☆☆☆☆☆☆☆☆☆☆☆☆☆☆☆☆
William Herschel,
"Catalogue of a Second Thousand of New Nebulae and Clusters of Stars; with a Few Introductory Remarks on the Construction of the Heavens"

[Read June 11, 1789 and published originally in the *Philosophical Transactions of the Royal Society*, 79 (1789), 212–55 and reprinted here from *The Scientific Papers of William Herschel*, ed. by J. L. E. Dreyer, vol. 1 (London, 1912), pp. 329–69.]

By the continuation of a review of the heavens with my twenty-feet reflector, I am now furnished with a second thousand of new Nebulae.

These curious objects, not only on account of their number, but also in consideration of their great consequence, as being no less than whole sidereal systems, we may hope, will in future engage the attention of Astronomers. With a view to induce them to undertake the necessary observations, I offer them the following catalogue, which, like my former one, of which it is a continuation, contains a short description of each nebula or cluster of stars, as well as its situation with respect to some known object.

The form of this work, it will be seen, is exactly that of the former part, the classes and numbers being continued, and the same letters used to express, in the shortest way, as many essential features of the objects as could possibly be crowded into so small a compass as that to which I thought it expedient to limit myself.

The method I have taken of *analyzing* the heavens, if I may so express myself, is perhaps the only one by which we can arrive at a knowledge of their construction. In the prosecution of so extensive an undertaking, it may well be supposed that many things must have been suggested, by the great variety in the order, the size, and the compression of the stars, as they presented themselves to my view, which it will not be improper to communicate.

To begin our investigation according to some order, let us depart from the objects immediately around us to the most remote that our telescopes, of the greatest *power to penetrate into space*, can reach. We shall touch but slightly on things that have already been remarked.

> In a paragraph omitted at this point, Herschel notes that whereas planets shine by reflected light, the sun and stars give off their own light. In this context, he stresses that every star "must likewise be a sun, shining by its own native brightness."

These suns, every one of which is probably of as much consequence to a system of planets, satellites, and comets, as our own sun, are now to be considered, in their turn, as the minute parts of a proportionally greater whole. I need not repeat that by my analysis it appears, that the heavens consist of regions where suns are gathered into separate systems, and that the catalogues I have given comprehend a list of such systems; but may we not hope that our knowledge will not stop short at the bare enumeration of phaenomena capable of giving us so much instruction? Why should we be less inquisitive than the natural philosopher, who sometimes, even from an inconsiderable number of specimens of a plant, or an animal, is enabled to present us with the history of its rise, progress, and decay? Let us then compare together, and class some of these numerous sidereal groups, that we may trace the operations of natural causes as far as we can perceive their agency. The most simple form, in which we can view a sidereal system, is that of being globular. This also, very favourably to our design, is that which has presented itself most frequently, and of which I have given the greatest collection.

But, first of all, it will be necessary to explain what is our idea of a cluster of stars, and by what means we have obtained it. For an instance, I shall take the phaenomenon which presents itself in many clusters: It is that of a number of lucid spots, of equal lustre, scattered over a circular space, in such a manner as to appear gradually more compressed towards the middle; and which compression, in the clusters to which I allude, is generally carried so far, as, by imperceptible degrees, to end in a luminous center, of a resolvable blaze of light. To solve this appearance, it may be conjectured, that stars of any given, very unequal magnitudes, may easily be so arranged, in scattered, much extended, irregular rows, as to produce the above described picture; or, that stars, scattered about almost promiscuously within the frustum of a given cone, may be assigned of such properly diversified magnitudes as also to form the same picture. But who, that is acquainted with the doctrine of chances, can seriously maintain such improbable conjectures? To consider this only in a very coarse way, let us suppose a cluster to consist of 5000 stars, and that each of them may be put into one of 5000 given places, and have one of 5000 assigned magnitudes. Then, without extending our calculation any further, we have five and twenty millions of chances, out of which only one will answer the above improbable conjecture, while all the rest are against it. When we now remark that this relates only to the given places within the frustum of a supposed cone, whereas these stars might have been scattered all over the visible space of the heavens; that they might have been scattered, even within the supposed cone, in a million of places different from the assumed ones, the chance of this apparent cluster's not being a real one, will be rendered so highly improbable that it ought to be intirely rejected.

MR. MICHELL computes, with respect to the six brightest stars of the Pleiades only, that the odds are near 500000 to 1 that no six stars, out of the number of those which are equal in splendour to the faintest of them, scattered at random in the whole heavens, would be within so small a distance from each other as the Pleiades are.[1]

Taking it then for granted that the stars which appear to be gathered together in a group are in reality thus accumulated, I proceed to prove also that they are nearly of an equal magnitude.

The cluster itself, on account of the small angle it subtends to the eye, we must suppose to be very far removed from us. For, were the stars which compose it at the same distance from one another as Sirius is from the sun; and supposing the cluster to be seen under an angle of 10 minutes, and to contain 50 stars in one of its diameters, we should have the mean distance of such stars twelve seconds; and therefore the distance of the cluster from us about seventeen thousand times greater than the distance of Sirius. Now, since the apparent magnitude of these stars is equal, and their distance from us is also equal,—because we may safely neglect the diameter of the cluster, which, if the center be seventeen thousand times the distance of Sirius from us, will give us seventeen thousand and twenty-five for the farthest and seventeen thousand wanting twenty-five for the nearest star of the cluster;—it follows that we must either give up the idea of a cluster, and recur to the above refuted supposition, or admit the equality of the stars that compose these clusters. It is to be remarked that we do not mean intirely to exclude all variety of size; for the very great distance, and

---

[1] *Phil. Trans.* vol. LVII. p. 246.

the consequent smallness of the component clustering stars, will not permit us to be extremely precise in the estimation of their magnitudes; though we have certainly seen enough of them to know that they are contained within pretty narrow limits; and do not, perhaps, exceed each other in magnitude more than in some such proportion as one full grown plant of a certain species may exceed another full-grown plant of the same species.

If we have drawn proper conclusions relating to the size of stars, we may with still greater safety speak of their relative situations, and affirm that in the same distances from the center an equal scattering takes place. If this were not the case, the appearance of a cluster could not be uniformly encreasing in brightness towards the middle, but would appear nebulous in those parts which were more crowded with stars; but, as far as we can distinguish, in the clusters of which we speak, every concentric circle maintains an equal degree of compression, as long as the stars are visible; and when they become too crowded to be distinguished, an equal brightness takes place, at equal distances from the center, which is the most luminous part.

The next step in my argument will be to shew that these clusters are of a globular form. This again we rest on the sound doctrine of chances. Here, by way of [adding] strength to our argument, we may be allowed to take in all round nebulae, though the reasons we have for believing that they consist of stars have not as yet been entered into. For, what I have to say concerning their spherical figure will equally hold good whether they be groups of stars or not. In my catalogues we have, I suppose, not less than one thousand of these round objects. Now, whatever may be the shape of a group of stars, or of a Nebula, which we would introduce instead of the spherical one, such as a cone, an ellipsis, a spheroid, a circle or a cylinder, it will be evident that out of a thousand situations, which the axes of such forms may have, there is but one that can answer the phaenomenon for which we want to account; and that is, when those axes are exactly in a line drawn from the object to the place of the observer. Here again we have a million of chances of which all but one are against any other hypothesis than that which we maintain, and which, for this reason, ought to be admitted.

> In a paragraph omitted at this point, Herschel formulates a geometrical argument to show "that these clusters of stars are more condensed towards the center than at the surface."

We may now venture to raise a superstructure upon the arguments that have been drawn from the appearance of clusters of stars and nebulae of the form I have been examining, which is that of which I have made mention in my "Theoretical view—Formation of Nebulae—Form I."[1] It is to be remarked that when I wrote the paragraph I refer to, I delineated nature as well as I do now; but, as I there gave only a general sketch, without referring to particular cases, what I then delivered may have been looked upon as little better than hypothetical reasoning, whereas in the present instance this objection is intirely removed, since actual and particular facts are brought to vouch for the truth of every inference.

Having then established that the clusters of stars of the 1st Form, and round nebulae, are of a spherical figure, I think myself plainly authorized to conclude that they are thus formed by the action of central powers.[2] To manifest the validity of this inference, the figure of the earth may be given as an instance; whose rotundity, setting aside small deviations, the causes of which are well known, is without hesitation allowed to be a phaenomenon decisively establishing a centripetal force. Nor do we stand in need of the revolving satellites of Jupiter, Saturn, and the Georgium Sidus,[3] to assure us that the same powers are likewise lodged in the masses of these planets. Their globular figure alone must be admitted as a sufficient argument to render this point uncontrovertible. We also apply this inference with equal propriety to the body of the sun, as well as to that of Mercury, Venus, Mars, and the Moon; as owing their spherical shape to the same cause. And how can we avoid inferring, that the

---

[1]*Phil. Trans.* vol. LXXV. p. 214.
[2][That is, by the action of gravitational forces. MJC]
[3][The planet Uranus, which Herschel had discovered and which he named after King George III, was until about 1850 frequently referred to by the name "Georgium Sidus" or as "Herschel." Presumably because this led to a rather odd sequence—Mars, Jupiter, Saturn, Herschel—the name "Uranus" gradually gained currency. MJC]

construction of the clusters of stars, and nebulae likewise, of which we have been speaking, is as evidently owing to central powers?

Besides, the step that I here make in my inference is in fact a very easy one, and such as ought freely to be granted. Have I not already shewn that these clusters cannot have come to their present formation by any random scattering of stars? The doctrine of chance, by exposing the very great odds against such hypotheses may be said to demonstrate that the stars are thus assembled by some power or other. Then, what do I attempt more than merely to lead the mind to the conditions under which this power is seen to act?

In a case of such consequence I may be permitted to be a little more diffuse, and draw additional arguments from the internal construction of spherical clusters and nebulae. If we find that there is not only a general form, which, as has been proved, is a sufficient manifestation of a centripetal force, what shall we say when the accumulated condensation, which every where follows a direction towards a center, is even visible to the very eye? Were we not already acquainted with attraction, this gradual condensation would point out a central power, by the remarkable disposition of the stars tending towards a center. In consequence of this visible accumulation, whether it may be owing to attraction only, or whether other powers may assist in the formation, we ought not to hesitate to ascribe the effect to such as are *central*; no phaenomena being more decisive in that particular, than those of which I am treating.

> In a paragraph omitted at this point, Herschel speculates that there may be central forces in addition to gravitational forces "concerned in the construction of the sidereal heavens."

I am fully aware of the consequences I shall draw upon myself in but mentioning other powers that might contribute to the formation of clusters. A mere hint of this kind, it will be expected, ought not to be given without sufficient foundation; but let it suffice at present to remark that my arguments cannot be affected by my terms: whether I am right to use the plural number,—central powers,—or whether I ought only to say,— the known central force of gravity,—my conclusions will be equally valid. I will however add, that the idea of other central powers being concerned in the construction of the sidereal heavens, is not one that has only lately occurred to me. Long ago I have entertained a certain theory of diversified central powers of attractions and repulsions; an exposition of which I have even delivered in the years 1780, and 1781, to the Philosophical Society then existing at Bath, in several mathematical papers upon that subject. I shall, however, set aside an explanation of this theory, which would not only exceed the intended limits of this paper, but is moreover not required for what remains at present to be added, and therefore may be given some other time, when I can enter more fully into the subject of the interior construction of sidereal systems.

To return, then, to the case immediately under our present consideration, it will be sufficient that I have abundantly proved that the formation of round clusters of stars and nebulae is either owing to central powers, or at least to one such force as refers to a center.

I shall now extend the weight of my argument, by taking in likewise every cluster of stars or nebula that shews a gradual condensation, or encreasing brightness, towards a center or certain point; whether the outward shape of such clusters or nebulae be round, extended, or of any other given form. What has been said with regard to the doctrine of chance, will of course apply to every cluster, and more especially to the extended and irregular shaped ones, on account of their greater size: It is among these that we find the largest assemblages of stars, and most diffusive nebulosities; and therefore the odds against such assemblages happening without some particular power to gather them, encrease exceedingly with the number of the stars that are taken together. But if the gradual accumulation either of stars or encreasing brightness has before been admitted as a direction to the seat of power, the same effect will equally point out the same cause in the cases now under consideration. There are besides some additional circumstances in the appearance of extended clusters and nebulae, that very much favour the idea of a power lodged in the brightest part. Although the form of them be not globular, it is plainly to be seen that there is a tendency towards sphericity, by the swell of the dimensions the nearer we draw towards the most luminous place, denoting as it were a course, or tide of stars, setting towards a center. And—if allegoral expressions may be allowed—it should seem as if the stars thus flocking towards the seat of power were stemmed by the crowd of those already assembled, and that while some of them are successful in forcing their predecessors sideways out of their places, others are themselves obliged to take up with lateral situations, while all of them seem equally to strive for a place in the central swelling, and generating spherical figure.

Since then almost all the nebulae and clusters of stars I have seen, the number of which is not less than three and twenty hundred, are more condensed and brighter in the middle; and since, from every form, it is now equally apparent that the central accumulation or brightness must be the result of central powers, we may venture to affirm that this theory is no longer an unfounded hypothesis, but is fully established on grounds which cannot be overturned.

Let us endeavour to make some use of this important view of the constructing cause, which can thus model sidereal systems. Perhaps, by placing before us the very extensive and varied collection of clusters, and nebulae furnished by my catalogues, we may be able to trace the progress of its operation, in the great laboratory of the Universe.

If these clusters and nebulae were all of the same shape, and had the same gradual condensation, we should make but little progress in this inquiry; but, as we find so great a variety in their appearances, we shall be much sooner at a loss how to account for such various phaenomena, than be in want of materials upon which to exercise our inquisitive endeavours.

Some of these round clusters consist of stars of a certain magnitude, and given degree of compression, while the whole cluster itself takes up a space of perhaps 10 minutes; others appear to be made up of stars that are much smaller, and much more compressed, when at the same time the cluster itself subtends a much smaller angle, such as 5 minutes. This diminution of the apparent size, and compression of stars, as well as diameter of the cluster to 4, 3, 2 minutes, may very consistently be ascribed to the different distances of these clusters from the place in which we observe them; in all which cases we may admit a general equality of the sizes, and compression of the stars that compose them, to take place. It is also highly probable that a continuation of such decreasing magnitudes, and encreasing compression, will justly account for the appearance of round, easily resolvable, nebulae; where there is almost a certainty of their being clusters of stars. And no Astronomer can hesitate to go still farther, and extend his surmises by imperceptible steps to other nebulae, that still preserve the same characteristics, with the only variations of vanishing brightness, and reduction of size.

Other clusters there are that, when they come to be compared with some of the former, seem to contain stars of an equal magnitude, while their compression appears to be considerably different. Here the supposition of their being at different distances will either not explain the apparently greater compression, or if admitted to do this, will convey to us a very instructive consequence: which is, that the stars which are thus supposed not to be more compressed than those in the former cluster, but only to appear so on account of their greater distance, must needs be proportionally larger, since they do not appear of less magnitude than the former. As therefore, one or other of these hypotheses must be true, it is not at all improbable but that, in some instances, the stars may be more compressed; and in others, of a greater magnitude. This variety of size, in different spherical clusters, I am however inclined to believe, may not go farther than the difference in size, found among the individuals belonging to the same species of plants, or animals, in their different states of age, or vegetation, after they are come to a certain degree of growth. A farther inquiry into the circumstance of the extent, both of condensation and variety of size, that may take place with the stars of different clusters, we shall postpone till other things have been previously discussed.

Let us then continue to turn our view to the power which is moulding the different assortments of stars into spherical clusters. Any force, that acts uninterruptedly, must produce effects proportional to the time of its action. Now, as it has been shewn that the spherical figure of a cluster of stars is owing to central powers, it follows that those clusters which, *ceteris paribus*, are the most compleat in this figure, must have been the longest exposed to the action of these causes. This will admit of various points of view. Suppose for instance that 5000 stars had been once in a certain scattered situation, and that other 5000 equal stars had been in the same situation, then that of the two clusters which had been longest exposed to the action of the modelling power, we suppose, would be most condensed, and more advanced to the maturity of its figure. An obvious consequence that may be drawn from this consideration is, that we are enabled to judge of the relative age, maturity, or climax of a sidereal system, from the disposition of its component parts; and, making the degrees of brightness in nebulae stand for the different accumulation of stars in clusters, the same conclusions will extend equally to them all. But we are not to conclude from what has been said that every spherical cluster is of an equal standing in regard to absolute duration, since one that is composed of a thousand stars only, must certainly arrive to the perfection of its form sooner than another, which takes in a range of a million. Youth and age are comparative expressions; and an oak of a certain age may be called very young, while a contemporary shrub is already on the verge of its decay. The method of judging with some assurance of the condition of any sidereal system may perhaps not improperly be drawn from the standard laid down; so that, for instance, a cluster or nebula which is very gradually more compressed and bright towards the middle, may be in the perfection of its growth, when another which approaches to the condition pointed out by a more equal compression, such as the nebulae I have called *Planetary* seem to present us with, may be looked upon as very aged, and drawing on towards a period of change, or dissolution. This has been before surmised, when, in a former paper, I considered the uncommon degree of compression that must prevail in a nebula to give it a planetary aspect; but the argument, which is now drawn from the powers that have collected the formerly scattered stars to the form we find they have assumed, must greatly corroborate that sentiment.

This method of viewing the heavens seems to throw them into a new kind of light. They now are seen to resemble a luxuriant garden, which contains the greatest variety of productions, in different flourishing beds; and one advantage we may at least reap from it is, that we can, as it were, extend the range of our experience to an immense duration. For, to continue the simile I have borrowed from the vegetable kingdom, is it not almost the same thing, whether we live successively to witness the germination, blooming, foliage, fecundity, fading, withering, and corruption of a plant, or whether a vast number of specimens, selected from every stage through which the plant passes in the course of its existence, be brought at once to our view?[1]

☆ ☆

Herschel's 1791 Paper
"On Nebulous Stars, Properly So Called"

Herschel's short 1791 paper reveals a major change in his thinking, particularly in regard to the resolvability of nebulae. An especially important observation was that recorded on 13 November 1790, when he observed what he called a "nebulous star." The object is now called a planetary nebula and is designated as NGC 1514 (nebula 1514 in the *New General Catalogue*). The term planetary nebula is applied to objects that have the appearance of a bright central spot surrounded by a hazy, more or less circular area. Herschel faced the question of how such an observation should be interpreted. Should the object be seen as a giant star surrounded by numerous small stars, which he could not, at least with the instruments available to him, resolve into individual stars? Or should it be viewed as an ordinary star surrounded by some glowing fluid or gas? Another question to keep in mind is: How does this paper represent a shift in the position Herschel advocated in his 1784 paper?

---

[1][The final portion of this paper consists of a listing of a thousand new nebulae that Herschel had detected. His listing has been omitted from this selection. MJC]

NGC 1514: Planetary Nebula Observed
by Herschel on 13 November 1790
(Lick Observatory Photograph)

☆☆☆☆☆☆☆☆☆☆☆☆☆☆☆☆☆☆☆☆☆☆☆☆☆☆☆☆
William Herschel
"On Nebulous Stars, Properly So Called"

[Read February 10, 1791 and published originally in the *Philosophical Transactions of the Royal Society*, 81 (1791), 71-88 and reprinted here from *The Scientific Papers of William Herschel*, ed. by J. L. E. Dreyer, vol. 1 (London, 1912), pp. 415-25.]

In one of my late examinations of a space in the heavens, which I had not reviewed before, I discovered *a star of about the 8th magnitude, surrounded with a faintly luminous atmosphere, of a considerable extent.* The phaenomenon was so striking that I could not help reflecting upon the circumstances that attended it, which appeared to me to be of a very instructive nature, and such as may lead to inferences which will throw a considerable light on some points relating to the construction of the heavens.

Cloudy or nebulous stars have been mentioned by several

astronomers; but this name ought not to be applied to the objects which they have pointed out as such; for, on examination, they proved to be either mere clusters of stars, plainly to be distinguished with my large instruments, or such nebulous appearances as might be reasonably supposed to be occasioned by a multitude of stars at a vast distance. The milky way itself, as I have shewn in some former Papers, consists intirely of stars, and by imperceptible degrees I have been led on from the most evident congeries of stars to other groups in which the lucid points were smaller, but still very plainly to be seen; and from them to such wherein they could but barely be suspected, till I arrived at last to spots in which no trace of a star was to be discerned. But then the gradations to these latter were by such well-connected steps as left no room for doubt but that all these phaenomena were equally occasioned by stars, variously dispersed in the immense expanse of the universe.

When I pursued these researches, I was in the situation of a natural philosopher who follows the various species of animals and insects from the height of their perfection down to the lowest ebb of life; when, arriving at the vegetable kingdom, he can scarcely point out to us the precise boundary where the animal ceases and the plant begins; and may even go so far as to suspect them not to be essentially different. But recollecting himself, he compares, for instance, one of the human species to a tree, and all doubt upon the subject vanishes before him. In the same manner we pass through gentle steps from a course cluster of stars, such as the Pleiades, the Praesepe, the milky way, the cluster in the Crab, the nebula in Hercules, that near the preceding hip of Bootes,[1] the 17th, 38th, 41st of the 7th class of my Catalogues, the 10th, 20th, 35th of the 6th class, the 33d, 48th, 213th of the 1st, the 12th, 150th, 756th of the 2d, and the 18th, 140th, 725th of the 3d, without any hesitation, till we find ourselves brought to an object such as the nebula in Orion, where we are still inclined to remain in the once adopted idea, of stars exceedingly remote,

---

[1][At this point and at numerous others in this paper, Herschel included footnotes in which he provided information on the right ascensions and polar distances of the objects mentioned. These footnotes have been omitted from the present selection. MJC]

and inconceivably crowded, as being the occasion of that remarkable appearance. It seems, therefore, to require a more dissimilar object to set us right again. A glance like that of the naturalist, who casts his eye from the perfect animal to the perfect vegetable, is wanting to remove the veil from the mind of the astronomer. The object I have mentioned above, is the phaenomenon that was wanting for this purpose. View, for instance, the 19th cluster of my 6th class, and afterwards cast your eye on this cloudy star, and the result will be no less decisive than that of the naturalist we have alluded to. Our judgement, I may venture to say, will be, that *the nebulosity about the star is not of a starry nature.*

But, that we may not be too precipitate in these new decisions, let us enter more at large into the various grounds which induced us formerly to surmise, that every visible object, in the extended and distant heavens, was of the starry kind, and collate them with those which now offer themselves for the contrary opinion.

It has been observed, on a former occasion, that all the smaller parts of other great systems, such as the planets, their rings and satellites, the comets, and such other bodies of the like nature as may belong to them, can never be perceived by us, on account of the faintness of light reflected from small, opaque objects; in my present remarks, therefore, all these are to be intirely set aside.

A well connected series of objects, such as we have mentioned above, has led us to infer, that all nebulae consist of stars. This being admitted, we were authorized to extend our analogical way of reasoning a little farther. Many of the nebulae had no other appearance than that whitish cloudiness, on the blue ground upon which they seemed to be projected; and why the same cause should not be assigned to explain the most extensive nebulosities, as well as those that amounted only to a few minutes of a degree in size, did not appear. It could not be inconsistent to call up a telescopic milky way, at an immense distance, to account for such phaenomena; and if any part of the nebulosity seemed detached from the rest, or contained a visible star or two, the probability of seeing a few near stars, apparently scattered over the far distant regions of myriads of sidereal collections, rendered nebulous by their distance, would also clear up these singularities.

In order to be more easily understood in my remarks on the comparative disposition of the heavenly bodies, I shall mention some of the particulars which introduced the ideas of *connection* and *disjunction*: for these, being properly founded upon an examination of objects that may be reviewed at any time, will be of considerable importance to the validity of what we may advance with regard to my lately discovered nebulous stars.

On June the 27th, 1786, I saw a beautiful cluster of very small stars of various sizes, about 15' in diameter, and very rich of stars. On viewing this object, it is impossible to withhold our assent to the idea which occurs, that these stars are connected so far one with another as to be gathered together, within a certain space, of little extent, when compared to the vast expanse of the heavens. As this phaenomenon has been repeatedly seen in a thousand cases, I may justly lay great stress on the idea of such stars being connected.

In the year 1779, the 9th of September, I discovered a very small star near ε Bootis. The question here occurring, whether it had any connection with ε or not, was determined in the negative; for, considering the number of stars scattered in a variety of places, it is very far from being uncommon, that a star at a great distance should happen to be nearly in a line drawn from the sun through ε, and thus constitute the observed double star.

The 7th of September, 1782, when I first saw the planetary nebula near ν Aquarii, I pronounced it to be a system whose parts were connected together. Without entering into any kind of calculation, it is evident, that a certain equal degree of light within a very small space, joined to the particular shape this object presents to us, which is nearly round, and even in its deviation consistent with regularity being a little elliptical, ought naturally to give us the idea of a conjunction in the things that produce it. And a considerable addition to this argument may be derived from a repetition of the same phaenomenon, in nine or ten more of a similar construction.

When I examined the cluster of stars, following the head of

the great dog, I found on the 19th of March, 1786, that there was within this cluster a round, resolvable nebula, of about two minutes in diameter, and nearly of an equal degree of light throughout. Here, considering that the cluster was free from nebulosity in other parts, and that many such clusters, as well as many such nebulae, exist in divers parts of the heavens, it appeared to me very probable, that the nebula was unconnected with the cluster; and that a similar reason would as easily account for this appearance as it had resolved the phaenomenon of the double star near ε Bootis; that is, a casual situation of our sun and the two other objects nearly in a line. And though it may be rather more remarkable, that this should happen with two compound systems, which are not by far so numerous as single stars, we have, to make up for this singularity, a much larger space in which it may take place, the cluster being of a very considerable extent.

On the 15th of February, 1786, I discovered that one of my planetary nebulae, had a spot in the center, which was more luminous than the rest, and with long attention, a very bright, round, well defined center became visible. I remained not a single moment in doubt, but that the bright center was connected with the rest of the apparent disk.

In the year 1785, the 6th of October, I found a very bright, round nebula, of about 1 1/2 minute in diameter. It has a large, bright nucleus in the middle, which is undoubtedly connected with the luminous parts about it. And though we must confess, that if this phaenomenon, and many more of the same nature, recorded in my catalogues of nebulae, consist of clustering stars, we find ourselves involved in some difficulty to account for the extraordinary condensation of them about the center; yet the idea of a connection between the outward parts and these very condensed ones within is by no means lessened on that account.

There is a telescopic milky way, which I have traced out in the heavens in many sweeps made from the year 1783 to 1789. It takes up a space of more than 60 square degrees of the heavens, and there are thousands of stars scattered over it: among others, four that form a trapezium, and are situated in the well known nebula of Orion, which is included in the above extent. All these stars, as well as the four I have mentioned, I take to be intirely unconnected with the nebulosity which involves them in appearance. Among them is also *d* Orionis, a cloudy star, improperly so called by former astronomers; but it does not seem to be connected with the milkiness any more than the rest.

I come now to some other phaenomena, that, from their singularity, merit undoubtedly a very full discussion. Among the reasons which induced us to embrace the opinion, that all very faint milky nebulosity ought to be ascribed to an assemblage of stars is, that we could not easily assign any other cause of sufficient importance for such luminous appearances, to reach us at the immense distance we must suppose ourselves to be from them. But if an argument of considerable force should now be brought forward, to shew the existence of a luminous matter, in a state of modification very different from the construction of a sun or star, all objections, drawn from our incapacity of accounting for new phaenomena upon old principles, will lose their validity.

Hitherto I have been shewing, by various instances in objects whose places are given, in what manner we may form the ideas of connection and its contrary by an attentive inspection of them only: I will now relate a series of observations, with remarks upon them as they are delivered, from which I shall afterwards draw a few simple conclusions, that seem to be of considerable importance.

To distinguish the observations from the remarks, the former are given in italics, and the date annexed is that on which the objects were discovered; but the descriptions are extracted from all the observations that have been made upon them.

October 16, 1784. *A star of about the 9th magnitude, surrounded by a milky nebulosity, or chevelure, of about 3 minutes in diameter. The nebulosity is very faint, and a little extended or elliptical, the extent being not far from the meridian, or a little from north preceding to south following. The chevelure involves a small star, which is about 1 1/2 minute north of the cloudy star; other stars of equal magnitude are perfectly free from this appearance.*

My present judgement concerning this remarkable object is,

that the nebulosity belongs to the star which is situated in its center. The small one, on the contrary, which is mentioned as involved, being one of many that are profusely scattered over this rich neighbourhood, I suppose to be quite unconnected with this phaenomenon. A circle of three minutes in diameter is sufficiently large to admit another small star, without any bias to the judgement I form concerning the one in question.

It must appear singular, that such an object should not have immediately suggested all the remarks contained in this Paper; but about things that appear new we ought not to form opinions too hastily, and my observations on the construction of the heavens were then but entered upon. In this case, therefore, it was the safest way to lay down a rule not to reason upon the phaenomena that might offer themselves, till I should be in possession of a sufficient stock of material to guide my researches.

October 16, 1784. *A small star of about the 11th or 12th magnitude, very faintly affected with milky nebulosity; other stars of the same magnitude are perfectly free from this appearance.*

February 23, 1786. *5 or 6 small stars within the space of 3 or 4', all very faintly affected in the same manner, and the nebulosity suspected to be a little stronger about each star.* But a third observation rather opposes this increase of the faintly luminous appearance.

Here the connection between the stars and the nebulosity is not so evident as to amount to conviction; for which reason we shall pass on to the next.

January the 6th, 1785. *A bright star with a considerable milky chevelure; a little extended, 4 or 5' in length, and near 4' broad; it loses itself insensibly. Other stars of equal magnitude are perfectly free from the chevelure.*

The connection between the star and the chevelure cannot be doubted, from the insensible gradation of its luminous appearance, decreasing as it receded from the center.

January 31, 1785. *A pretty considerable star, with a very faint, and very small, irregular, milky chevelure; other stars of the same size are perfectly free from such appearance.*

I can have no doubt of the connection between the star and its chevelure.

> A section in which Herschel describes twelve more observations, made between 1785 and 1790, of stars associated with nebulosity has been omitted at this point. Then comes an observation that was crucial for his interpretation of such objects.

November 13, 1790. A most singular phaenomenon! A star of about the 8th magnitude, with a faint luminous atmosphere, of a circular form, and of about 3' in diameter. The star is perfectly in the center, and the atmosphere is so diluted, faint, and equal throughout, that there can be no surmise of its consisting of stars; nor can there be a doubt of the evident connection between the atmosphere and the star. Another star not much less in brightness, and in the same field with the above, was perfectly free from any such appearance.

This last object is so decisive in every particular, that we need not hesitate to admit it as a pattern, from which we are authorized to draw the following important consequences.

Supposing the connection between the star and its surrounding nebulosity to be allowed, we argue, that one of the two following cases must necessarily be admitted. In the first place, if the nebulosity consist of stars that are very remote, which appear nebulous on account of the small angles their mutual distances subtend at the eye, whereby they will not only, as it were, run into one another, but also appear extremely faint and diluted; then, what must be the enormous size of the central point, which outshines all the rest in so superlative a degree as to admit of no comparison? In the next place, if the star be no bigger than common, how very small and compressed must be those other luminous points that are the occasion of the nebulosity which surrounds the central one? As, by the former supposition, the luminous central point must far exceed the standard of what we call a star, so, in the latter, the shining matter about the center will be much too small to come under the same denomination; we therefore either have a central body which is not a star, or have a star which is involved in a shining fluid, of a nature totally unknown to us.

I can adopt no other sentiment than the latter, since the probability is certainly not for the existence of so enormous a body as would be required to shine like a star of the 8th magnitude, at a distance sufficiently great to cause a vast system of stars to put on the appearance of a very diluted, milky nebulosity.

But what a field of novelty is here opened to our conceptions! A shining fluid, of a brightness sufficient to reach us from the remote regions of a star of the 8th, 9th, 10th, 11th, or 12th magnitude, and of an extent so considerable as to take up 3, 4, 5, or 6 minutes in diameter! Can we compare it to the coruscations of the electrical fluid in the aurora borealis? Or to the more magnificent cone of the zodiacal light as we see it in spring or autumn? The latter, not withstanding I have observed it to reach at least 90 degrees from the sun, is yet of so little extent and brightness as probably not to be perceived even by the inhabitants of Saturn or the Georgian planet, and must be utterly invisible at the remoteness of the nearest fixed star.

More extensive views may be derived from this proof of the existence of a shining matter. Perhaps it has been too hastily surmised that all milky nebulosity, of which there is so much in the heavens, is owing to starlight only. These nebulous stars may serve as a clue to unravel other mysterious phaenomena. If the shining fluid that surrounds them is not so essentially connected with these nebulous stars but that it can also exist without them, which seems to be sufficiently probable, and will be examined hereafter, we may with great facility explain that very extensive, telescopic nebulosity, which, as I mentioned before, is expanded over more than sixty degrees of the heavens, about the constellation of Orion; a luminous matter accounting much better for it than clustering stars at a distance. In this case we may also pretty nearly guess at its situation, which must commence somewhere about the range of the stars of the 7th magnitude, or a little farther from us, and extend unequally in some places perhaps to the regions of those of the 9th, 10th, 11th, and 12th. The foundation for this surmise is, that, not unlikely, some of the stars that happen to be situated in a more condensed part of it, or that perhaps by their own attraction draw together some quantity of this fluid greater than what they are intitled to by their situation in it, will, of course, assume the appearance of cloudy stars; and many of those I have named are either in this stratum of luminous matter, or very near it.

We have said above, that in nebulous stars the existence of the shining fluid does not seem to be so essentially connected with the central points that it might not also exist without them. For this opinion we may assign several reasons. One of them is the great resemblance between the chevelure of these stars and the diffused extensive nebulosity mentioned before, which renders it highly probable that they are of the same nature. Now, if this be admitted, the separate existence of the luminous matter, or its independence on a central star, is fully proved. We may also judge, very confidently, that the light of this shining fluid is no kind of reflection from the star in the center; for, as we have already observed, reflected light could never reach us at the great distance we are from such objects. Besides, how impenetrable would be an atmosphere of a sufficient density to reflect so great a quantity of light? And yet we observe, that the outward parts of the chevelure are nearly as bright as those that are close to the star; so that this supposed atmosphere ought to give no obstruction to the passage of the central rays. If, therefore, this matter is self-luminous, it seems more fit to produce a star by its condensation than to depend on the star for its existence.

Many other diffused nebulosities, besides that about the constellation of Orion, have been observed or suspected; but some of them are probably very distant, and run out far into space. For instance, about 5 minutes in time preceding ξ Cygni, I suspect as much of it as covers near four square degrees; and much about the same quantity 44' preceding the 125 Tauri. A space of almost 8 square degrees, 6' preceding α Trianguli, seems to be tinged with milky nebulosity. Three minutes preceding the 46 Eridani, strong, milky nebulosity is expanded over more than two square degrees. 54' preceding the 13th Canum venaticorum, and again 48' preceding the same star, I found the field of view affected with whitish nebulosity throughout the whole breadth of the sweep, which was 2° 39'. 4' following the 57 Cygni, a considerable space is filled with faint, milky nebulosity which is pretty bright in some places, and contains the 37th nebula of my Vth class, in the brightest part of it. In the neighbourhood of the 44th Piscium, very faint nebulosity appears to be diffused over more than 9 square degrees of the heavens. Now, all these phaenomena, as we have already seen, will admit of a much easier explanation by a luminous fluid than by stars at an immense distance.

The nature of planetary nebulae, which has hitherto been involved in much darkness, may now be explained with some degree of satisfaction, since the uniform and very considerable brightness of their apparent disk accord remarkably well with a much condensed, luminous fluid; whereas to suppose them to consist of clustering stars will not so completely account for the milkiness or soft tint of their light, to produce which it would be required that the condensation of the stars should be carried to an almost inconceivable degree of accumulation. The surmise of the regeneration of stars, by means of planetary nebulae, expressed in a former Paper, will become more probable, as all the luminous matter contained in one of them, when gathered together into a body of the size of a star, would have nearly such a quantity of light as we find the planetary nebulae to give. To prove this experimentally, we may view them with a telescope that does not magnify sufficiently to shew their extent, by which means we shall gather all their light together into a point, when they will be found to assume the appearance of small stars; that is, of stars at the distance of those which we call of the 8th, 9th, or 10th magnitude. Indeed this idea is greatly supported by the discovery of a well defined, lucid point, resembling a star, in the center of one of them: for the argument which has been used, in the case of nebulous stars, to shew the probability of the existence of a luminous matter, which rested upon the disparity between a bright point and its surrounding shining fluid, may here be alleged with equal justice. If the point be a generating star, the further accumulation of the already much condensed, luminous matter, may complete it in time.

How far the light that is perpetually emitted from millions of suns may be concerned in this shining fluid, it might be presumptuous to attempt to determine; but, notwithstanding the unconceivable subtlety of the particles of light, when the number of the emitting bodies is almost infinitely great, and the time of the continual emission indefinitely long, the quantity of emitted particles may well become adequate to the constitution of a shining fluid, or luminous matter, provided a cause can be found that may retain them from flying off, or reunite them. But such a cause cannot be difficult to guess at, when we know that light is so easily reflected, retracted, inflected, and deflected; and that, in the immense range of its course, it must pass through innumerable systems, where it cannot but frequently meet with many obstacles to its rectilinear progression. Not to mention the great counteraction of the united attractive force of whole sidereal systems, which must be continually exerting their power upon the particles while they are endeavouring to fly off. However, we shall lay no stress upon a surmise of this kind, as the means of verifying it are wanting: nor is it of any immediate consequence to us to know the origin of the luminous matter. Let it suffice, that its existence is rendered evident, by means of nebulous stars.

I hope it will be found, that in what has been said I have not launched out into hypothetical reasonings; and that facts have all along been kept sufficiently in view. But, in order to give every one a fair opportunity to follow me in the reflections I have been led into, the place of every object from which I have argued has been purposely added, that the validity of what I have advanced might be put to the proof by those who are inclined, and furnished with the necessary instruments to undertake an attentive and repeated inspection of the same phaenomena.

☆☆☆☆☆☆☆☆☆☆☆☆☆☆☆☆☆☆☆☆☆☆☆☆☆☆☆☆☆☆

Herschel's 1811 Paper
"Astronomical Observations Relating to the Construction of the Heavens"

William Herschel's enthusiasm for research in stellar astronomy continued into the last years of his long life. One of his most interesting papers, that of 1811, was published when he was seventy-three. Nor was this paper the last of his publications. In fact, he published four more papers before his death in 1822. His 1811 paper is long, but its main points are contained in the following selection from it. The first part of this selection not only contains a description of the direction of his thought at this time, but also begins with a statement on the "ultimate object" of his observational work. This statement can be taken as applying to his researches over the previous decades.

In reading this selection from his 1811 paper, it will be helpful to concentrate on understanding the two pages of diagrams included in it. What was Herschel attempting to represent by these diagrams? How do they represent the dramatic change in his thought that had occurred since the mid–1780s? And, in turn, what has happened to the hypothesis that nebulae are island universes?

☆☆☆☆☆☆☆☆☆☆☆☆☆☆☆☆☆☆☆☆☆☆☆☆☆☆☆☆☆
William Herschel,
"Astronomical Observations Relating to the Construction of the Heavens, Arranged for the Purpose of a Critical Examination, the Result of Which Appears to Throw Some New Light upon the Organization of the Celestial Bodies"

[Read June 20, 1811 and published originally in the *Philosophical Transactions of the Royal Society* (1811), 269–336 and reprinted here from *The Scientific Papers of William Herschel*, ed. by J. L. E. Dreyer, vol. 2 (London, 1912), pp. 459–97.]

A knowledge of the construction of the heavens has always been the ultimate object of my observations, and having been many years engaged in applying my forty, twenty, and large ten feet telescopes, on account of their great space-penetrating power to review the most interesting objects discovered in my sweeps, as well as those which had before been communicated to the public in the *Connoissance des Temps*, for 1784, I find that by arranging these objects in a certain successive regular order, they may be viewed in a new light, and, if I am not mistaken, an examination of them will lead to consequences which cannot be indifferent to an inquiring mind.

138 Modern Theories of the Universe

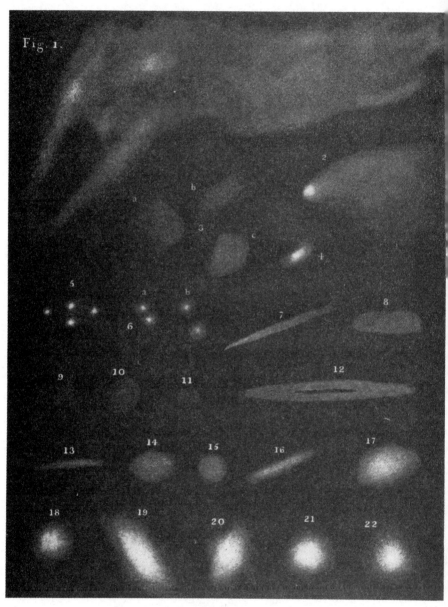

Plate I from Herschel's 1811 Paper

Plate II from Herschel's 1811 Paper

If it should be remarked that in this new arrangement I am not entirely consistent with what I have already in former papers said on the nature of some objects that have come under my observation, I must freely confess that by continuing my sweeps of the heavens my opinion of the arrangement of the stars and their magnitudes, and of some other particulars, has undergone a gradual change; and indeed when the novelty of the subject is considered, we cannot be surprised that many things formerly taken for granted, should on examination prove to be different from what they were generally, but incautiously, supposed to be.

For instance, an equal scattering of the stars may be admitted in certain calculations; but when we examine the milky way, or the closely compressed clusters of stars, of which my catalogues have recorded so many instances, this supposed equality of scattering must be given up. We may also have surmised nebulae to be no other than than clusters of stars disguised by their very great distance, but a longer experience and better acquaintance with the nature of nebulae, will not allow a general admission of such a principle, although undoubtedly a cluster of stars may assume a nebulous appearance when it is too remote for us to discern the stars of which it is composed.

Impressed with an idea that nebulae properly speaking were clusters of stars, I used to call the nebulosity of which some were composed, when it was of a certain appearance, *resolvable*; but when I perceived that additional light, so far from resolving these nebulae into stars, seemed to prove that their nebulosity was not different from what I had called milky, this conception was set aside as erroneous. In consequence of this, such nebulae as afterwards were suspected to consist of stars, or in which a few might be seen, were called *easily resolvable*; but even this expression must be received with caution, because an object may not only contain stars, but also nebulosity not composed of them.

It will be necessary to explain the spirit of the method of arranging the observed astronomical objects under consideration in such a manner, that one shall assist us to understand the nature and construction of the other. This end I propose to obtain by assorting them into as many classes as will be required to produce the most gradual affinity between the individuals contained in any one class with those contained in that which precedes and that which follows it: and it will certainly contribute to the perfection of this method, if this connection between the various classes can be made to appear so clearly as not to admit of a doubt. This consideration will be a sufficient apology for the great number of assortments into which I have thrown the objects under consideration; and it will be found that those contained in one article, are so closely allied to those in the next, that there is perhaps not so much difference between them, if I may use the comparison, as there would be in an annual description of the human figure, were it given from the birth of a child till he comes to be a man in his prime.

The similarity of the objects contained in each class will seldom require the description of more than one of them, and for this purpose, out of the number referred to, the selected one will be that which has been most circumstantially observed; however, those who wish either to review any other of the objects, or to read a short description of them, will find their place in the heavens, or the account of their appearance either in the catalogues I have given of them in the *Philos. Trans.* or in the *Connoissance des Temps* for 1784, to which in every article proper references will be given for the objects under consideration.

If the description I give should sometimes differ a little from that which belongs to some number referred to, it must be remembered that objects which had been observed many times, could not be so particularly and comprehensively detailed in the confined space of the catalogues as I now may describe them: additional observations have also now and then given me a better view of the objects than I had before. This remark will always apply to the numbers which refer to the *Connoissance des Temps*; for the nebulae and clusters of stars are there so imperfectly described, that my own observation of them with large instruments may well be supposed to differ entirely from what is said of them. But if any astronomer should review them, with such high space-penetrating-powers, as are absolutely required, it will be found that I have classed them very properly.

It will be necessary to mention that the nebulous delineations in the figures are not intended to represent any of the individuals of the objects which are described otherwise than in the circumstances which are common to the nebulae of each assortment: the irregularity of a figure, for instance, must stand for every other irregularity; and the delineated size for every other size. It will however, be seen, that in the figure referred to there is a sufficient resemblance to the described nebula to show the essential features of shape and brightness then under consideration.

> Keeping in mind what Herschel has said up to this point, especially his stress on nebulae as evolving structures, examine the two diagrams that accompany this paper. The idea of the paper is that by arranging numerous nebulae into a sequence, Herschel could speculate on the ways that nebulae change. In the diagrams, figure 1 should be seen as a nebula very early in its history, whereas figure 36 represents a nebula much later in time with the nebulae from figures 2 to 35 representing intermediate stages. The suggestion that Herschel is making is that nebulae (some? all?) are condensing to form stars. Rather than being immense galaxies comparable to the Milky Way and, like it, composed of millions of stars, some nebulae on this view are patches of gradually condensing "shining fluid." After discussing this idea in some detail, Herschel in section 22 goes so far as to comment:

22. *Of Nebulae that have a Cometic appearance.*

Among the numerous nebulae I have seen, there are many that have the appearance of telescopic comets. The following are of that sort.

I. 4 is "A pretty large cometic nebula of considerable brightness; it is much brighter in the middle, and the very faint chevelure is pretty extensive." Fig. 22.

By the appellation of cometic, it was my intention to express a gradual and strong increase of brightness towards the center of a nebulous object of a round figure; having also a faint chevelure or coma of some extent, beyond the faintest part of the light, gradually decreasing from the center.

It seems that this species of nebulae contains a somewhat greater degree of condensation than that of the round nebulae of that last article, and might perhaps not very improperly have been included in their number. Their great resemblance to telescopic comets, however, is very apt to suggest the idea, that possibly such small telescopic comets as often visit our neighbourhood may be composed of nebulous matter, or may in fact be such highly condensed nebulae.

> Herschel's paper concludes with the following remarks as well a long tabular synopsis of its contents and a postscript. The synopsis has been omitted from this selection.

35. *Concluding Remarks.*

The total dissimilitude between the appearance of a diffusion of the nebulous matter and of a star, is so striking, that an idea of the conversion of the one into the other can hardly occur to any one who has not before him the result of the critical examination of the nebulous system which has been displayed in this paper. The end I have had in view, by arranging my observations in the order in which they have been placed, has been to shew, that the above mentioned extremes may be connected by such nearly allied intermediate steps, as will make it highly probable that every succeeding state of the nebulous matter is the result of the action of gravitation upon it while in a foregoing one, and by such steps the successive condensation of it has been brought up to the planetary condition. From this the transit to the stellar form, it has been shown, requires but a very small additional compression of the nebulous matter, and several instances have been given which connect the planetary to the stellar appearance.

The faint stellar nebulae have also been well connected with all sorts of faint nebulae of a larger size; and in a number of the smaller sort, their approach to the starry appearance is so advanced, that in my observations of many of them it became doubtful whether they were not stars already.

It must have been noticed, that I have confined myself in every one of the preceding articles to a few remarks upon the appearance of the nebulous matter in the state in which my observations represented it; they seemed to be the natural result of the observations under consideration, and were not given with a view to establish a systematic opinion, such as will admit of complete demonstration. The observations themselves are arranged so conveniently that any astronomer, chemist, or philosopher, after having considered my critical remarks, may form what judgment appears most probable to him. At all events, the subject is of such a nature as cannot fail to attract the notice of every inquisitive mind to a contemplation of the stupendous construction of the heavens; and what I have said may at least serve to throw some new light upon the organization of the celestial bodies.

POSTSCRIPT.

It will be seen that in this paper I have only considered the nebulous part of the construction of the heavens, and have taken a star for the limit of my researches. The rich collection of clusters of stars contained in the 6th, 7th, and 8th classes of my Catalogues, and many of the *Connoissance des Temps*, have as yet been left unnoticed. Several other objects, in which stars and nebulosity are mixed, such as nebulous stars, nebulae containing stars, or suspected clusters of stars which yet may be nebulae, have not been introduced, as they appeared to belong to the sidereal part of the construction of the heavens, into a critical examination of which it is was not my intention to enter in this Paper.

☆☆☆☆☆☆☆☆☆☆☆☆☆☆☆☆☆☆☆☆☆☆☆☆☆☆☆☆☆☆

Concluding Comments
on Herschel's Contribution to Stellar Astronomy

When Herschel died in 1822, he left the most important questions in stellar astronomy still awaiting solution. Although he had become convinced in the 1780s that all or nearly all nebulae are resolvable and hence probably island universes comparable to the Milky Way, his confidence in this conclusion began to erode with his observations of the objects discussed in his 1791 paper, and especially of the object he first saw on 13 November 1790. By the time of his 1811 paper, he seems to have believed that most nebulae are simply stars in the process of formation. This may very well appear as an anticlimactic conclusion to his research, but this is where his decades of research had brought him. Anticlimactic though it may be, it would be a serious mistake to undervalue Herschel's contribution to stellar astronomy. Herschel had, after all, increased the number of known nebulae by a factor of 25 and had also begun to establish a classification of them. Most importantly, he had formulated a number of questions that established the research program for stellar astronomy for the next hundred years or more.

Note: For a laboratory exercise designed to accompany and illuminate this chapter, see the appendix.

Chapter Four

From William Herschel to 1860

Introduction

As we enter the period between William Herschel and 1860, it is vital to keep in mind the pioneering character of Herschel's work: he virtually began the in-depth study of stellar astronomy. The great majority of eighteenth-century astronomers did not view stellar astronomy as part, or at least as a significant part, of astronomy, which in their minds was concerned primarily with objects in our solar system, especially the exact determination and prediction of the positions of those objects. This point is effectively illustrated by the title of an 1983 essay by M. E. W. Williams: "Was There Such a Thing as Stellar Astronomy in the Eighteenth Century?"[1] The point of her essay is to suggest that, except for Herschel and a very few others, eighteenth-century astronomers paid little attention to stellar astronomy. Part of the reason for this lack of interest is that almost no astronomers possessed telescopes comparable to those that enriched Herschel's work.

Interest in stellar astronomy and in Herschel's investigations did not immediately increase as the nineteenth century began. We shall see that progress was gradually attained, but for decades stellar astronomy continued to be viewed as an esoteric, somewhat overly speculative speciality. Consequently, most astronomy textbooks of the early nineteenth century gave little or no attention to stellar astronomy. A striking illustration of this point is that when in 1833 John Herschel, William Herschel's son, published his *Treatise on Astronomy*, he devoted only one chapter—38 pages in a 418 page book—to stellar astronomy. This overall point is also exemplified in the views concerning astronomy presented by Auguste Comte, a leading French philosopher of the period.

---

[1]M. E. W. Williams, "Was There Such a Thing as Stellar Astronomy in the Eighteenth Century?" *History of Science*, 21 (1983), 369–85.

Auguste Comte

Among the most influential philosophers of the first half of the nineteenth century was Auguste Comte (1798–1857), who is often described as the founder of Positivism. One of the central tenets of that philosophy is that scientists should not waste their energies in attempts to answer unanswerable questions. In his *Cours de philosophie positive* (1830–42), Comte wrote:

> Astronomy up till now has been the only branch of natural philosophy in which the human mind has been set free from every theological and metaphysical influence, direct or indirect; it is this that makes it particularly easy to bring out its real philosophic character. But to attain a true idea of the nature and composition of this science, it is indispensable to set aside the vague definitions of it that are still being given, and to mark the boundaries of the positive knowledge that we are able to gain of the stars.
> Of the three senses that reveal the existence of distant bodies, that of sight is clearly the only one applicable to the celestial bodies. There could be no astronomy for any blind species, however intelligent one might imagine it to be; and for ourselves, the invisible stars, possibly more numerous than the visible ones, are excluded from study, and their existence can be suspected only by induction. Any research that is irreducible to actual visual observation is necessarily excluded in regard to the stars, which are thus of all natural entities those that we know under the least varying aspects. We conceive the possibility of determining their forms, their distances, their magnitudes, and their movements, but we can never by any means investigate their chemical composition or mineralogical structure, still less the nature of the organic beings that live on their surface, etc. In short, to put the matter in scientific terms, the positive knowledge we can have of the stars is limited solely to their geometrical and mechanical phenomena, and can never be extended by physical, chemical, physiological, and social research, such as can be expended on entities accessible to all our diverse means of observation.[1]

---

[1] Auguste Comte, *Cours de philosophie positive*, vol. 2 (Paris, 1924), pp. 1–2.

Comte not only stressed that astronomers should avoid useless speculations on the physical makeup of celestial bodies, he also urged that they should forsake stellar astronomy, which he viewed as both useless and incapable of attaining certain knowledge. He wrote:

> It is therefore necessary that we separate more completely than is commonly done the solar from the universal point of view, the idea of the world from that of the universe: the first is the highest we can actually attain, and it is also the only one that truly interests us.
>
> Thus, without entirely renouncing hope of some knowledge of the stellar, we must conceive of positive astronomy as consisting essentially in the geometrical and mechanical study of a small number of celestial bodes composing the *world* of which we are part. It is only within such limits that astronomy merits by its perfection the supreme rank that it occupies today among the natural sciences. As for those innumerable stars scattered in the sky, they have scarcely any interest for astronomy other than as markers in our observations, their position being regarded as fixed relative to the movements internal to our system, which alone concern us.[1]

Comte pressed the same point in his *Traité philosophique d'astronomie populaire* (1844), where he warned:

> It is then in vain that for half a century it has been endeavoured to distinguish two astronomies, the one solar, the other sidereal. In the eyes of those for whom science consists of real laws and not of incoherent facts, the second exists only in the name, and the first alone constitutes the true astronomy; and I am not afraid to assert that it will always be so.[2]

When Comte wrote the above, he must have known of William Herschel's researches. Yet his conviction was that stellar astronomy consisted of nothing more than an accumulation of incoherent empirical information and fruitless speculations. Comte's position raises the question whether

---

[1] Comte, *Cours*, pp. 6–7.

[2] Auguste Comte, *Traité philosophique d'astronomie populaire* (Paris, 1844), p. 114.

stellar astronomy had as yet attained a level that merited the term *scientific* or the attention of serious astronomers.

Olbers's Paradox

One scientist who certainly would not have shared Comte's views was H. W. M. Olbers (1758–1840), a German ophthalmologist and astronomer, who in 1823 presented a paper entitled "On the Transparency of Space," which sets out what has come to be known as "Olbers's Paradox." This idea, in which Olbers had been to some extent anticipated in the eighteenth century by Halley and by de Chéseaux, has received extensive discussion by cosmologists from Olbers's time to the present. In fact, concerning Olbers's paper, the prominent cosmologist H. Bondi remarked in 1958 that "we can put a precise date to the moment when [cosmology] became a scientific subject and left the road of philosophical speculation. This date is 1826 [when] Olbers published a little investigation which, although I doubt whether he realised it, made cosmology a science."[1] This is overstatement, but it suggests the importance of Olbers's short paper.

In his publication, Olbers presented astronomers with an elementary mathematical argument to the effect that either (1) the universe must be finite or (2) the sky must be completely filled with stars and be, night or day, blazing bright, or (3) space must not be completely transparent.

Olbers's Argument

Assume that the stars are more or less uniformly distributed throughout the heavens and that space is infinite. Let 1 be taken as the average distance of stars of the first magnitude. Assume also that the stars can be divided into groups in such a manner that every star is associated with a spherical shell of radius R, where the values of R range over the distances 1, 2, 3, 4, ... To determine which shell is associated with an

---

[1] As quoted in Stanley L. Jaki, *The Paradox of Olbers' Paradox* (New York: Herder and Herder, 1969), p. 233. This book is a useful study of the history of Olbers's Paradox.

individual star, simply determine which number in the sequence 1, 2, 3, 4, ..., best approximates its distance from earth. Set the average radius of a star associated with the first shell as seen from earth equal to r. We shall calculate the area in the sky as seen from earth that is taken up by these stars.

For the first shell, that with radius $R = 1$, the average area for the stars associated with it will be πr^2. If n stars exist at that distance, their total surface area will be $n\pi r^2$.

For a spherical shell of radius 2, the number of stars will be $4n$. The reason is that the surface area of a sphere of radius R is $4\pi R^2$. This means that for a doubled radius, the surface area of the sphere will be four times greater. Because of the assumed uniform distribution of stars, this will entail that there are 4 times more stars there. Because these stars are twice as distant, their apparent widths will be half as great. Hence their radii will appear to be $r/2$. Consequently at this distance, the stars will take up an area equal to $4n\pi(r/2)^2 = n\pi r^2$.

For distance 3, comparable calculations give the number of stars as $9n$ and their average radii as $r/3$. Thus the area will be $9n\pi(r/3)^2 = n\pi r^2$.

As these cases suggest, at whatever distance we take our shell of stars, the stars from that shell will cover an area in the sky equal to $n\pi r^2$. This is because the increase in the number of stars in a more distant shell will exactly offset the decrease in the area taken up by those stars.

If we assume that space is infinite, then we can construct an infinite number of such shells. If we add up the amount of sky covered by the stars of each shell, these quantities in each case being $n\pi r^2$, we see that the sky will eventually be entirely covered. Physically, this means that the sky should be blazing bright, even at night!

Obviously, this is not the case. Olbers's chief proposal for a way around this paradox was to suggest that the sky is not transparent, that matter exists in space blocking some of the light from reaching us. This was an unattractive idea because it implied that astronomers are barred from having a clear view of distant celestial objects.

Josef Fraunhofer and the Refractor

The argument in Olbers's paper did not depend on the telescopic technology of the early nineteenth century, whereas other advances did. The quality of refracting telescopes dramatically improved in the opening decades of the century, due in good measure to Josef Fraunhofer (1787–1826), one of the most gifted optical instrument makers of all time. We shall later discuss his contribution to spectroscopy, but at present it is important to stress his signal contribution to telescope construction. Around 1799, Pierre Guinand developed a method of making clear flint glass lenses larger than 5 inches, which had been the previous practical limit on the size of objective lenses for refracting telescopes. Based on this method, an optical institute was created in Munich, Fraunhofer becoming its director. During Fraunhofer's life as well as later, this institute produced refracting telescopes of unrivaled excellence. This was partly due to the Guinand method, but also to Fraunhofer's skill in constructing telescopes with superior mountings that enhanced the capacity of such telescopes for precise positional determinations. By 1850, most of the major research telescopes were refractors and the best of these were Munich-made. Listed below are four of the most famous telescopes that came from the Munich optical works:
(1) In 1824, Wilhelm Struve, then at the observatory at Dorpat in Russia, received a 9.5-inch-aperture refractor.
(2) In 1829, Friedrich Bessel acquired a 6.25-inch-aperture refractor for the Königsberg Observatory in Prussia.
(3) In 1839, Struve, who by then had become director of Pulkova Observatory, which is located near St. Petersburg in Russia, acquired a 15-inch-aperture refractor.
(4) In 1847, William Cranch Bond of Harvard College Observatory received a 15-inch refractor comparable in quality to Struve's Pulkova telescope.
Some of the most important astronomical advances of the mid-nineteenth century were achieved with these instruments.

The Continuing Search for Stellar Parallax

As noted previously, Robert Hooke had claimed in 1669 that he had measured a stellar parallax of about 15" in the star gamma Draconis. In the 1690s, John Flamsteed, England's Astronomer Royal, and Olaus Roemer, a prominent Danish astronomer, also reported results that seemed to indicate a stellar parallax. Flamsteed's observations were of Polaris, for which he announced a parallax of 40". These results, which seem to have been viewed with substantial skepticism even at the time when they were announced, were recognized as definitely erroneous as a result of James Bradley's unsuccessful search for parallax, which culminated in his discovery of the aberration of starlight. In fact, the aberrational shift was probably the major factor in producing the misleading results reported by Hooke and Flamsteed. Bradley had also claimed with considerable authority that were a parallactic shift as large as 1" of arc present in gamma Draconis, he would have been able to detect it.

Even as the truth of the Copernican theory became ever more widely accepted and consequently less in need of the proof that the determination of parallax would provide, interest in determining parallax remained strong because it would reveal the distance of at least the nearer stars.

Around 1780, William Herschel had become very interested in determining a parallax, hoping to do this by observing double stars. The reason that Herschel selected double stars (i.e., stars very near each other in the sky) for observation was that he realized that if one of the double stars were near and the other very remote, then the motions of the near star would show up by comparison with the essentially fixed position of the more distant star, which, being so distant, would not show a parallactic shift. This method had the advantage that refraction, aberration, etc., being the same for the two stars, would not distort the results. Herschel consequently set about cataloguing a large number of double stars. One result that came from Herschel's double star observations was his determination, which he announced around 1802, that some double stars consist of physical doubles, that is, two stars that have orbits around their common center of gravity. This was an ironic result because it presented the problem that any double star pairs that are physical doubles, being immediately adjacent to each other, are not suitable for a parallactic measurement. The reason for this is because neither of the two stars, being located at essentially the same distance from the earth and thus having the same parallax, can be used as a fixed reference point for determining the parallactic motion of the other. In any case, Herschel never claimed to have detected a parallax.

Around 1808, Giuseppe Piazzi, a prominent Italian astronomer, who was also a Theatine monk, announced that he had detected a parallactic shift in three stars: for Aldeberan, he had measured a parallax of 1.6"; for Sirius, 4"; and for Procyon, 5.7". Another Italian astronomer, Calendrelli, also at this time claimed to have found a parallactic shift of 4.4" in Vega. Moreover, in 1814, Ireland's Astronomer Royal, John Brinkley, published a paper reporting his determination of parallaxes of 1.0" for the star Vega, 2.7" for Altair, 1.1" for Arcturus, and 1.0" for Deneb. His results were challenged by John Pond, England's Astronomer Royal, leading to a controversy that lasted for more than a decade. In an 1822 paper in the dispute, Pond asserted:

> The history of annual parallax appears to me to be this: in proportion as instruments have been imperfect in their construction, they have misled observers into the belief of the existence of sensible parallax. This has happened in Italy to astronomers of the very first reputation. The Dublin instrument [Brinkley's] is superior to any of similar construction on the Continent; and accordingly it shows a much less parallax than the Italian astronomers imagined they had detected. Conceiving that I have established, beyond a doubt, that the Greenwich instrument approaches still nearer to perfection, I can come to no other conclusion than that this is the reason why it discovers no parallax.[1]

Pond went on to assert that his observations assured him that the parallax of Vega could not be as large as 0.1" (in fact, Vega has a parallax of 0.12"). Pond's conclusion that no parallax

---

[1] As quoted in J. D. Fernie, "The Historical Search for Stellar Parallax," *Journal of the Royal Astronomical Society of Canada*, 69 (1975), 226.

had ever been measured seems to have been widely accepted.

Bessel and the Detection of Stellar Parallax

The first successful detection of stellar parallax was that announced in 1838 by Friedrich Wilhelm Bessel (1784–1846). Sir John Herschel was among those who applauded this discovery, describing it as "the greatest and most glorious triumph which practical astronomy has ever witnessed."[1]

Bessel had not originally planned on a career in astronomy; in fact, at age fifteen he became an unpaid apprentice in a leading mercantile firm in Bremen, Germany. In accord with this work, he began to study geography, whence he turned to navigation in hopes of becoming the cargo officer on a merchant vessel. From navigation he proceeded to study astronomy, making such progress that by age twenty, he had completed a paper on observations of Halley's comet that led Olbers to take an interest in the young man. Later Olbers remarked that his greatest contribution to astronomy was his recognition of Bessel's extraordinary abilities. Olbers helped Bessel secure a position in the observatory of Johannes Schröter, where Bessel progressed with such success that in 1809 he was named director of the new Königsberg Observatory in Prussia. One of the first important publications that Bessel completed appeared in 1818 as his *Fundamenta astronomiae pro anno 1755*. This was based not primarily on his own observations, but rather observations of 3,222 stars made by James Bradley between 1750 and 1762, but left in an unreduced state. Bessel reduced these observations, correcting them for such factors as refraction, aberration, and the instrumental errors that Bessel had found in Bradley's instruments. Thus Bessel provided very precise positional data for these stars as they appeared in the year 1755. Such information was valuable for many reasons, not least, the detection of proper motions.

Bessel had a long-standing interest in determining stellar parallax. He as well as other astronomers faced the problem of deciding which stars to select for observation, the problem

---

[1] John Herschel, "An Address ... on Presenting the Honorary Medal to M. Bessel," *Royal Astronomical Society Memoirs*, 12 (1842), 453.

being to select a star that is near enough to manifest a parallax. At least two rough criteria for deciding which stars are nearest were available. The first of these was that one could select a very bright star, assuming that its brightness was due to its proximity. The difficulty in this is that a star's apparent brightness is a function not only of how near it is, but also how luminous it is. Consequently, a star that appears very bright may in fact be remote, but producing vast quantities of light. The second criterion was to select a star with a large proper motion, this because it was recognized that the nearer any given star is, the more evident must be whatever motions it has. This approach was beset by the problem that a star's proper motion is a function not only of its distance, but also of its absolute motion and the direction of that motion—a star moving directly toward or away from us would show no proper motion, however great its actual velocity. In 1812, Bessel urged the advantages of the proper-motion criterion and in 1815 attempted to observe a parallax for the star 61 Cygni, which had the largest proper motion (5.2" per year) of any star for which Bessel had data. Bessel's 1815 attempt failed as did a second attempt conducted in 1818, the latter attempt being made not in his telescope's field of view but in Bradley's highly reliable observational data.

In 1829, Bessel acquired what he referred to in his 1838 discovery paper as his "great Fraunhofer Heliometer." This was a telescope made by Fraunhofer with a 6.25-inch main lens. It was called a heliometer because it was constructed in such manner as to make it especially useful for measuring the widths of the sun and planets. In particular, the main lens consisted of two semi-circular halves, which could be shifted relative to each other. Were the image of Venus, for example, seen in the two halves of this lens and then one lens displaced so that the left side of one image coincided with the right side of the other, the amount of displacement could be measured and the apparent width of the planet easily determined.

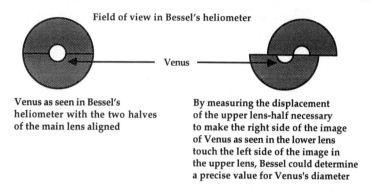

| Venus as seen in Bessel's heliometer with the two halves of the main lens aligned | By measuring the displacement of the upper lens-half necessary to make the right side of the image of Venus as seen in the lower lens touch the left side of the image in the upper lens, Bessel could determine a precise value for Venus's diameter |

This same method could also be used to determine the precise distance between two nearby stars. The acquisition of this telescope gave Bessel one of the premier telescopes in the world and correspondingly enhanced his chances of detecting a parallax. In assessing the importance of this telescope for Bessel's work, one should keep in mind the following comment made by Bessel: "Every instrument ... is made twice, once in the workshop of the artisan, in brass and steel, but then again by the astronomer on paper, by means of the list of necessary corrections, which he determines by his investigations."[1] Bessel's point was to stress that the astronomer can measure more precisely than any instrument maker can build.

In 1837, Wilhelm Struve, who also had the advantage of a Fraunhofer telescope, announced that he had determined a parallax of 0.125" for the star Vega (modern value = 0.123"). This might seem a sufficient basis for crediting Struve with the first detection of stellar parallax. Struve, however, put forth his result in a very tentative manner. Moreover, in 1839, he published a revised value: 0.2619" ±0.0254"! Bessel, seeing Struve's 1837 result, began again to observe 61 Cygni, which is a star of 5.6 magnitude, i.e., it is nearly at the limit of naked eye visibility. Using as reference two adjacent far dimmer and presumably more distant stars, Bessel announced in October, 1838 that he had detected a parallactic shift of 0.3136" ±0.0202" (modern value = 0.292" ±0.004"). And this was

---

[1] F. W. Bessel, "Ueber die Verbindung der astronomischen Beobachtungen mit der Astronomie" in *Populäre Vorlesungen* (Hamburg, 1848), p. 432.

widely accepted as a reliable determination.

To achieve an understanding of Bessel's determination, recall that a parallax of 0.3" for 61 Cygni would mean that 61 Cygni in the course of a year would move through an ellipse (call it a circle) of arc radius 0.3". The geometry of this can be represented by a right angled triangle with one very short side. Imagine yourself standing at one end of that short side and then moving to the other end.

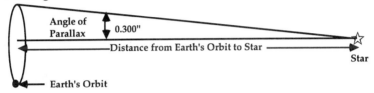

Having determined the angle and knowing the radius of the earth's orbit, Bessel could proceed to find the length of the long side of the triangle, i.e., the distance of the star.

To keep the figures comprehensible at first, imagine a person standing atop a tall building and looking at a distant object. Suppose also that the person moves two feet to the left and notices that as a result of this, the object has shifted by .300". How far is the distant object? The formula to calculate this distance (R) is given below with d representing the distance through which the person has moved (2 ft.).

$$\frac{d=2\text{ft.}}{2\pi R} = \frac{0.300"}{360 \cdot 60 \cdot 60"};$$

hence $R = \dfrac{360 \cdot 60 \cdot 60" \cdot 2\text{ft.}}{0.300" \cdot 2\pi} = 1.38 \cdot 10^6 \text{ft.} = 260 \text{ miles.}$

Thus Bessel's measurement was equivalent to detecting a change in the position of an object moving around an ellipse with a major axis of 2 feet, the object being positioned 260 miles distant.

From the above formula along with the fact that the radius of the earth's orbit is 93 million miles, one can calculate the distance of 61 Cygni. In his 1838 paper, Bessel stated: "we find the distance of the star 61 Cygni from the sun [is] 657,700 mean distances of the earth from the sun...." And he added that "light employs 10.3 years to traverse this distance."[1] A moment's reflection on the result of this calculation will suggest why astronomers avoid using feet and miles as measures of stellar distances. They find it more convenient to use either light-years or parsecs to indicate distances. A parsec is the distance of an object that produces a parallax of one second of arc, or $1.92 \bullet 10^{13}$ miles. A light-year is the distance that light travels in one year, or $5.88 \bullet 10^{12}$ miles. Thus 1 parsec = 3.26 light-years. You can check your answer by using the fact that 61 Cygni is 3.46 parsecs distant.

Bessel's choice of 61 Cygni was fortunate; it ranks 12th in the list of nearest stars. Thomas Henderson detected a parallax in the star alpha Centauri around the same time as Bessel's work, basing his analysis on observations that he had made earlier in the decade, but had not found time to reduce. Henderson's choice was even more fortunate because alpha Centauri is the nearest star at 1.31 parsecs. Henderson's observations were, however, made from South Africa, whereas Bessel, whose observatory was located in Prussia, could not observe alpha Centauri. The questions of interstellar travel and of UFOs from other stars are frequently discussed. You can get an insight into this by calculating how long it would take for a rocket ship with a speed of 10,000 mph to travel to the nearest star, alpha Centauri.

Wilhelm Struve

Another astronomer who had the advantage of Munich-made telescopes was Friedrich Georg Wilhelm Struve (1793–1864), who in 1824 while at Dorpat Observatory in Russia secured a 9-inch Fraunhofer refractor and in 1839, by which time he had become director of the Pulkova Observatory, received a 15-inch aperture refractor. Struve presented some of his main conclusions concerning stellar astronomy in his *Études d'astronomie stellaire* (1847). In that volume, he provided evidence for interstellar obscuring matter by showing that stars

---

[1] F. W. Bessel, "The Parallax of 61 Cygni" in Harlow Shapley and Helen Howarth (eds.), *A Source Book in Astronomy* (New York: McGraw-Hill, 1929), p. 219.

grow dimmer more rapidly than is reconcilable with the assumption that they are homogeneously distributed in the plane of the Milky Way. In other words, we see fewer eighth-magnitude stars than would be expected were the stars all equally bright and homogeneously distributed. This implied that space contains some sort of matter that progressively dims the light of stars at greater distances. Struve attempted to quantify this effect, stating that "light, in passing through a distance equal to that of a first-magnitude star, is subject to extinction by nearly one hundredth—in particular, it loses 1/107th of its intensity."[1] More distant stars, according to this analysis, would be diminished in brightness to an ever greater extent. Struve in his 1847 book also analyzed a number of William Herschel's papers, urging, for example, the conclusion *"that the [Milky Way] system of Herschel, which he proposed in 1785, breaks down completely in light of the author's later researches; and that Herschel himself has entirely abandoned it."*[2] Subsequently Struve, claiming that our telescopes probably do not reach the extreme portions of the Milky Way, added that *"if we consider all the fixed stars which surround the sun, as forming a great system, that of the Milky Way, we are in complete ignorance as to its extent, and that we have not the least idea as to the external form of that immense system."*[3]

Sir John Herschel

In the second sentence of his *Origin of Species* (1859), Charles Darwin admitted that in his book he would discuss "the origin of species—that mystery of mysteries as it has been called by one of our greatest philosophers." By that accolade, Darwin referred to Sir John Herschel (1792–1871), the only child of William Herschel. The younger Herschel's eminence was also evidenced some years earlier when John Pringle Nichol, a Glasgow astronomer, defended some astronomical ideas presented in his *Views of the Architecture of the*

---

[1] F. G. W. Struve, *Études d'astronomie stellaire* (St. Petersburg, 1847), p. 87.
[2] Struve, *Études*, p. 34.
[3] Struve, *Études*, p. 63.

Heavens by ascribing them to "Sir John Herschel—a mere hint from whom is of authority." John Herschel's scientific reputation among Victorians was such that it has been compared to that attained in the twentieth century by Einstein.

William Herschel's son came to astronomy by a very different route from that taken by his autodidact father. John Herschel not only graduated from the University of Cambridge in 1813, but also proved himself the most mathematically capable of that year's degree candidates by capturing first places in both the Wrangler and the Smith Prize competitions. A founder with Babbage (pioneer of the computer) of the Analytical Society, which urged the introduction of continental mathematics at traditionally Euclidean and Newtonian Cambridge, Herschel by 1813 had also achieved election to the Royal Society. Concerned as to what career to follow, he pondered the ministry at his father's suggestion, studied law for over a year, nearly won election to the professorship of chemistry at Cambridge, but by 1816 had decided on "going under my father's direction to take up stargazing." In 1820, two years before William Herschel's death, the father and son constructed a new twenty-foot focal length reflector and by 1825 John Herschel had won a gold medal from England's Astronomical Society for a catalogue of double stars he compiled in collaboration with James South. Writing to his aunt Caroline Herschel concerning nebulae in 1825, he stated: "These curious objects I shall now take into my especial charge...." Between that time and 1833, in which year he published his *Treatise on Astronomy* and completed a new catalogue of nebulae, adding 525 that he had himself

Sir John Herschel

discovered, he had also published a major study of the new wave theory of light and an analysis of the methodology of science, his *Preliminary Discourse on the Study of Natural Philosophy* (1830), which both Darwin and John Stuart Mill warmly admired. He had also married in 1829 and was knighted two years later.

On 13 November 1833, Herschel, with his wife and their three small children and his large telescope, sailed for the Cape of Good Hope in South Africa, planning to extend his father's researches to the southern celestial hemisphere.

**Sir John Herschel's 20' Reflector,
Erected at Feldhausen in South Africa**

By the time he returned to England in 1838, he had discovered hundreds of new double stars, 1,269 new nebulae, made a map of the Magellanic Clouds showing 1,163 objects, and expanded his family by three, six more children being born after his return. In 1847, he published his *Results of Astronomical Observations Made ... at the Cape of Good Hope,* having in the interim made a number of advances in photography, a then new technology to which he contributed the terms "positive" and "negative" as well as much else. In 1849, he published his *Outlines of Astronomy,* which for decades was deemed the most authoritative astronomy text in English and which was translated into Chinese, Russian, and possibly Arabic. Serving from 1850 to 1855 as Master of the Mint, a position earlier held by Newton, he died in 1871 and was buried next to Newton in Westminster Abbey, his last years having been devoted to publishing a consolidated catalogue of 5,079 nebulae (1864), a translation of Homer's *Iliad* (1866), and a catalogue of 10,300 double stars, left incomplete at his death.

Turning now to John Herschel's theoretical views in stellar astronomy, it is noteworthy that in regard to the structure of the Milky Way, in 1835 he wrote William Rowan Hamilton, an Irish astronomer and mathematician, that he had concluded that "the Milky Way is not a mere stratum but an annulus...."[1] Fourteen years later in his *Outlines of Astronomy*, he expressed his view in more general terms, writing that "beyond the obvious conclusion that its form must be, generally speaking, *flat*, and of a thickness small in comparison with its area in length and breadth, the laws of perspective afford us little further assistance in the inquiry."[2]

On the nature of the nebulae, he stated in 1825 that they are probably composed of "a self-luminous or phosphorescent substance, gradually subsiding into stars and sidereal systems." In his *Treatise on Astronomy* (1833), he noted that the nebulae furnish

> ... an inexhaustible field of speculation and conjecture. That by far the larger share of them consist of stars there can be little doubt; and in the interminable range of system upon system, ... the imagination is bewildered and lost. On the other hand, if it be true, as, to say the very least, it seems extremely probable, that a phosphorescent or self-luminous matter also exists, disseminated through extensive regions of space, in the manner of a cloud or fog—now assuming capricious shapes, like actual clouds drifted by the wind, and now concentrating itself like a cometic atmosphere around particular stars;—what ... is the nature and destination of this nebulous matter? Is it absorbed by the stars in whose

---

[1] As given in Michael Hoskin, "Astronomical Correspondence of William Rowan Hamilton," *Journal for the History of Astronomy*, 15 (1984), p. 72. See also M. Hoskin, "John Herschel's Cosmology," *Journal for the History of Astronomy*, 18 (1988), 1–34.
[2] John Herschel, *Outlines of Astronomy*, 3rd ed. (London: Longman, 1850), p. 532.

nebulous matter? Is it absorbed by the stars in whose neighborhood it is found, to furnish by its condensation, their supply of light and heat? or is it progressively concentrating itself by the effect of its own gravity into masses, and so laying the foundation of new sidereal systems or of insulated stars? It is easier to propound such questions than to offer any probable reply to them.[1]

Nonetheless, in his *Results...* (1847), he labeled his father's shining fluid "purely hypothetical" and added concerning the resolvabilty of nebulae:

> The distinction between nebulae properly so called, and those which we are to consider as certainly or very probably clusters of stars, resting, as it must do, on the merely temporary and conventional ground of the capacity or incapacity of our telescopes, wholly or partially to resolve them, can never become a permanent ground of classification, since every new improvement in the powers of telescopes will cause more and more nebulae to pass into the class of clusters.[2]

He put this more succinctly in the 1849 first edition of his *Outlines of Astronomy*, where he stated: "it may very reasonably be doubted whether there be really any essential physical distinction between nebulae and clusters of stars, at least in the nature of the matter of which they consist...."[3]

John Herschel also expressed views concerning the location of the nebulae; in *Results...*, he stated that "the nebulous system is distinct from the sidereal, though involving, and to a certain extent intermixed with the latter."[4] It is very important to understand the empirical basis Herschel provided for this conclusion; in that book, he included a diagram (see p. 165) of the heavens on which he plotted the number of known nebulae in each region. In this diagram, the outermost circle corresponds to the celestial equator, whereas the dashed line

---

[1] John Herschel, *Treatise on Astronomy* (London: Longman, 1833), pp. 406–7.
[2] John Herschel, *Results of Astronomical Observations . . . at the Cape of Good Hope* (London: Smith, Elder, and Co., 1847), p. 139.
[3] Herschel, *Outlines*, p. 598.
[4] Herschel, *Results*, p. 136.

running across the circles represents the plane of the Milky Way. An inspection of this diagram reveals that very few nebulae lie in the plane of the Milky Way; in fact, the great majority lie toward its poles. A natural conclusion to draw from this is that the nebulae cannot be "island universes," because were this to be the case, they should be more or less randomly distributed over the heavens rather than oriented in relation to the plane of the Milky Way. Although John Herschel did not press this point, later authors, for example, R. A. Proctor, did do so; in fact, this became one of the most convincing arguments against belief in island universes.

John Herschel's analysis in his 1847 book of the larger and smaller Magellanic Clouds (which he called the "Nebeculae") was also seen as relevant to the question of the resolvability of nebulae. Concerning these two large objects, which lie fairly near the south celestial pole and hence beyond the visibility of most persons in the northern hemisphere, he stated:

> The Nubecula Major, like the Minor, consists partly of large tracts and ill-defined patches of irresolvable nebula, and of nebulosity in every stage of resolution, up to perfectly resolved stars like the Milky Way, as also of regular and irregular nebulae properly so called, of globular clusters in every stage of resolvability, and of clustering groups sufficiently insulated and condensed to come under the designation of "clusters of stars" .... [1]

This report was later used by William Whewell, as we shall see shortly, as another argument against the resolvability of all nebulae.

As a final point, it is significant that whereas John Herschel in his *Treatise on Astronomy* (1833) had devoted only one chapter of 38 pages to stellar astronomy, in the 1849 first edition of his *Outlines of Astronomy*, which was to a large extent based on the earlier *Treatise,* he assigned that subject three chapters or 102 pages. In short, stellar astronomy, due in good measure to the efforts of William and John Herschel, was coming to assume an ever larger role in astronomy, a tendency that has continued into the contemporary period.

---

[1] Herschel, *Results*, p. 146.

From William Herschel to 1860

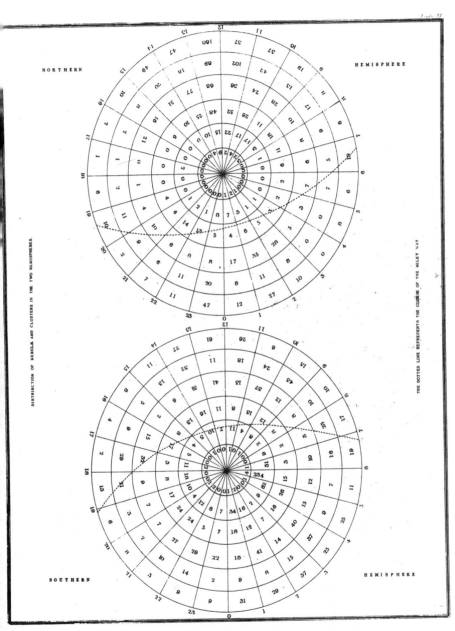

John Herschel's Diagram of the Distribution of the Nebulae

Lord Rosse

By 1840, William Herschel's giant 48-inch reflector had long since ceased to be functional. Consequently, on 1 January 1840, John Herschel assembled his family in the tube of that telescope, read an obituarial poem concerning it that he had composed for the occasion, and laid it to rest. The base of the telescope now decorates the courtyard at Royal Greenwich Observatory.

Although the younger Herschel did not display his father's desire for building ever larger reflectors, this passion had passed to William Parsons, the Third Earl of Rosse (1800–1867). Lord Rosse was an Irish nobleman, who after studies at Trinity College, Dublin and at Oxford University presided over the family estates from Birr Castle in central Ireland. In 1840, Rosse erected a 36-inch-aperture reflector.

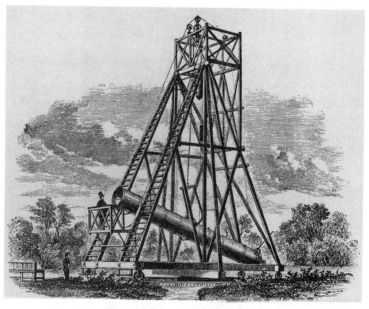

Lord Rosse's 36"-Aperture Reflecting Telescope

By 1845, he had completed "the Leviathan of Parsonstown" (see next picture), a reflector of 72-inch aperture and 54 feet in focal length. Until 1917, it remained the largest telescope ever constructed. Possessing more than double the light-gathering power of William Herschel's 48-inch reflector, it was mounted in the plane of the meridian between thick brick walls nearly sixty feet high. Its copper and tin mirrors, which required six weeks to cool after casting, weighed over four tons. Rosse's reflector was ideal for studying nebulae, although administrative and political matters, especially his efforts to alleviate the problems caused in the 1840s by the Irish potato famine, limited the time he could devote to astronomy. Nonetheless, important new results began to issue from his observatory.

Lord Rosse's 72"-Aperture Reflecting Telescope

Concerning the resolvability of Orion, Rosse in March, 1846, expressed "little if any doubt" that it could be resolved. He saw it as "abounding with stars, and exhibiting the characteristics of resolvability strongly marked." Another Irish astronomer, T. R. Robinson, reported in 1848 that Rosse had observed 50 of the brightest nebulae in John Herschel's 1833 catalogue of nebulae and they "were all resolved without exception." By 1849, John Herschel had cautiously stated that all nebulae are in principle resolvable, and in 1850, J. P. Nichol, a Glasgow astronomer, asserted: "Every shred of that evidence which induced us to accept as a reality, accumulations in the heavens of matter not stellar, is for ever and hopelessly destroyed."

In 1845, Rosse discovered the spiral structure of the nebula M51. This was the first known case of a large class of nebulae that are known as *spiral nebulae*. By 1850, he had announced the detection of 14 other spiral nebulae (it is now known that not all of these are actually spirals). The next reading consists of a selection from his 1850 paper presenting his discovery of spiral nebulae. This was Rosse's most important discovery with his telescope, which instrument turned out to be less productive than many had hoped. Although Rosse rivaled William Herschel's telescope-making skills, he lacked the latter's genius for creative astronomical work.

☆☆☆☆☆☆☆☆☆☆☆☆☆☆☆☆☆☆☆☆☆☆☆☆☆☆☆
Selection from Rosse's "Observations on the Nebulae," *Philosophical Transactions of the Royal Society*, 140 (1850), 499–514:504–6.

It will be at once remarked, that the spiral arrangement so strongly developed in Plate XXXV.H.1622, 51 Messier, fig.1, is traceable, more or less distinctly, in several of the sketches. More frequently indeed there is a nearer approach to a kind of irregular interrupted annular disposition of the luminous material than to the regularity so striking in 51 Messier; but it can scarcely be doubted that these nebulae are systems of a very similar nature, seen more or less perfectly, and variously placed to the line of sight. In general the details which characterize objects of this class are extremely faint, scarcely perhaps to be seen with certainty on a moderately good night with less than the full aperture of 6 feet: in 51 Messier, however, and perhaps a few more, it is not so. A 6-feet aperture so strikingly brings out the characteristic features of 51 Messier, that I think considerably less power would suffice, on a very fine night, to bring out the principal convolutions. This nebula has been seen by a great many visitors, and its general resemblance to the sketch at once recognized even by unpractised eyes. Messier describes this object as a double nebula without stars; Sir William Herschel as a bright round nebula, surrounded by a halo or glory at a distance from it, and accompanied by a companion; and Sir John Herschel observed the partial subdivision of the *s.f.* limb of the ring into two branches. Taking Sir J. Herschel's figure, and placing it as it would be if seen with a Newtonian telescope, we shall at once recognize the bright convolutions of the spiral, which were seen by him as a divided ring. We thus observe, that with each successive increase of optical power, the structure has become more complicated and more unlike anything which we could picture to ourselves as the result of any form of dynamical law, of which we find a counterpart in our system. The connection of the companion with the greater nebula, of which there is not the least doubt, and in the way represented in the sketch, adds, as it appears to me, if possible, to the difficulty of forming any conceivable hypothesis. That such a system should exist, without internal movement, seems to be in the highest degree improbable: we may possibly aid our conceptions by coupling with the idea of motion that of a resisting medium; but we cannot regard such a system in any way as a case of mere statical equilibrium. Measurements therefore are of the highest interest, but unfortunately they are attended with great difficulties. Measurements of the points of maximum brightness in the mottling of the different convolutions must necessarily be very loose; for although on the finest nights we see them breaking up into stars, the exceedingly minute stars cannot be seen steadily, and to identify one in each case would be impossible with our present means. The nebula itself, however, is pretty well studded with stars, which can be distinctly seen of various sizes, and of a few of these, with reference to the principal nucleus, measurements were taken by my assistant, Mr. Johnstone Stoney, in the spring of 1849, during my absence in London; for some time before the weather had been continually cloudy. These measurements have been again repeated by him this year, 1850, during the months of April and May. Just as was the case last year, in February and March the sky was almost constantly overcast. He has also taken some measures from the centre of the principal nucleus to the apparent boundary of the coils, in different angles of position. The micrometer employed was furnished with broad lines formed of a coil of silver wire in the way I have described, seen without illumination. Some of the stars in the nebula are so bright, I have little doubt they would bear illumination; if so, their positions with respect to some one star might be obtained with great accuracy of course by employing spiders' lines; this season however it is too late to make the attempt. Several of these stars are no doubt within the reach of the great instruments at Pulkova and at Cambridge, U.S., and I hope the distinguished astronomers who have charge of them will consider the subject worthy of their attention. Their better climate gives them many advantages, of which not the least is the opportunity of devoting time to measurements without any serious interruption to other work. I need perhaps hardly add, that measurements taken from the estimated centre of a nucleus, and still more from the estimated termination of nebulosity, are but the roughest approximations; they are however the only measurements nebulosity admits of, and if sufficiently numerous, I think they will bring to light any considerable change of place, or form, which may occur.

The spiral arrangement of 51 Messier was detected in the spring of 1845. In the following spring an arrangement, also spiral but of a different character, was detected in 99 Messier, Plate XXXV. fig. 2. This object is also easily seen, and probably a smaller instrument, under favourable circumstances, would show everything in the sketch. Numbers 3239 and 2370 of Herschel's Southern Catalogue are very probably objects of a similar character, and as the same instrument does not seem to have revealed any trace of the form of 99 Messier, they are no doubt much more conspicuous. It is not therefore unreasonable to hope, that whenever the southern hemisphere shall be re-examined with instruments of great power, these two remarkable nebulae will yield some interesting result.

**Lord Rosse's Drawings of
Two Spiral Nebulae (M51 at top, M99 below)**

The other spiral nebulae discovered up to the present time are comparatively difficult to be seen, and the full power of the instrument is required, at least in our climate, to bring out the details. It should be observed that we are in the habit of calling all objects spirals in which we have detected a curvilinear arrangement not consisting of regular re-entering curves; it is convenient to class them under a common name, though we have not the means of proving that they are similar systems. They at present amount to fourteen, four of which have been discovered this spring; there are besides other nebulae in which indications of the same character have been observed, but they are still marked doubtful in our working list, having been seen when the air was not very transparent; 51 Messier, Plate XXXV. fig.1, is the most conspicuous object of that class.

☆☆☆☆☆☆☆☆☆☆☆☆☆☆☆☆☆☆☆☆☆☆☆☆☆☆☆☆☆☆☆☆

Bond, the Harvard Refractor, and the Resolution of Orion

Harvard 15" Refractor

In 1847, Harvard College, where William Cranch Bond (1789–1859) was the leading astronomical observer, acquired a magnificent 15-inch-aperture Munich-made refracting telescope, very similar to Wilhelm Struve's famous Pulkova telescope. Persons interested in the new telescope pressed Bond for evidence of its excellences. A natural question was whether this instrument was at all comparable in quality to Lord Rosse's giant reflector. Very little time passed before Bond began to announce discoveries with this powerful instrument. On 22 September 1847, he wrote Harvard's President Everett to reveal that with the newly acquired instrument, he had achieved, like Lord Rosse, the resolution of the Orion Nebula, an accomplishment that many had unsuccessfully sought before, not least because of its bearing on the questions of the nature of the nebulae and of the existence of island universes.

Bond's enthusiastic letter, which received wide circulation

through its publication in the *American Journal of Science and Arts*, is the next reading.

☆☆☆☆☆☆☆☆☆☆☆☆☆☆☆☆☆☆☆☆☆☆☆☆☆☆☆☆☆
William Cranch Bond, "[Letter to President Everett]," *American Journal of Science and Arts*, Ser. 2, 4 (1847), 427.

Cambridge Observatory, Sept. 22d, 1847
Dear Sir—

You will rejoice with me that the great nebula in Orion has yielded to the power of our incomparable telescope.

This morning the atmosphere being in a favorable condition, at about 3 o'clock the telescope was set upon the Trapezium in the great nebula of Orion.—Under a power of 200, the 5th star was immediately conspicuous; but our attention was directly absorbed with the splendid revelations made in its immediate neighborhood. This part of the nebula was resolved into bright points of light. The number of stars was too great to attempt counting them; many were however readily located and mapped. The double character of the brightest star of the Trapezium was readily recognized with a power of 600.—This is "Struve's 6th star;" and certain of the stars composing the nebula were seen as double stars under this power.

It should be borne in mind that this nebula and that of Andromeda have been the last strong-hold of the nebular theory; that is, the idea first thrown out by the elder Herschel, of masses of nebulous matter in process of condensation into systems. The nebula in Orion yielded not to the unrivaled skill of both the Herschels, armed with their excellent Reflectors.

It even defied the power of Lord Rosse's three-foot mirrors, giving "not the slightest trace of resolvability," or separation into a number of *single* sparkling points.

And even when, for the first time, Lord Rosse's grand Reflector of six-feet speculum was directed to this object, "not the veriest trace of a star was to be seen." Subsequently his Lordship communicated the result of his farther examination of Orion as follows:—

"I think I may safely say, that there can be little if any doubt as the resolvability of the nebula.—We could plainly see that all about the Trapezium is a mass of stars; the rest of the nebula also abounding in stars, and exhibiting the characteristics of resolvability strongly marked."

This has hitherto been considered as the greatest effort of the largest reflecting telescope in the world;—and this our own telescope has accomplished.

I feel deeply sensible of the odiousness of comparisons;—but innumerable applications have been made to me for evidence of the excellence of the instrument, and I can see no other way in which the public are to be made acquainted with its merits.

With sincere respect and esteem, I remain, Sir, your obedient servant,

W. C. Bond

☆ ☆

Stephen Alexander: Spiral Nebulae and the Milky Way

One of the most interesting offshoots of Rosse's discovery of spiral nebulae came in 1852 in a paper by Stephen Alexander, a professor of astronomy and mathematics at the College of New Jersey (now Princeton University). Alexander published a long paper on the classification of nebulae in the second volume of a new American journal called *The Astronomical Journal*. From our present perspective, the most striking idea in that paper is contained in a section titled: "The Milky Way—A Spiral." In that section, Alexander stated: "The following coincidences are consistent with the supposition that the Milky Way and the stars within it together constitute a spiral with several (it may be *four*) branches, and a central (probably spheroidal) cluster...."[1] It is interesting to reflect on the strengths and weaknesses of Alexander's bold speculation. For example, one may ask whether it fits with any observations already available in the 1850s?

---

[1] Stephen Alexander, "On the Origin of the Forms and the Present Condition of Some of the Clusters of Stars and Several of the Nebulae," *Astronomical Journal*, 2 (1852), p. 101.

Whewell, Spencer, and Struve: Some Dissenting Views

The resolvability of nebulae pointed to the idea of a large cosmos, and for some persons supported belief in "island universes," to use a term introduced in 1845 by Alexander von Humboldt. Not everyone, however, saw the situation in this way. William Whewell (1794–1866), although not a professional astronomer, was among the most brilliant and broadly learned persons of his day. In 1853, he published *Of the Plurality of Worlds: An Essay*, in which he challenged the then widespread belief that intelligent life exists throughout the universe. In analyzing nebulae as possible island universes, he noted John Herschel's statement (discussed previously) on the Magellanic Clouds, drawing from it the inference that if nebular material occurs in conjunction with stars in these structures, then nebulae must be no more distant than these stars. If this were so, it implied the rejection of the view of nebulae as immensely distant and as "island universes." Moreover, calling attention to the similarity in appearance of spiral nebulae and cometary materials, he suggested that the spirals are solar systems in formation and hence nearby. He concluded his chapter on nebulae by stating: "Thus we appear to have good reason to believe nebulae are vast masses of incoherent or gaseous matter, of immense tenuity and [thus] to have made it certain that *these* celestial objects at least are not inhabited."

Dissenting views also came from Herbert Spencer (1820–1903), a prominent English philosopher and proponent of evolutionary ideas. In an 1858 essay on the nebular hypothesis of Laplace, who around 1800 had suggested that our solar system was formed from the contraction of a rotating nebular mass, Spencer stated in regard to nebulae: "The hypothesis now commonly entertained is, that all nebulae are galaxies of stars more or less like in nature to that immediately surrounding us; but that they are so inconceivably remote, as to look through an ordinary telescope like small faint spots." This view, which Spencer described as "of late years idly repeated and uncritically received," he found to be "totally irreconcileable with the facts."[1] Spencer's own convictions were "that the nebulae are not further off from us than parts of our own sidereal system, of which they must be considered members; and that when they are resolvable into discrete masses, these masses cannot be considered as stars in anything like the ordinary sense of that word."[2] Spencer marshaled a number of arguments in support of this conclusion. First, he noted that nearly all nebulae lie off the plane of the Milky Way and toward its poles. This effect is so prevalent that it cannot be mere coincidence; it must be seen as indicating a "physical connexion" between nebulae and our stellar system, for otherwise we cannot explain why nebulae, if they are remote galaxies, avoid the plane of the Milky Way. Second, Spencer argued, if our telescopes can barely distinguish the fainter stars in our own system, the Milky Way, how is it that they can resolve these distant galaxies into individual stars? They must consequently be no more distant than the extreme parts of our own system. How are we to explain the fact, he continued, that some of the largest nebulae (Andromeda) cannot be resolved, whereas some small ones can be, if we assume the island universe doctrine? Finally, he restated Whewell's objection based on John Herschel's description of the Magellanic Clouds.

Wilhelm Struve also dissented from prevailing views on the resolvability of nebulae, basing his argument primarily on reports that some nebulae had changed. William Herschel, for example, had stated in 1802: "The changes I have observed in the great milky nebulosity of Orion, 23 years ago, and which have also been noticed by other astronomers, cannot permit us to look upon this phenomenon as arising from immensely distant regions of fixed stars." In 1853, Struve stated:

> Recently astronomers have believed they have found, in the resolution into stars of a small number of nebulae with the help of fine modern telescopes, a complete refutation of the ideas of Sir William Herschel; although to some extent the alleged miracles of resolution, for example of the Orion and Andromeda nebulae, are nothing but illusions. The study of

---

[1] As quoted in Stanley L. Jaki, *The Milky Way: An Elusive Road for Science* (New York: Science History, 1972), p. 267.
[2] As quoted in Jaki, *Milky Way*, p. 268.

a small number of nebulae will by no means settle the question of the nature of these objects. To do this we must examine and compare the 6000 nebulae now known, or at least all those visible from a given place. Meanwhile, we believe that the discoveries of M. Liapounov have given, to the contrary, a direct proof of the truth of William Herschel's view. For since in certain nebulae there are major rapid changes, these nebulae are not clusters of stars; for the starry heavens, even in the regions nearest to us, present only changes of position that are extremely slow and comparatively small.[1]

The materials presented in this chapter shed further light on our three central questions:
(1) What size, shape, and structure does the Milky Way have?
(2) What are the nebulae and are they resolvable?
(3) Are the nebulae "island universes"?
Nonetheless, these questions remained some distance from resolution.

---

[1] As quoted in Michael Hoskin, "Apparatus and Ideas in Mid-nineteenth-century Cosmology," *Vistas in Astronomy*, 9 (1967), 83.

Chapter Five

The New Astronomy

Some Instrumental Advances

In the middle third of the nineteenth century, astronomers gained access to a number of new instrumental techniques, in particular, photography and spectroscopy. Photography has proved important in at least three ways:
(1) Most notably, because photographic exposures can be made over extended periods of time, photography gives telescopes increased light-gathering power. Dim objects by being exposed for many minutes (or hours) are made visible, whereas they are not visible to the naked eye, which is unable to accumulate light to any degree.
(2) Photography permits astronomers to record fleeting events, e.g., solar flares or eclipses.
(3) Photography is objective; as they say, "the camera does not lie." Astronomers can record what their telescopes "see" without individual peculiarities (of astronomers, not of telescopes) being a factor.

As important as photography has been, spectroscopy has proven of even greater significance. Indeed, so important was spectroscopy that it led astronomers to speak of the "New Astronomy." In effect, it enabled the astronomer to function as an astrophysicist and astrochemist. Josef Fraunhofer was one of its pioneers. It was, however, Wilhelm Bunsen and Gustav Kirchhoff who in 1859 were the first fully to understand the significance of spectroscopy. William Huggins was the leading pioneer of spectroscopy as applied to astronomy.

This chapter presents two readings from Huggins. The information provided in the following chronology, especially that associated with 1859, provides a sound basis for understanding the selections from Huggins.

Highlights in the Development of Spectroscopy as Related to Astronomy

1666 Isaac Newton discovered the heterogeneous nature of white light; that is, by passing sunlight through a prism (see diagram), he found that it is composed of an array of colors. He had thereby produced a spectrum.

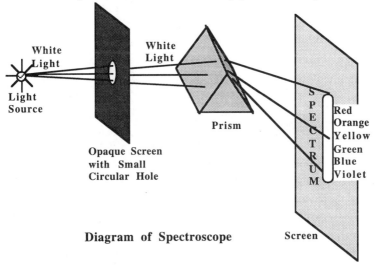

Diagram of Spectroscope

1800 William Herschel, in measuring the temperature increase in a thermometer placed at various positions in the spectrum, discovered the invisible infrared region and showed that infrared rays are reflected and refracted according to the same laws as light.

1801 J. W. Ritter discovered ultraviolet rays, that is, rays beyond the violet end of the spectrum. He did this by showing that such rays produce chemical effects.

1802 W. H. Wollaston discovered 7 dark lines in the solar spectrum; such a spectrum is known as a dark-line spectrum.

1827 John Herschel, as a result of studying spectra of flames impregnated with various metallic salts, stated: "The colours thus communicated by the different bases to flames afford, in many cases, a ready and neat way of detecting extremely minute quantities of them...."

By 1828 Josef Fraunhofer had published papers in which he
(1) described over 500 dark lines in the solar spectrum;
(2) showed that light from the planets or moon exhibits the same pattern of dark lines as sunlight;
(3) found that the spectra of some stars differ from the solar spectrum;
(4) discussed bright-line spectra, i.e., spectra consisting of an array of thin bright lines with large dark spaces between the bright lines; he suggested that some bright lines correspond to some dark lines in the solar spectrum;
(5) developed the diffraction grating, a device that can be used in place of the prism in experiments designed to produce spectra. One simple form of diffraction grating is a plate of glass or a mirror with numerous parallel lines or grooves cut into it. Fraunhofer made one diffraction grating of the reflecting type with 3,625 lines per centimeter.

1832 Sir David Brewster showed that when a continuous spectrum is passed through a gas a dark-line spectrum results.

1835 Auguste Comte in his *Cours de philosophie positive* stated concerning astronomical bodies that "we can never by any means investigate their chemical composition or mineralogical structure ... [nor] true mean temperatures."

1847 John Draper determined that, as he put it, "an ignited body will give a continuous spectrum, or one devoid of fine lines; an ignited gas will give a discontinuous spectrum, one broken up by lines or bands or spaces." He also noted: "as the temperature of an incandescent body rises, it emits rays of light of an increasing refrangibility."

1859   Wilhelm Bunsen and Gustav Robert Kirchhoff established spectrum analysis on a firm basis; they stated that
(1) glowing solids and liquids produce a continuous spectrum;
(2) glowing gases produce a bright-line spectrum;
(3) continuous spectra, when passed through a gas, become dark-line spectra.
(4) The line positions in both dark- and bright-line spectra are characteristic of the chemical constitution of the matter producing the lines; each element and compound has its characteristic spectrum; the element absorbs or emits light only of particular wavelengths. Consequently, the determination of the spectrum produced by an unknown substance will permit its chemical identification, if an identical spectrum is known to result from a body of already identified chemical composition.

1859   Kirchhoff, on the basis of an analysis of the spectral lines in the light from the sun, published a study of the chemical constitution of the sun.

1861   Kirchhoff and Bunsen spectroscopically discovered two new chemical elements, Cesium and Rubidium. Later many more new elements were discovered by means of spectroscopic analysis.

1864   Sir William Huggins, using a spectroscope in conjunction with his telescope (see drawing), showed (what no telescope by itself could have shown) that some nebulae are glowing gases. He did this by discovering that they produce bright-line spectra, which can only be produced by glowing gases. Such spectra also indicate that the object is at a very high temperature.

**Sir William Huggins's Observatory and 8" Refractor
with a Spectroscope Attached**

1864   Giovanni Battista Donati used the spectroscope to show that comets must be at least partly self-luminous because they give bright-line spectra.

1868   Huggins stated that according to Doppler's principle, spectral lines produced by a celestial object moving in such a manner that a component of its motion is in the observer's line of sight should appear shifted in position, the amount of shift being a function of the speed of the object toward or away from the observer. Thus he found a way to measure the radial component of the velocity of celestial objects, that is, how rapidly such objects are approaching or receding.

By 1868 Father Angelo Secchi published a spectroscopic classification of stellar spectra into four spectral types: (1) white to bluish, (2) yellow, (3) red, and (4) faint fiery red stars producing a spectrum somewhat different from his third type.

1868 Through observing a solar eclipse, Jules Janssen and J. Norman Lockyer identified lines in the solar spectrum that corresponded to no chemical substance known at that time; eventually (1895) this element (helium) was discovered on the earth. Thus helium was first discovered in the sun.

1872 Henry Draper (son of J. W. Draper) made the first spectrograph of a star, that is, a photograph of the star's spectrum.

1882 Death of Henry Draper; his widow gave his instruments and a quantity of money to Harvard to be used for the study of stellar spectra.

1890 Publication of the first Henry Draper Catalogue, which consisted of a catalogue of the spectra of 10,000 stars divided into seven types: A, B, F, G, K, M, N. A, B correspond to Secchi's first type; F, G, K to his second; M to his third; and N to his fourth. The Harvard astronomers mainly responsible for the Draper catalogue and this classification were Antonia Maury and Annie Jump Cannon. In 1918–24, a more complete Draper catalogue appeared, classifying 225,000 stars spectroscopically.

1895 Henry Rowland, using diffraction gratings, produced a table of the solar spectrum showing 20,000 lines.

Sir William Huggins

The leading pioneer of astronomical spectroscopy was the Englishman Sir William Huggins (1824–1910). Although unable as a young man to attend Cambridge because of family financial needs, Huggins by 1854 had sold the family business so as to devote his time to science. In 1858, he acquired and erected at his home near London an excellent 8-inch-aperture refractor made by Alvan Clark of Boston. The following quotations as well as the two later readings from Huggins have been taken from an 1897 paper entitled "The New Astronomy: A Personal Retrospect," which he published in the journal *Nineteenth Century* and later included in his *Scientific Papers*. It consists of a survey of his nearly four decades of work up to that time in astronomical spectroscopy. Referring to the period immediately after 1858, Huggins stated in that essay:

> I soon became a little dissatisfied with the routine character of ordinary astronomical work, and in a vague way sought about in my mind for the possibility of research upon the heavens in a new direction or by new methods. It was just at this time . . . that the news reached me of Kirchhoff's great discovery of the true nature of the chemical constitution of the sun from his interpretation of the Fraunhofer lines.
>
> This news was to me like a coming upon a spring of water in a dry and thirsty land. Here at last presented itself the very order of work for which in an indefinite way I was looking— namely, to extend his novel methods of research upon the sun to the other heavenly bodies.[1]

The task Huggins took up was not easily accomplished; he noted, for example, that "The light received at the earth from a first magnitude star, as Vega, is only about one forty thousand millionth part of that received from the sun."[2] He overcame this and other difficulties, collaborating until 1864 with the chemist William Miller, and later, after his marriage in 1875, with his talented wife. One striking result of the new techniques he pioneered was that they transformed the nature of the astronomical observatory; as he stated:

> Then it was that an astronomical observatory began, for the first time, to take on the appearance of a laboratory.

---

[1] William Huggins, "The New Astronomy: A Personal Retrospect," *The Nineteenth Century*, 41 (1897), 911.
[2] Huggins, "New Astronomy," p. 912.

> Primary batteries, giving forth noxious gases, were arranged outside one of the windows; a large induction coil stood mounted on a stand on wheels so as to follow the positions of the eye-end of the telescope, together with a battery of several Leyden jars; shelves with Bunsen burners, vacuum tubes, and bottles of chemicals . . . lined its walls.
>
> The observatory became a meeting place where terrestrial chemistry was brought into direct touch with celestial chemistry. . . .
>
> This time was, indeed, one . . . of scientific exaltation for the astronomer, almost without parallel; for nearly every observation revealed a new fact, and almost every night's work was red-lettered by some discovery.[1]

Huggins and Miller made numerous spectroscopic observations of stars and planets, achieving many interesting results. The discovery of most immediate interest for us concerned nebulae and occurred in 1864. It is described in the following selection from an essay Huggins published in the *Nineteenth Century Review* in 1897, summarizing some of the results that he had obtained through the use of the spectroscope.

☆☆☆☆☆☆☆☆☆☆☆☆☆☆☆☆☆☆☆☆☆☆☆☆☆☆☆
From Sir William Huggins, *The Scientific Papers of Sir William Huggins*, ed. by Sir William Huggins and Lady Huggins (London, 1909), pp. 105–8.

Soon after the completion of the joint work of Dr. Miller and myself, and then working alone, I was fortunate in the early autumn of the same year, 1864, to begin some observations in a region hitherto unexplored; and which, to this day, remain associated in my memory with the profound awe which I felt on looking for the first time at that which no eye of man had seen, and which even the scientific imagination could not foreshow.

The attempt seemed almost hopeless. For not only are the nebulae very faintly luminous—as Marius put it, "like a rushlight shining through a horn"—but their feeble shining cannot be increased in brightness, as can be that of the stars, neither to the eye nor in the spectroscope, by any optic tube, however great.

---

[1]Huggins, "New Astronomy," p. 913.

Shortly after making the observations of which I am about to speak, I dined at Greenwich, Otto Struve being also a guest; when, on telling of my recent work on the nebulae, Sir George Airy said, "It seems to me a case of 'Eyes and No Eyes.'" Such work indeed it was, as we shall see, on certain of the nebulae.

The nature of these mysterious bodies was still an unread riddle. Towards the end of the last century the elder Herschel, from his observations at Slough, came very near suggesting what is doubtless the true nature, and place in the Cosmos, of the nebulae. I will let him speak in his own words:

> A shining fluid of a nature unknown to us.
> What a field of novelty is here opened to our conceptions! . . . We may now explain that very extensive nebulosity, expanded over more than sixty degrees of the heavens, about the constellation of Orion; a luminous matter accounting much better for it than clustering stars at a distance. . . .
> If this matter is self-luminous, it seems more fit to produce a star by its condensation, than to depend on the star for its existence.

This view of the nebulae as parts of a fiery mist out of which the heavens had been slowly fashioned, began, a little before the middle of the present century, at least in many minds, to give way before the revelations of the giant telescopes which had come into use, and especially of the telescope, six feet in diameter, constructed by the late Earl of Rosse at a cost of not less than £12,000.

Nebula after nebula yielded, being resolved apparently into innumerable stars, as the optical power was increased; and so the opinion began to gain ground that all nebulae may be capable of resolution into stars. According to this view, nebulae would have to be regarded, not as early stages of an evolutional progress, but rather as stellar galaxies already formed, external to our system—cosmical "sandheaps" too remote to be separated into their component stars. Lord Rosse himself was careful to point out that it would be unsafe from his observations to conclude that all nebulosity is but the glare of stars too remote to be resolved by our instruments. In 1858 Herbert Spencer showed clearly that, notwithstanding the

Parsonstown revelations, the evidence from the observation of nebulae up to that time was really in favour of their being early stages of an evolutional progression.

On the evening of August 29, 1864, I directed the telescope for the first time to a planetary nebula in Draco. The reader may now be able to picture to himself to some extent the feeling of excited suspense, mingled with a degree of awe, with which, after a few moments of hesitation, I put my eye to the spectroscope. Was I not about to look into a secret place of creation?

I looked into the spectroscope. No spectrum such as I expected! A single bright line only! At first I suspected some displacement of the prism, and that I was looking at a reflection of the illuminated slit from one of its faces. This thought was scarcely more than momentary; then the true interpretation flashed upon me. The light of the nebula was monochromatic, and so, unlike any other light I had as yet subjected to prismatic examination, could not be extended out to form a complete spectrum. After passing through the two prisms it remained concentrated into a single bright line, having a width corresponding to the width of the slit, and occupying in the instrument a position at that part of the spectrum to which its light belongs in refrangibility. A little closer looking showed two other bright lines on the side towards the blue, all the three lines being separated by intervals relatively dark.

The riddle of the nebulae was solved. The answer, which had come to us in the light itself, read: Not an aggregation of stars, but a luminous gas. Stars after the order of our own sun, and of the brighter stars, would give a different spectrum; the light of this nebula had clearly been emitted by a luminous gas. With an excess of caution, at the moment I did not venture to go further than to point out that we had here to do with bodies of an order quite different from that of the stars. Further observations soon convinced me that, though the short span of human life is far too minute relatively to cosmical events for us to expect to see in succession any distinct steps in so august a process, the probability is indeed overwhelming in favour of an evolution in the past, and still going on, of the heavenly hosts.

A time surely existed when the matter now condensed into the sun and planets filled the whole space occupied by the solar system, in the condition of gas, which then appeared as a glowing nebula, after the order, it may be, of some now existing in the heavens. There remained no room for doubt that the nebulae, which our telescopes reveal to us, are the early stages of long processions of cosmical events, which correspond broadly to those required by the nebular hypothesis in one or other of its forms.[1]

Not, indeed, that the philosophical astronomer would venture to dogmatise in matters of detail, or profess to be able to tell you pat off by heart exactly how everything has taken place in the universe, with the flippant tongue of a Lady Constance after reading "The Revelations of Chaos":

"It shows you exactly how a star is formed; nothing could be so pretty. A cluster of vapour—the cream of the Milky Way; a sort of celestial cheese churned into light."

It is necessary to bear distinctly in mind that the old view which made the matter of the nebulae to consist of an original fiery mist—in the words of the poet:

> ... a tumultuous cloud
> Instinct with fire and nitre—

could no longer hold its place after Helmholtz had shown, in 1854, that such an originally fiery condition of the nebulous stuff was quite unnecessary, since in the mutual gravitation of widely separated matter we have a store of potential energy sufficient to generate the high temperature of the sun and stars.

The solution of the primary riddle of the nebulae left pending some secondary questions. What chemical substances are represented by the newly found bright lines? Is solar matter common to the nebulae as well as to the stars? What are the

---

[1][The two main forms of the nebular hypothesis were proposed by William Herschel and Pierre Simon Laplace. Herschel's nebular hypothesis was presented in his 1811 paper (see Chapter Three). A few years earlier Laplace had suggested that the solar system may have formed from the condensation and rotation of a glowing nebulous mass. MJC]

physical conditions of the nebulous matter?

....

Are the nebulae very hot, or comparatively cool? The spectroscope indicates a high temperature: that is to say, that the individual molecules or atoms, which by their encounters are luminous, have motions corresponding to a very high temperature, and in this sense are very hot. On account of the great extent of the nebulae, however, a comparatively small number of luminous molecules might be sufficient to make them as bright as they appear to us: taking this view, their mean temperature, if they can be said to have one, might be low, and so correspond with what we might expect to find in gaseous masses at an early stage of condensation.

In the nebulae I had as yet examined, the condensation of nearly all the light into a few bright lines made the observations of their spectra less difficult than I feared would be the case. It became, indeed, a case of "Eyes and No Eyes" when a few days later I turned the telescope to the Great Nebula in Andromeda. Its light was distributed throughout the spectrum, and consequently extremely faint. The brighter middle part only could be seen, though I have since proved, as I at first suggested might be the case, that the blue and the red ends are really not absent, but are not seen on account of their feebler effect upon the eye. Though continuous, the spectrum did not look uniform in brightness, but its extreme feebleness made it uncertain whether the irregularities were due to certain parts being enhanced by bright lines, or the other parts enfeebled by dark lines.

Out of sixty of the brighter nebulae and clusters, I found about one-third, including the planetary nebulae and that of Orion, to give the bright-line spectrum. It would be altogether out of place here to follow the results of my further observations along the same lines of research, which occupied the two years immediately succeeding.

☆ ☆

The Discovery by Huggins of a Method for Measuring Radial Motions

By 1868, Huggins had attained yet another spectroscopic discovery. In that year, he announced a method of measuring the radial motion of any celestial object from which he could obtain a spectrum. Radial motion is motion in the line of sight, either toward or away from the observer. Hence, he could measure the speed of an object moving toward or away from him, but could derive no information about an object moving in a direction perpendicular to the line between him and the object. The method depended on the Doppler principle, presented in 1842 by Christian Doppler. Consider a stationary source S of waves (see diagram at left) that emits waves of wavelength λ, doing this in a periodic fashion. Wave 1 was emitted first, wave 2 second, wave 3 third. Consider now the situation (see diagram at the right) that represents the case where the source moves at constant velocity from position S_1 to position S_2 and finally to position S_3. An observer at A will perceive these waves as separated by a different although fixed wavelength; moreover, that observer will experience the waves as arriving at shorter intervals, that is, at increased frequency compared to the case for a stationary source. An observer at B will experience the opposite effect; the wavelengths will appear longer and the frequency (number of waves per time) will decrease.

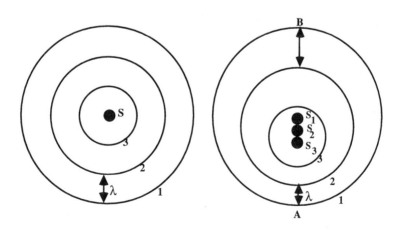

The Doppler effect is well known in regard to sound. We hear horns or sirens from cars coming toward us at shortened wavelength, i.e., at an elevated pitch. Similarly, we hear sounds from sources moving away from us at longer wavelength and lower pitch. Doppler's principle, Huggins realized, can be applied to light, including light from celestial bodies. Such application reveals that light received from, say, a star moving away from us, should appear to have increased wavelength, that is, the light should be shifted toward the red end of the spectrum, red having the longest wavelength. Hence celestial objects moving away from us are said to be redshifted; those moving toward us are described as blueshifted. What allows astronomers to measure this effect is the property of the spectral lines that they precisely mark particular wavelengths. The way the astronomer measures this effect is to compare the positions of the spectral lines of some element as produced from a stationary source with the positions of the lines for that element coming from a moving celestial object. This difference allows the determination of the velocity. The specific formula used is:

$$\frac{\Delta \lambda}{\lambda} = \frac{v}{c}$$

In this equation, $\Delta \lambda$ is the change in the wavelength, λ is the wavelength emitted by the object, v is the relative velocity of the object with respect to the observer, and c is the speed of light. It is important to realize that this method tells the astronomer nothing at all either about the motion of the body in a direction perpendicular to the line from observer to object or about the distance of the object; nonetheless, the ability to determine the radial velocity of celestial objects is extremely important. As Huggins predicts in the following report on this discovery, it proved to be of ever increasing usefulness.

☆☆☆☆☆☆☆☆☆☆☆☆☆☆☆☆☆☆☆☆☆☆☆☆☆☆☆☆
From Sir William Huggins, *The Scientific Papers of Sir William Huggins*, ed. by Sir William Huggins and Lady Huggins (London: William Wesley and Son, 1909), pp. 195–7.

From the beginning of our work upon the spectra of the stars, I saw in vision the application of the new knowledge to the creation of a great method of astronomical observation which could not fail in future to have a powerful influence on the progress of astronomy; indeed, in some respects greater than the more direct one of the investigation of the chemical nature and the relative physical conditions of the stars.

It was the opprobrium of the older astronomy—though indeed one which involved no disgrace, for *à l'impossible nul n'est tenu*[1]—that only that part of the motions of the stars which is across the line of sight could be seen and directly measured. The direct observation of the other component in the line of sight, since it caused no change of place and, from the great distance of the stars, no appreciable change of size or of brightness within an observer's lifetime, seemed to lie hopelessly quite outside the limits of man's powers. Still, it was only too clear that, so long as we were unable to ascertain directly those components of the stars' motions which lie in the line of sight, the speed and direction of the solar motion in space, and many of the great problems of the constitution of the heavens, must remain more or less imperfectly known.

Now as the colour of a given kind of light, and the exact position it would take up in a spectrum, depends directly upon the length of the waves, or, to put it differently, upon the number of waves which would pass into the eye in a second of time, it seemed more than probable that motion between the source of the light and the observer must change the apparent length of the waves to him, and the number reaching his eye in a second. To a swimmer striking out from the shore each wave is shorter, and the number he goes through in a given time is greater than would be the case if he had stood still in the water. Such a change of wave-length would transform any given kind of light, so that it would take a new place in the spectrum, and from the amount of this change to a higher or to a lower place, we could determine the velocity per second of the relative motion between the star and the earth.

The notion that the propagation of light is not instantaneous, though rapid far beyond the appreciation of our senses, is due, not as is sometimes stated to Francis, but to Roger

---

[1] ['no one should be expected to do the impossible.' MJC]

Bacon. ...

The discovery of its actual velocity was made by Roemer in 1675, from observations of the satellites of Jupiter. Now, though the effect of motion in the line of sight upon the apparent velocity of light underlies Roemer's determinations, the idea of a change of colour in light from motion between the source of light and the observer was announced for the first time by Doppler in 1841. Later, various experiments were made in connection with this view by Ballot, Sestini, Klinkerfues, Clerk Maxwell, and Fizeau. But no attempts had been made, nor were indeed possible, to discover by this principle the motions of the heavenly bodies in the line of sight. For, to learn whether any change in the light had taken place from motion in the line of sight, it was clearly necessary to know the original wave-length of the light before it left the star.

As soon as our observations had shown that certain earthly substances were present in the stars, the original wave-lengths of their lines became known, and any small want of coincidence of the stellar lines with the same lines produced upon the earth might safely be interpreted as revealing the velocity of approach or of recession between the star and the earth.

These considerations were present to my mind from the first, and helped me to bear up under many toilsome disappointments: "Studio fallente laborem."[1] It was not until 1866 that I found time to construct a spectroscope of greater power for this research. It would be scarcely possible, even with greater space, to convey to the reader any true conception of the difficulties which presented themselves in this work, from various instrumental causes, and of the extreme care and caution which were needful to distinguish spurious instrumental shifts of a line from a true shift due to the star's motion.

At last, in 1868, I felt able to announce, in a paper printed in the *Transactions of the Royal Society* for that year, the foundation of this new method of research, which, transcending the wildest dreams of an earlier time, enables the astronomer to measure off directly in terrestrial units the invisible motions

---

[1] ['The effort falls short of the task.' MJC]

in the line of sight of the heavenly bodies.

To pure astronomers the method came before its time, since they were then unfamiliar with Spectrum Analysis, which lay completely outside the routine work of an observatory. It would be easy to mention the names of men well known, to whom I was "as a very lovely song of one that hath a pleasant voice." They heard my words, but for a time were very slow to avail themselves of this new power of research. My observations were, however, shortly afterwards confirmed by Vogel in Germany; and by others the principle was soon applied to solar phenomena.

....

It would be scarcely possible, without the appearance of great exaggeration, to attempt to sketch out even in broad outline the many glorious achievements which doubtless lie before this method of research in the future.

☆☆☆☆☆☆☆☆☆☆☆☆☆☆☆☆☆☆☆☆☆☆☆☆☆☆☆☆

Chapter Six

From the New Astronomy to Henrietta Leavitt

The Milky Way and Island Universes: 1860–1912

As noted previously, the number of known nebulae increased from 103 in 1781, when Messier published his catalogue, to 5,709, when in 1865 Sir John Herschel's new cumulative catalogue of nebulae appeared. In 1887, J. L. E. Dreyer published his *New General Catalogue,* a revision of John Herschel's catalogue, in which Dreyer listed a total of 7,840 nebulae. A few years later Dreyer added 1,529 in a supplement. Large as these numbers of nebulae were, it seemed highly probable that more could be found. Around 1899, J. E. Keeler of Lick Observatory in California, using a 36" photographic reflector to study nebulae, reached the conclusion that 120,000 nebulae could in principle be detected with that instrument.

Concerning the question of "island universes," as the end of the century approached, more and more astronomers became convinced that the claim that "island universes" exist would have to be given up. A few supporters of the earlier view did remain; one of these was Joseph Plassmann of Freiburg University, who in an 1898 book accepted the existence of at least some island universes, for example, Andromeda. This was, however, decidedly a minority view. One representative of the majority was Richard A. Proctor, an English astronomical writer, who was strongly influenced by the fact that nebulae (except for gaseous and planetary nebulae) lie off the plane of the Milky Way. In 1874, Proctor stated:

> The sidereal system is altogether more complicated, altogether more varied in structure, than has hitherto been supposed. Within one and same region co-exist stars of many orders of real magnitude, the greatest being thousands of times larger than the least. All the nebulae hitherto discovered, whether gaseous or stellar, irregular, planetary, ring-formed, or elliptic, exist within the limits of the sidereal system. They all form part and parcel of that wonderful

system whose nearer and brighter parts constitute the glories of our nocturnal heavens.[1]

Moreover, Proctor argued for the rejection of the disk theory of the Milky Way. He urged that such phenomena as the unevenness and branches of the Milky Way as well as such dark areas as the Coal Sack can best be explained by assuming that the Milky Way has a ring-like structure. Part of Proctor's evidence came from a careful analysis of the writings of the two Herschels. In this, Proctor attempted to show that their mature views more or less corresponded with his own position. Proctor was not alone in arguing for a ring structure for the Milky Way; the Italian astronomer Giovanni Celoria in the 1870s and Cornelius Easton of Holland in the 1890s argued for a single or double ring theory of the Milky Way's structure.

The most prominent American astronomer of the late nineteenth century was Simon Newcomb. In 1882, basing his arguments primarily on the clustering of the nebulae at the poles of the Milky Way, Newcomb concluded that the nebulae cannot be island universes. In this context, he labeled speculations about island universes as idle talk. He also presented a diagram of the Milky Way that represented the nebulae as far smaller structures than it and as existing directly above and below the Milky Way so as to form a sort of halo.

Agnes M. Clerke (1842–1907), whose name is pronounced like Clark, was an English astronomer, who wrote the finest history of nineteenth-century astronomy, as well as contributing directly to that science. In the 1887 edition of that history, she stated:

> There is no maintaining nebulae to be simply remote worlds of stars in the face of an agglomeration like the Nubecula Major [the larger Magellanic Cloud], containing in its (certainly capacious) bosom *both* stars and nebulae. Add the evidence of the spectroscope to the effect that a large proportion of these perplexing objects are gaseous, with the facts of their distribution telling of an intimate relation between the mode of their scattering and the lie of the Milky Way, and it becomes impossible to resist the conclusion that

---

[1]R. A. Proctor, *The Universe and the Coming Transits* (London, 1874), pp. 201–2.

both nebular and stellar systems are parts of a single scheme.[1]

Herbert Spencer in 1899 re-edited his 1858 paper, some sections of which were discussed previously. Whereas in 1858 he had stated that truly extragalactic nebulae are "next to impossible," Spencer in his 1899 revision dropped the words "next to," leaving only the unqualified adjective "impossible."[2]

Astronomers outside the English-speaking world were no less disinclined to believe in island universes. In 1907, the German astronomer Max Wolf argued for a more or less spherical Milky Way, stating:

> By taking everything into account, we seem to be justified in considering it likely that the nebulae and clusters of stars represent an essential part of our star-island and perhaps lie relatively close to us. They all form, together with the stars of the Milky Way, an organic whole. Distant, isolated Milky Ways have never been sighted by man.[3]

Thus the consensus picture of the cosmos around 1900 included the notions that the Milky Way is essentially the entire visible universe and that it is about 7,000 to possibly 30,000 light-years in diameter with the sun located near its center. Nebulae are basically of three types: (1) gaseous nebulae, which almost invariably are found in the plane of the Milky Way; (2) globular clusters, which lie off but near that plane, generally in the direction of the Milky Way toward Sagittarius; and (3) spiral nebulae, sometimes called the "white nebulae," which also avoid the plane of the Milky Way, generally clustering toward its poles.

---

[1] Agnes Clerke, *A Popular History of Astronomy during the Nineteenth Century*, 2nd ed. (Edinburgh, 1887), pp. 456–7.
[2] As quoted in Stanley L. Jaki, *The Milky Way: An Elusive Road for Science* (New York: Science History, 1972), p. 275.
[3] As quoted in Jaki, *Milky Way*, p. 278.

Large Magellanic Cloud ↑   ↓ Small Magellanic Cloud

Alfred Russel Wallace on "Man's Place in the Universe"

The questions whether the Milky Way is a ring or disk or has some other more complicated structure and whether nebulae should be seen as island universes may seem to have little bearing on such issues as the nature of humanity and the relation of God to the universe. That such is not the case is suggested by a paper published in 1903 by Alfred Russel Wallace (1823–1913), a naturalist who became famous as the co-discoverer with Charles Darwin of the theory of evolution by natural selection. Wallace's paper, although a striking exemplification of some of the views discussed above, is atypical in the conclusions its author drew from it. Entitled "Man's Place in the Universe," Wallace's essay appeared in both the *Fortnightly Review* and the *Independent*. By late 1903, Wallace had expanded it into a book, which carried the same title as his paper. The paper and book were so controversial that they generated over fifty published responses. Although Wallace's paper was not a significant contribution to stellar astronomy, it merits consideration as a striking example of the larger human significance seen in developments in stellar astronomy and as a vivid example of the degree to which the island universe theory had fallen into disfavor by the beginning of the twentieth century.

The background for the paper is that at the end of the nineteenth century, Wallace, in writing his *Wonderful Century*, in which he surveyed the development of science in the nineteenth century, had made a review of recent advances in astronomy and detected therein some patterns that he believed to be emerging. For example, the observational fact that as telescopes reach farther and farther into space, they detect relatively fewer stars as well as what was in effect Olbers's paradox, seemed to Wallace to lead to the conclusion that the universe is finite. Moreover, various indications pointed, as noted previously, toward a ring theory of the Milky Way. As Wallace stated, the evidence indicates that

> the Galaxy is a vast annular agglomeration of stars forming a great circle round the heavens, although in places very irregular.... [And Wallace added:] "But what is more

> important is, that we must be situated not in any part of [the Milky Way] as was once supposed, but at or near the very central point in the plane of the ring, that is, nearly equally distant from every part of it. This must be the case, because from any other position the ring would not appear to us so symmetrical as it does.[1]

After arguing that the sun is, moreover, in the main plane of the Milky Way and enclosed in a cluster of stars, Wallace summarized this view by stating that recent astronomical investigations have shown "that our Sun is one of the central orbs of a globular star-cluster, and that this star-cluster occupies a nearly central position in the exact plane of the Milky Way." (p. 405) Wallace then added an inference: "But I am not aware that any writer has taken the next step, and combining these two conclusions, has stated definitely that our Sun is thus shown to occupy a position very near to, if not actually at, the centre of the whole visible universe, and therefore, in all probability, in the *centre* of the *whole material universe.*" (p. 405)

Wallace also noted that recent planetary studies had pointed to the conclusion that within our solar system, only our earth possesses the conditions necessary for the development of life. As he stated, "All the evidence at our command goes to assure us that our earth alone in the Solar System has been from its very origin adapted to be the theatre for the development of organised and intelligent life." (p. 409) At the end of his paper Wallace drew its various parts together as a basis for his dramatic concluding statement:

> The three startling facts—that we *are* in the centre of a cluster of suns, and that that cluster *is* situated not only precisely in the *plane* of the Galaxy, but also *centrally* in that plane, can hardly now be looked upon as chance coincidences without any significance in relation to the culminating fact that the planet so situated *has* developed humanity.
>
> Of course the relation here pointed out *may* be a true relation of cause and effect, and yet have arisen as the result of one in a thousand million chances occurring during almost

---

[1] A. R. Wallace, "Man's Place in the Universe," *Fortnightly Review*, 73 (1903), 395–411:403.

infinite time. But, on the other hand, those thinkers may be right who, holding that the universe is a manifestation of Mind, and that the orderly development of Living Souls supplies an adequate reason why such an universe should have been called into existence, believe that we ourselves are its sole and sufficient result, and that nowhere else than near the central position in the universe which we occupy, could that result have been attained. (p. 411)

At another point in the paper, the famous evolutionist (who had broken with Christianity many years earlier) stated the same point in a somewhat different manner by claiming that observational results attained by astronomers in the last quarter of the nineteenth century

tend to show that our position in the material universe is special and probably unique, and that it is such as to lend support to the view, held by many great thinkers and writers to-day, that the supreme end and purpose of this vast universe was the production and development of the living soul in the perishable body of man. (p. 396)

Introduction to the Reading from Agnes Clerke

The next reading is from the second edition (1905) of Agnes Mary Clerke's *System of the Stars*. This selection from Clerke's volume provides a soundly researched, elegantly written, and historically influential survey of the state of astronomical thought at the beginning of the twentieth century concerning the status and structure of the Milky Way and the nature of the nebulae.

Agnes Mary Clerke

☆☆☆☆☆☆☆☆☆☆☆☆☆☆☆☆☆☆☆☆☆☆☆☆☆☆☆
Agnes Clerke, Selection from her *The System of the Stars*, 2nd ed. (London: Adam and Charles Black, 1905), pp. 333–59.

CHAPTER XXV

THE MILKY WAY

The Milky Way shows to the naked eye as a vast, zone-shaped nebula; but is resolved, with very slight optical assistance, into innumerable small stars. Its stellar constitution, already conjectured by Democritus, was one of Galileo's earliest telescopic discoveries. The general course of the formation, however, can only be traced through the perception of its cloudy effect; and this is impaired by the application even of an opera-glass. Rendered the more arduous by this very circumstance, its detailed study demands exceptional eyesight improved by assiduous practice in catching fine gradations of light. Our situation, too, close to the galactic plane is the most disadvantageous possible for purposes of survey. Groups behind groups, systems upon systems, streams, sheets, lines, knots of stars, indefinitely far apart in space, may all be projected without distinction upon the same sky-ground. Unawares, our visual ray sounds endless depths, and brings back only simultaneous information about the successive objects met with. We are thus presented with a flat picture totally devoid of perspective-indications. Only by a long series of inductions (if at all) can we hope to arrange the features of the landscape according to their proper relations. . . .[1]

The bright spaces of the galactic zone are commonly surrounded and set off by dark winding channels, and the rapid alternation of amazingly rich with poor, or almost vacant patches of sky, is a constantly recurring phenomenon,[2] associated by Mr. Maunder[3] with slow processes of stellar

---

[1][Five paragraphs of description of the Milky Way have been omitted at this point.]
[2]Herschel, *Outlines*, arts. 790, 797.
[3]*Knowledge*, vol. xviii. p. 36.

agglomeration. The most remarkable instance occurs in the Southern Cross, the brilliant gems of which emblazon a broad galactic mass very singularly interrupted by a pear-shaped black opening eight degrees long by five wide, named by early navigators the "Coal-sack." This yawning excavation figures in Australian folk-lore as the embodiment of evil in the shape of an Emu, who lies in wait at the foot of a tree represented by the stars of the Cross, for an opossum driven by his persecutions to take refuge among its branches.[1] The denudation of the Coal-sack is, however, shown by Mr. H. C. Russell's photographs to be complete only towards its northern end.[2] To the south, a considerable invasion of small stars modifies the contrasting darkness.

Partial galactic vacuities, evidently of the same nature with the southern Coal-sack, occur elsewhere, notably in Cygnus; but they are inconspicuous to casual observers.[3]

Sir William Herschel was perfectly satisfied that, with his 20-foot reflector (equivalent to a modern refractor about 14 inches in aperture), the Milky Way was, in general, "fathomable." The stars composing it, that is to say, were of definite numbers, and appeared projected upon a perfectly black sky. But this was not so everywhere; certain parts completely baffled the penetrative faculty of his instrument. One such was met with in Cepheus, where he found the small stars to become "gradually less till they escape the eye, so that appearances here favour the idea of a succeeding, more distant clustering part." And he remarked, in exploring between Sagitta and Aquila, that "the end of the stratum cannot be seen."[4] Again, in the galactic branch traversing Ophiuchus, Sir John Herschel encountered "large milky nebulous irregular patches and banks, with few stars of visible magnitudes"; he described "a very large space" of the Milky Way in Sagittarius as "completely nebulous like the diffused nebulosity of the Magellanic

---

[1]Macpherson, *Journ. R. Soc. N.S. Wales*, 1861, p. 72.
[2]*Monthly Notices*, vol. li. p. 40.
[3][Two and one half paragraphs of description of the Milky Way have been omitted at this point.]
[4]*Phil. Trans.* vol. cvii, p. 326.

Cloud";[1] and observed a similar spot in Scorpio, "where, through the hollows and deep recesses of its complicated structure, we behold what has all the appearance of a wide and indefinitely prolonged area strewed over with discontinuous masses and clouds of stars which the telescope at length refuses to analyse."[2]

Even with the best telescopes of recent construction, this perplexing and indeterminate aspect cannot be altogether got rid of. Professor Holden could obtain with the 36-inch Lick achromatic directed to the Milky Way "no final resolution of its finer parts into stars. There is always the background of unresolved nebulosity on which hundreds and thousands of stars are studded—each a bright, sharp, separate point."[3] The lingering nebulosity was strongly indicated to be of stellar nature, but whether it was due to the presence of innumerable small stars mixed up in the same region with larger ones, or to the indefinite extension into outer space of galactic agglomerations could not be pronounced off-hand.

The explanations attempted of these complicated phenomena may be divided into disc-theories, ring-theories, and spiral-theories. The "disc-theory" of the Milky Way was first propounded by Thomas Wright of Durham in 1750.[4] He supposed all the stars to be distributed in a comparatively shallow layer, producing an annular effect by its enormous lateral spread. Irregularities, he thought, were partly due to our eccentric position within the stratum, partly to "the diversity of motion that may naturally be conceived amongst the stars themselves, which may, here and there, in different parts of the heavens, occasion a cloudy knot of stars."[5]

To this view Sir William Herschel gave wide currency and apparent stability by the application to its support of his ingenious method of "star-gauges." By counting the stars

---

[1] *Cape Observations*, p. 389.
[2] *Outlines of Astr.* art. 798.
[3] *Sid. Mess.* August 1888, p. 298.
[4] [Clerke is mistaken in this statement. As noted previously, Wright put forth three theories of the Milky Way—but not a disk theory.]
[5] *An Original Theory of the Universe*, p. 63.

simultaneously visible with his great reflector in various portions of the sky, he showed that their paucity or abundance depended upon the situation of the gauge-fields relative to the Milky Way. In its neighbourhood, stars were copiously, away from it, they were sparsely distributed. And this by a regular progression of density from the galactic poles to the galactic equator, the latter region being on an average thirty times richer than the former. Now, if we were to admit, as Herschel did, a nearly equable scattering of stars in space, there would be no alternative but to suppose the sidereal system extended in any direction proportionately to the number of stars seen in that direction. Their crowding should, *on that hypothesis*, be purely optical—the effect of the indefinite spreading out in the line of sight of their evenly serried ranks. Sounding the star-depths upon this principle, Herschel measured the length of his line by their seeming populousness, and constructed, from the numerical data thus obtained, the "cloven disc" model, long accepted as representing the true form of the stellar universe.

But his own observations at the very moment of enouncing this theory, fatally undermined it. Already, in 1785, he remarked that two or three hundred "beginning, or gathering clusters," might be pointed out in the galactic system, and he surmised its eventual separation, "after numbers of ages," into so many distinct "nebulae."[1] "Equable scattering," then, was an ideal state of things long since abolished by the "ravages of time." The conviction that such was the case grew with his experience. "The immense starry aggregation" of the Milky Way, he wrote in 1802[2] "is by no means uniform. The stars of which it is composed are very unequally scattered, and show evident marks of clustering together into many separate allotments." Nor did he fail to perceive, from the gradual increase of brightness towards the centres of these "allotments," that they tended to assume a spherical form, and thus suggested "the breaking-up of the Milky Way, in all its minute parts, as the unavoidable consequence of the clustering power arising out of those preponderating attractions which

---

[1] *Phil. Trans.* vol. lxxv. p. 255.
[2] *Ibid.* vol. xcii. p. 495.

have been shown to be everywhere existing in its compass."[1] The formal announcement of this conviction "that the Milky Way itself consists of stars very differently scattered from those which are immediately about us,"[2] amounted to a recantation of the principle of star-gauging.

With it disappeared from Herschel's mind the conception of an optically-produced galaxy. In his ultimate opinion the actual corresponded very closely with the apparent structure: it was composed, that is to say, mainly, if not wholly, of real clouds of stars. Credit was thus restored to the early impression of Galileo, who in 1610 described the Milky Way as "nothing else but a mass of innumerable stars planted together in clusters."[3]

Wilhelm Struve's[4] effort towards the reorganisation of the stratum-theory, though aided by all the resources of his great ability and address, could scarcely be counted as a step in advance. Substituting for the hypothesis of equable distribution that of concentration in parallel planes, he imagined the average interval of space between the stars to diminish regularly with approach to the central horizon of the system. The swarming aspect of the Milky Way was hence interpreted as agreeing with fact, but the annular appearance as being illusory. Of illimitable dimensions, the system was conceived to stretch away, still preserving its specific character, to an infinite, or at least unimaginable remoteness, comparatively narrow *visual* bounds being set to it by a supposed extinction of light.

But the quasi-geometrical regularity of Struve's galaxy is belied by innumerable details of the original. The swell of the tide of stars towards the galactic plane is neither uniformly progressive,[5] nor does it proceed without conspicuous interruptions. Thus, the region near the horns of Taurus, although close to the Milky Way, is absolutely the poorest in

---

[1] *Phil. Trans.* vol. civ. p. 282.
[2] *Ibid.* vol. xcii. p. 480.
[3] *Sidereus Nuncius*, trans. by E. S. Carlos, p. 42.
[4] *Études d'Astronomie Stellaire*, 1847.
[5] C. S. Peirce, *Harvard Annals*, vol. ix. p. 174.

the northern hemisphere;[1] and it is matched in the southern[2] by an almost clean-swept space in Scorpio, on meeting which Sir William Herschel exclaimed in amazement, "Hier ist wahrhaftig ein Loch im Himmel!" But it is the openings in the formation itself which most decisively negative the stratum-theory in any of its modifications. Is it credible that a boundlessly extended layer of stars should be pierced, in many of its densest portions, by *tunnels* converging directly upon our situation within it?[3] No sane mind, we venture to say, realising all that such an affirmation implies, can assent to it. But, indeed, the entire conformation of the Milky Way—its streaming offsets, convoluted windings, promontories, and sharply bounded inlets, no less than its breaches of continuity—is absolutely irreconcilable with the hypothesis of optical creation out of star-materials equably distributed.

We seem then led to the alternative belief that it is a definite structure, at a definite distance from ourselves—a belief forced upon Sir John Herschel by his Cape experiences, notwithstanding his natural reluctance to drift far away from the position originally taken up by his father. The shape suggested by him for the galaxy was that of "a flat ring, or some other re-entering form of immense and irregular breadth and thickness."[4] Expanded indefinitely along the central plane, the new model scarcely differed from the old except in so far as the idea of homogeneous construction was given up. The disc remained, but with its centre scooped out. The solar system was located in an enormous space of relative vacuity.

The Milky Way, thus regarded, appeared to consist of an indefinite number of stellar collections "brought by projection into nearly the same visual line"—to represent the foreshortened effect (more especially at a particular spot in Sagittarius) of "a vast and illimitable area scattered over with discontinuous masses and aggregates of stars in the manner

---

[1] Argelander, *Bonner Beob.* Bd. v. *Einleitung;* Proctor, *Universe of Stars,* p. 32.
[2] Thome, *Cordoba Durchmusterung,* Introduction.
[3] Proctor, *loc. cit.* p. 15.
[4] *Outlines,* art. 788.

of the cumuli of a mackerel-sky."[1] But in an assemblage of this nature seen edgewise, a "Coal-sack" would be a phenomenon as anomalous as in a uniform stratum; nor could it, without violent improbability, be conceived of as rent by the colossal fractures dividing the actual Milky Way in Argo and Ophiuchus.

To remedy these inconveniences, Professor Stephen Alexander devised in 1852,[2] upon the model of the wheel-shaped nebula in Virgo (M 99), a spiral galaxy with four curvilinear branches diverging from a central cluster formed by the sun and lucid stars. By properly adjusting the mode of projection of these radiating star-streams, the effects of rifts and coal-sacks were duly produced; but the arrangement, however admired for ingenuity, gave no persuasion of reality, and quickly dropped out of remembrance. Essentially different, although with some features in common, was that by which Mr. Proctor replaced it in 1869.[3] Rather than a "spiral," indeed, the new design resembled a bent and broken ring, with long, riband-like ends, looped back on either side of an opening, accommodated to the shape of the gap in the visible structure in Argo. One of these loops, by the apparent intercrossing of its near with its remoter branch, was supposed to generate the Coal-sack in Crux; while the other end, trailing lengthily backward, afforded a deceptive effect of bifurcation. Excessive distance was invoked, as in Professor Alexander's scheme, to explain the cessation of nebulous light in Ophiuchus.

Of the manifold objections to which this hypothesis is liable,[4] only two need here be mentioned. In the first place, it involves a wholly inadmissible rationale of the opening seen in the Milky Way. If these were due to the interlacing by perspective of branches really far apart in space, the enclosing luminous formation should be markedly fainter on one side than on the other. But this is not so. The borders of the southern Coal-sack are approximately of the same brightness all round. A single vivid mass has obviously been the scene of what, in

---

[1] *Cape Observations*, p. 389.
[2] *Astr. Journ.* vol. ii. p. 101.
[3] *Monthly Notices*, vol. xxx. p. 50.
[4] See Mr. J. R. Sutton's remarks, *Illustrated Science Monthly*, vol. ii. pp. 63, 199; *Knowledge*, vol. xiv. p. 41; Easton, *ibid.* vol. xxi. p. 12.

the absence of better knowledge, may be described as an excavatory process.

Again, on the spiral theory, the great rift in the Milky Way should represent the interval between branches mutually disconnected except through the optical effect of projection. But their mutual dependence is manifest. They strike apart gradually, and as the result of changing conditions. Premonitory cavities seem to announce their impending separation; and even after it has become definitive, abortive efforts towards reunion are indicated by the correspondence of opposite projections and by the occasional bridging of the fissure. The bifurcation is beyond question a physical reality.

The prevalence of spiral forms among the nebulae, emphasised by Keeler's photographic explorations, adds weight to the arguments for assigning a more or less similar structure to the great firmamental zone. M. Easton has illustrated the subject by a plan-sketch, copied in Fig. [1].

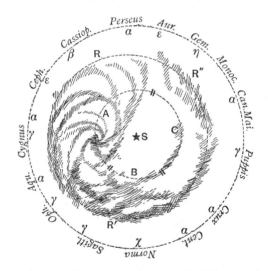

Figure [1].—The Galactic Spiral (Easton)

It does not profess to give "even an approximate representation of the Milky Way, seen from a point in space situated on its axis,"[1] but merely to indicate a possible mode of distribution of

---

[1]*Astroph. Journ.* vol. xii. p. 157

stellar accumulations by which the observed phenomenon might be produced. A remarkable bright patch of the Milky Way in Cygnus serves in it as the nucleus of the convolutions. The sun's position, though central, is detached from them.

Some such design strongly tempts thought. It is recommended by analogy; it is recommended also by its adaptability. A multitude of facts, at first sight incongruous, are combined by it into plausible unity. The ring-theory may be said to have broken down when Celoria[1] found it necessary to establish a double annular system. His researches, together with those of Plassmann, Seeliger, and Easton, leave no doubt that the various sections of the starry girdle differ prodigiously in distance. Moreover, it seems impossible so to place the sun in such an enclosing structure as to get a reasonable explanation of these differences without having recourse to further complexities of artificial arrangement. True, no design corresponding in any degree with what appears, can be other than intricate. Were it permissible to adopt the opinion that the Milky Way *is* very much what it *seems*, we should describe it as a ring with streaming appendages extending from the main body in all possible directions, some nearly straight towards, or away from us, others at every imaginable angle with our line of sight. The results in perspective foreshortening would evidently, under these circumstances, be highly complex; the eye being presented with groups and streams of stars, immensely various in remoteness, but all projected indiscriminately upon the same zone of the heavens. Thus, while some branches, pursued along their outward course, fade at last into dim nebulosity, other Milky Way groups may be distinguished as bright separate stars, because much nearer to us than the generality of their associates. Closed rings, however, are beginning to appear alien to the cosmic plan of structure. Nebulae presenting that aspect are, perhaps without exception, resolvable into helical or spiral figures. And it would be hazardous to assert that the Milky Way lies outside the mysterious law imposed upon minor aggregations.

Its internal organisation is of baffling intricacy. It collects

---

[1]*Memorie del R. Istituto Lombardo*, t. xiv. p. 82.

within its ample round, there is every reason to suppose, an absolutely endless variety of separate systems. A multitudinous aggregate of individual clusters, it is composed moreover as a whole very much like one single cluster on a colossal scale. Its fringed edges, its rifts and vacuities, are, as we have seen, reproduced in miniature in numberless star-groups. "Rings," and "sprays," and "streams" of stars are unmistakably common to the two orders of formation; and the stellar constituents of both are frequently involved with gaseous nebulae in a way showing most intimate association by origin and development. The laws then governing stellar aggregation in the one case govern it also in the other; and so, from this direction independently, we again reach Herschel's conclusion that the Milky Way "consists of stars very differently scattered from those which are immediately about us."

But are these stars *suns*, co-ordinate with our own? or must we regard them as comparatively insignificant bodies, sharing a sun-like nature, indeed, but on a far lower level of power and splendour? The question is equivalent to this other perpetually recurring one, What is their average distance from ourselves? In what portion of space do the true galactic condensations occur? How far outward should we have to travel before finding ourselves actually in the midst of the crowded objects producing, to terrestrial observers, the "milky" effect of a nebulous stratum?

Certain knowledge in this matter is not to be had; indications and probable estimates have to take its place. Professor Newcomb, as the upshot of a discussion embracing all the available facts, arrives at the conclusion that the Milky Way, in most of its sections, is no nearer to us than would be signified by a parallax of one-thousandth of a second,[1] corresponding to a light-journey of 3200 years. Now our sun, if thus unimaginably removed into space, would shrink to a star of the fifteenth magnitude; it would seem just one of the grains of shining sand coagulated into heaps out near the confines of the sidereal world. Presumably, then, a large proportion of the

---

[1] *The Stars*, p. 317.

lustrous specks forming, in their unnumbered aggregate, the nebulous arch amid the constellations are really suns of the scale of our own. They are, however, pretty evidently intermingled with many smaller globes, and with some vastly larger. M. Easton has succeeded,[1] by detailed comparisons, in establishing a correlation between galactic structures and the stars, from about the sixth to the fifteenth magnitudes, enumerated by Argelander and others. The inference is thus rendered compulsory that a percentage of these comparatively bright orbs are genuine constituents of the clusters with which they are collineated. There are besides strong grounds for the belief that many, if not most, of the bluish brilliants giving helium spectra dominate these comprehensive groups. Mr. Ranyard pointed out in 1891,[2] from the evidence of Mr. H. C. Russell's photographs, that the chief luminary of the Southern Cross is centrally situated in a curiously symmetrical little cluster of excessively faint stars; and the irresistible conclusion of its being physically related to them may be safely applied to a number of other lustrous objects of the same spectral class. The range of actual size and splendour, then, among the components of galactic star-drifts is astonishingly great—much greater, in Professor Newcomb's opinion, than in ordinary detached clusters, such as the Pleiades, or the double Sword-handle group. The helium stars of the zone are, indeed, veritable tritons amidst shoals of minnows; they are frequently of a magnitude transcending the powers of imagination to realise, while their dwarfed associates may well be of average sun-like stature.

The Milky Way clouds are not condensed from the general contents of the sidereal heavens; they are markedly distinct. Their spectral peculiarities make this clear. They are built up, essentially and fundamentally, out of Sirian stars;[3] those of the solar and Antarian types seem to be totally absent from them. They include, however, nearly all the helium and bright-line stars that exist; but they are relatively few; they scarcely

---

[1] *Distribution de la Lumière Galactique*, p. 24; Newcomb, *The Stars*, p. 273.
[2] *Knowledge*, vol. xiv. p. 112.
[3] Pickering, *Harvard Annals*, vol. lvi. p. 25.

count as ingredients; they are *rari nantes in gurgite vasto*.[1] Many galactic tracts, too, are suffused with a phosphorescent glow; they harbour nebulous formations which only the photographic camera, through its faculty of persistent gazing, has been able to actualise and define. This surprising characteristic affords an additional proof that cosmic conditions of a special kind prevail in the enigmatical girdle which enclasps the mystery of the universe.

From a most careful study of the Milky Way at Cordoba, where it was seen to peculiar advantage, Dr. Gould inclined to regard it as the product of two or more superposed galaxies.[2] The fact of the two narrowest and brightest, and the two most diffused parts lying in pairs opposite to each other is certainly remarkable, and lends some countenance to the surmise that the "necks" in Cassiopeia and Crux really represent the intersections of the two crossed rings visibly diverging in Ophiuchus. M. Celoria, too, as we have seen, adopted the hypothesis of a compound Milky Way, but of such a form as to allow the possibility of one of its constituent annuli being comprehended by the other. The transition to a true spiral shape was thence easily effected, and a wider range of facts was rendered capable of theoretical accommodation. What is unmistakable is that the entire formation, single or compound, while individual and specific, is yet no isolated phenomenon. The contents of the firmament are arranged mainly with reference to it. It is a large part of a larger design exceeding the compass of finite minds to grasp in its entirety.

---

[1] ["swimming far apart in a wide expanse." Virgil, *Aeneid*, Bk. I, l. 118. MJC]

[2] *Uran. Argentina*, p. 381.

CHAPTER XXVI

STATUS OF THE NEBULAE

The question whether nebulae are external galaxies hardly any longer needs discussion. It has been answered by the progress of research. No competent thinker, with the whole of the available evidence before him, can now, it is safe to say, maintain any single nebula to be a star system of co-ordinate rank with the Milky Way. A practical certainty has been attained that the entire contents, stellar and nebular, of the sphere belong to one mighty aggregation, and stand in ordered mutual relations within the limits of one all-embracing scheme. All-embracing, that is to say, so far as our capacities of knowledge extend. With the infinite possibilities beyond, science has no concern.

The chief reasons justifying the assertion that the status of the nebulae is intra-galactic are of three kinds. They depend, first, upon the nature of the bodies themselves; secondly, upon their individual stellar associations; thirdly, upon their systematic arrangement as compared with the systematic arrangement of the stars.

The detection of gaseous nebulae not only directly demonstrated the non-stellar nature of a large number of these objects, but afforded a rational presumption that the others, however composed, were on a commensurate scale of size, and situated at commensurable distances. It may indeed turn out that gaseous and non-gaseous nebulae form an unbroken series rather than two distinct classes separated by an impassable barrier. Their spectra have perhaps more in common than would, at first, be supposed. For the vivid rays of green nebulae are superimposed upon a gauzy background of continuous light, which appears to be resolvable into a multitude of bright lines in juxtaposition;[1] and the spectra of white nebulae show, not a smooth, prismatic gradation, but slight inequalities in the flow of light, indicating effects of absorption, of emission, or of both combined. Before indeed any settled opinion can be formed as to

---

[1] Palmer, *Lick Bulletin*, No. 35.

whether these analogies have really the transitional meaning we might be inclined to attribute to them, nebular spectroscopy must be a good deal further advanced than it is at present. But apart from this question, relationship between the various orders of nebulae is manifest. The tendency of all to assume spiral forms demonstrates, in itself, their close affinity; so that to admit some to membership of the sidereal system while excluding others would be a palpable absurdity. And since those of a gaseous constitution must be so admitted, the rest follow inevitably.

Of the physical connection of nebulae with particular stars, fresh and incontrovertible proofs accumulate day by day. Nothing can be more certain than that objects of each kind coexist in the same parts of space, and are bound together by most intimate mutual ties. To argue the matter seems, as the French say, like "battering in an open door." We need only recall the stars of the Pleiades, photographically shown to be intermixed with nebulae, and those in Orion still bearing in their spectra traces of their recent origin from the curdling masses around. The nuclear positions so frequently occupied in nebulae by stars single and multiple, reiterate the same assertion of kinship, emphasised still further by the phenomena of stellar outbursts *in* nebulae. The scenes of these *must*, as the late Mr. Proctor insisted, lie within the circuit of the Milky Way, unless we are prepared to assume the occurrence, in extra-sidereal space, of conflagrations on a scale outraging all probability. It has been calculated that if the Andromeda nebula were a universe apart of the same real extent as the Galaxy, it should be situated, in order to reduce it to its present apparent dimensions, at a minimum distance of twenty-five galactic diameters.[1] And a galactic diameter being estimated by the same authority at thirteen thousand light-years, it follows that, on the supposition in question, light would require 325,000 years to reach us from the nebula. The seventh-magnitude star then which suddenly shone out in the midst of it in August 1885 should have been an absolutely portentous orb. In real light it should have been equivalent to

---

[1] Weisse, *Schriften Wiener Vereins*, Bd. v. p. 318.

762,000 stars like Sirius, or to sixteen million such suns as our own! But even this extravagant result inadequately represents the real improbability of the hypothesis it depends upon; since the Andromeda nebula, if an external galaxy, would almost certainly be at a far greater remoteness from a sister-galaxy than would be represented by twenty-five of its own diameters.

Just as the Milky Way might be described as a great compound cluster made up of innumerable subordinate clusters, so the greater Magellanic Cloud seems to be a gigantic nebula combining into some kind of systemic unity multitudes of separate nebulae. To the naked eye it shows vaguely a brighter axis spreading at the extremities so as to produce a resemblance to the "Dumb-bell" nebula; photographic exposures bring out unequivocal traces of a spiral conformation; either way it shows signs of definite organisation as a coherent whole; and it includes, strangely enough, among its inmates, a miniature of itself (N.G.C. 1978), but of much greater intensity and distinctness. Sir John Herschel's enumeration in 1847, of the contents of the "Cloud," gave conclusive evidence of the interstellar situation of nebulae—evidence, the full import of which Dr. Whewell was the first to perceive. Over an area of forty-two square degrees, 278 nebular objects (stars being copiously interspersed) are distributed with the elsewhere unparalleled density of 6 1/2 to the square degree. "The Nubecula Major," Herschel wrote, "like the Minor, consists partly of large tracts and ill-defined patches of irresolvable nebula and of nebulosity in every stage of resolution, up to perfectly resolved stars like the Milky Way, as also of regular and irregular nebulae properly so called, of globular clusters in every stage of resolvability, and of clustering groups sufficiently insulated and condensed to come under the designation of clusters of stars."[1]

Here then we find—in a system certainly, as Herschel said, "*sui generis*," yet none the less, on that account, instructive as to cosmical relationships—undoubted stars and undoubted nebulae at the same general distance from the earth. Some of the nebulae may indeed very well be placed actually nearer to us

---

[1] *Cape Results*, p. 146.

than some of the stars; and the extreme possible difference of their remoteness cannot, if the Cloud be approximately globular, exceed one-tenth of the interval between its hither edge and ourselves. We learn, too, the plain lesson that distance is only one factor in the production of irresolvability. For stars in every stage of crowding, from loose groups to the veriest dust-streaks, globular clusters coarse and fine, nebulae of all kinds and species, range side by side in this extraordinary collection, proving beyond question that differences of aggregation subsist in themselves, and need no additional abysses of space to account for them.

Even, however, if all these mutually confirmatory arguments could be dismissed as invalid, the mode of scattering of nebulae on the sky-surface would alone suffice to demonstrate their association with the sidereal system. Sir William Herschel was early struck with the occurrence of beds of these objects, preceded and followed by spaces void of stars. So familiar to him was this sequence of phenomena that he would sometimes warn his assistant "to prepare, since he expected in a few minutes to come at a stratum of the nebulae, finding himself already on nebulous ground."[1] The relation, after a century of partial oblivion, was photographically rediscovered by Dr. Max Wolf. It is curiously exemplified in the star-denuded nidus of the America nebula[2] (see Plate XVIII)[3], which seems to have absorbed or repelled from its immediate neighbourhood the dense galactic clouds impending towards it. Similarly, a comparatively vacant region encompasses the Orion nebula, within which, nevertheless, stars swarm and flicker after the manner of the components of globular clusters. The replacement of stars by nebulous matter is again conspicuous in Plate [I], which reproduces, by Dr. Wolf's kind permission, a photograph taken by him, July 10, 1904, with an exposure of four hours. The depicted nebula, which had been discovered ten years previously, is about 10' in diameter, of a round shape, and a complex structure. "It is placed centrally," Dr. Wolf writes, "in

---

[1]*Phil. Trans.* vol. lxxiv. 449.
[2]*Publicationen des Observatoriums Königstuhl-Heidelberg,* Bd. i. p. 181 (Kopff).
[3][Plate XVIII is not included in this selection.]

a very fine lacuna void of faint stars, which surrounds the luminous cloud like a trench."[1] Moreover, this negative "halo forms the end of a long channel, running eastward from the western nebulous clouds and their lacunae, to a length of more than two degrees." The coexistence in the same sidereal district of nebulae and stars could not well be asserted with stronger emphasis than by the clearing of a dark fosse for the accommodation of the cocoon-like object in Plate [I].

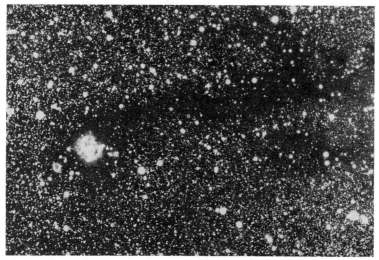

Plate [I]. The Cocoon Nebula in Cygnus
Photographed by Dr. Max Wolf, July 10, 1904

The larger plan of nebular distribution, as being the inverse to that of stars, partially revealed itself to the elder Herschel; but Sir John first brought into clear view the distinct and striking division of the nebulae into "two chief strata, separated by the Galaxy." Taking the circle of the Milky Way as a horizon, he remarked that the accumulation of them in Virgo and Coma Berenices "forms, as it were, a canopy occupying the zenith, and descending thence to a considerable distance on all sides, but chiefly on that towards which the (celestial) north pole lies."[2]

---

[1] *Monthly Notices*, vol. lxiv, p. 839.
[2] *Cape Results*, p. 137.

This crowding about the galactic pole is less marked in the southern hemisphere, though here too there is a "chief nebular region" approximately corresponding to that in Virgo. The distribution is, however, on the whole much more uniform than in the northern hemisphere, or rather, more uniformly *patchy*, rich districts alternating with more or less ample vacuities. One of these extends about fifteen degrees all round the south pole, the Lesser Cloud marking its edge. The remarkable fact, too, was noticed by Sir John Herschel that the larger nubecula seems "to terminate something approaching to a zone of connected patches of nebulae," reaching across Dorado, Eridanus, and Cetus to the equator, where it unites with the "nebular region of Pisces." A similar line of communication is less conspicuously kept open with the minor nubecula, and this feature of "streams" of nebulae with terminal aggregations was considered by Proctor to be distinctive of southern skies.[1] He adverted besides to the coincidence of two of them with stellar "streams" in Eridanus and Aquarius, and was struck with a significant deficiency of bright stars over the intervals between nebular groups.

Ampler acquaintance with this class of objects has, on the whole, served to ratify earlier conclusions relative to their mode of scattering. Their overwhelming tendency to congregate about the north galactic pole is accentuated by the results of Dr. Max Wolf's photographic survey;[2] while the presence of a secondary focus of aggregation in Perseus and Andromeda, remarked by Waters,[3] was rendered unmistakable by the subsequent investigations of Stratonoff.[4] M. Easton, discussing the subject with his accustomed thoroughness in 1904, was especially struck with the disturbing influence of the Magellanic Clouds. Hence possibly so fundamental a diversity in the modes of nebular distribution on either side of the Milky Way, that he considers it as "very probable that the structure of the southern galactic sky with regard to the nebulae differs

---

[1]*Monthly Notices*, xxix, p. 340.
[2]*Report of Königstuhl Observatory* for 1902; *Astr. Nach.* No. 3812.
[3]*Monthly Notices*, vol. liv. p. 527.
[4]Ristenpart, *V. J. S. Astr. Ges.* Jahrg. xxxvii. p. 356.

entirely from that of the northern galactic sky."[1]

The leading facts of nebular distribution were correctly described by Herbert Spencer in 1854. "In that zone," he wrote, "of celestial space where stars are excessively abundant, nebulae are rare; while in the two opposite celestial spaces that are furthest removed from this zone nebulae are abundant. Scarcely any nebulae lie near the galactic circle, and the great mass of them lie round the galactic poles. Can this be mere coincidence? When to the fact that the general mass of nebulae are antithetical in position to the general mass of stars, we add the fact that local regions of nebulae are regions where stars are scarce, and the further fact that single nebulae are habitually found in comparatively starless spots, does not the proof of a physical connection become overwhelming?"[2]

Accompanying, but considerably overlapping the Milky Way along its entire round, is a "zone of nebular dispersion" (as Proctor called it)—a wide track of denudation, so far as these objects are concerned. The nebular multitude shrinks, as it were, from association with the congregated galactic stars. A relation of avoidance is strongly accentuated. But withdrawal implies recognition. It implies the subordination of stars and nebulae alike to a single idea embodied in a single scheme. The range of our possible acquaintance is accordingly restricted to one "island universe"—that within whose boundaries our temporal lot is cast, and from whose shores we gaze wistfully into infinitude.

Dismissing, then, the grandiose but misleading notion that nebulae are systems of equal rank with the galaxy, we may turn our attention to the problems presented by the peculiarities of their interior situation. When these are subjected to a detailed examination, distinctions become evident between the different classes of nebulae. Distinctions so marked as to lead almost to their separation.

The "relation of avoidance" to the Milky Way just adverted to prevails *only* among the "unresolved" nebulae. These, it is true, are the great majority of the entire, so that the

---

[1] *Proc. Acad. of Sciences*, Amsterdam, June 25, 1904; *Astr. Nach.* No. 3969.
[2] *The Nebular Hypothesis* (with Addenda), p. 112.

conclusion of nebular crowding away from that zone remains unimpeachable. For certain classes of minor numerical, but high cosmical importance, the relation is precisely inverted. Over gaseous nebulae and clusters, the Milky Way seems to exercise an attractive influence equally strong with its repulsive effect upon nebulae of other kinds.

Forty out of one hundred and eleven globular clusters belong to the galactic zone,[1] which is hence twice as richly furnished as the rest of the sky with these wonderful objects. And the excess rises to twenty-five times for irregular or nondescript clusters, 434 out of 535 of which—that is, 81 per cent—are located in, or close to, the Milky Way. Many clusters, indeed, obviously form an integral part of the formation itself; of others, it is difficult to decide whether they should be ranked as distinct, or simply as intensifications of ordinary galactic star-groupings. To the latter category almost certainly belongs a collection (M 24) visible to the naked eye as a dim cloudlet near μ Sagittarii, and named by Father Secchi "Delle Caustiche," from the peculiar arrangement of its stars in rays, arches, caustic curves, and intertwined spirals. This again is included in a great oval condensation of galactic stars, obviously endowed with some degree of structural independence.

Gaseous nebulae, like gaseous stars, are nearly exclusive in their galactic affinities.[2] Very few planetaries can be found at any considerable distance from the favoured zone; the spectroscopic search for stellar nebulae is fruitless unless within its borders; and they embrace—with one exception—*all* the irregular nebulae. This single exception is a most significant one. It is that of the "great looped nebula" (see Plate XV),[3] an important constituent of the greater

---

[1]Taken as of the uniform width of thirty degrees, and covering 1/4.5 of the sphere. Major Markwick (*Journal Liv. Astr. Soc.* vol. vii. p. 182) finds the proportionate area of the Milky Way in the northern hemisphere to be 1/4.37, in the southern 1/3. Pickering (*Harvard Annals,* vol. xlviii. p. 165) estimates the galactic area at 15,612 square degrees, or 1/2.7 of the sphere.

[2]Bauschinger, *V. J. S. Astr. Ges.* Jahrg. xxiv. p. 43.

[3][Plate XV is not included in this selection.]

Magellanic Cloud. Plainly, then, the conditions allowing primitive cosmical matter to remain uncondensed in galactic regions, prevail also in the nubecula. The individuality of its organisation has been strongly accentuated by the discovery, lately made at Harvard College, that very many of the stars contained in it fluctuate rapidly in light.[1] Miss Leavitt's examination of the Arequipa plates yielded at once a harvest of 152 variables; and more doubtless await recognition.

Gaseous and white nebulae meet on equal terms only in this comprehensive assemblage. The Milky Way is more exclusive. It favours the former class largely at the expense of the latter. Within its precincts only one in sixteen of those dim, often fantastically-shaped, objects is met with, the analysed light of which gives no indication of gaseity, while their even mixture, under the highest telescopic powers, suggests no approach to the stage of breaking-up into stars. What then is their nature? Is the difference separating them in appearance from the resolvable aggregations of star-dust, crowding the Milky Way, a difference of distance solely? Are they too clusters beyond the reach, through remoteness, of effective scrutiny? There is nothing in their aspect to preclude this supposition. So far as observation can tell they may be of stellar composition. Only it is not easy to understand why nebulae situated near the galactic poles should be immensely and consistently more distant than nebulae thronging the vicinity of the galactic equator.

Mr. Cleveland Abbe[2] sought to overcome the difficulty by imagining the nebulae to be equably distributed over the surface of a "prolate ellipsoid," its longer axis coinciding approximately with the axis of the Milky Way; and this arrangement would undoubtedly give an appearance of crowding in the observed directions, since to an eye placed near

The adjacent prolate ellipsoid portrays the figure that Clerke probably had in mind. The ellipse at the center of the figure represents the Milky Way.

---

[1] *Harvard Circular*, No. 82; *Astr. Nach.* No. 3965.
[2] *Monthly Notices*, vol. xxvii. p. 262.

the centre of such an oval figure, objects uniformly scattered over its surface would produce, by perspective, the effect of running together near its pointed ends. But this highly artificial contrivance was scarcely realisable. Our minds demand from a theory not barely that it "cover the phenomena," but also that it show itself congruous with the general plan of operations upon which we can see that nature works. Besides, the local distribution of nebulae is so far from uniform, that antecedent probability is in favour of their general distribution being also marked by striking irregularities. The "canopy" of nebulae in Virgo is then, we may rest assured, as genuine an accumulation in its own way as the spherical assemblage in the Magellanic Cloud.

But if there be no systematic difference of distance between the nebular classes occupying contrasted situations as regards the lines of galactic structure, there must be a systematic difference of constitution.[1] The parts of those objects crowding towards the poles must be comparatively small and close together. We have indeed already found reason to believe that clusters do, in point of fact, merge insensibly into nebulae—that groups of genuine suns at wide intervals stand at the summit of an unbroken gradation of systems with smaller and closer constituents, down to accumulations of what is almost literally "star-dust." Resolvability is hence a question of constitution quite as much as of distance, and we are brought to the conclusion that, while galactic nebulae are of what we may roughly describe as stellar composition, non-galactic nebulae are more or less pulverulent. We cannot of course pretend to account for this remarkable distinction. All that can be said is that it *appears* to be actually existent. The irresolvable "polar" nebulae perhaps escaped influences powerful over the "equatorial" ones. Their development, at any rate, seems to have taken a different course.

The spectroscope informs us that nebulae possess radial velocities of the same order as those of stars. And their apparent fixity to visual observation unquestionably results, in part from their extreme remoteness, in part from their

---

[1]Proctor, *Monthly Notices*, vol. xxix. p. 342.

evasiveness of accurate measurement. The singularities of nebular distribution afford besides indirect evidence of movement. "Streaminess," if it mean anything, implies that the bodies affected by it advance in common towards a common goal. Aggregation at the end of a stream prompts the conjecture that a motion of advance was at a certain point, by some supervening attraction, swayed into a motion of revolution. A hint as to the origin of the Magellanic Clouds may hence be derived. They represent in some sort vessels filled through long pipes from a vast reservoir. And since the pipes are still there, the flow may be conceived to be still in progress. Were it to cease, the connection of the nubeculae with the main nebular body would eventually be interrupted, and their insulation would become complete.

The fidelity with which gaseous nebulae and clusters adhere to the Milky Way as seen projected upon the sphere, warrants the inference that their distribution in space is of a similar character. It would be unreasonable to disconnect them from a formation of which they so closely follow the lines. We can scarcely err in supposing that they lie in general *within*, not behind, or in front of it. Thus, the globular clusters richly strewn over the branch of the Milky Way from Scorpio to Ophiuchus, but withdrawn from the conterminous dark rift, plainly belong to the cloud-like stellar masses, owing to the absence of which the fissure seems black, although densely stocked with stars to the tenth magnitude. Other compressed groups stand out from a curtain of apparently still more remote stars, representing possibly a divergent galactic ramification.

Such ramifications must in many cases be greatly foreshortened as viewed from our nearly central position; in some, they may appear only as brilliant knots upon the "trunk" of the Milky Way. Possibly, the double cluster in Perseus may partake of this optical character. It may be the termination of a branch spreading inward, and seen nearly end-on.

Like the Perseus clusters, the Orion nebula gives indications of greater proximity than the main galactic accumulation, to which it is nevertheless beyond doubt structurally related. For a winding nebulous extension from the Milky Way can be traced past α Orionis through the belt and sword, the bright stars marking which are demonstrably associated with the nebula. The inference then presents itself that the whole mixed system, or series of systems, is placed upon an obliquely directed offset from the galactic zone. Reasoning of the same kind may perhaps apply to the combined nebula and cluster M 8. It occurs as a premonitory outlier of the leading division of the fissured Milky Way, from which it lies a little apart; and it seemed to Sir John Herschel only an "intense exaggeration" of the stellar collections in its neighbourhood.[1]

Summarising our conclusions, we find the unity of the stellar and nebular systems to be fully ascertained. They are bound together by relations of agreement and contrast scarcely less visibly intimate than those severally connecting individual members of each order. The general plan of nebular distribution is into two vast assemblages, one on either side of the Milky Way; but while this is, comparatively speaking, avoided by the unresolved crowd, it is densely thronged with clusters and gaseous nebulae. The conditions of aggregation within the zone may hence be inferred to differ from those prevailing outside it; but how and why they differ remains inscrutable. As to the distances of the nebulae, we know nothing positive; they no doubt vary extensively; nor can either fineness of grain, or faintness of light (both of which may be inherent qualities) serve to distinguish between those nearest to, and those further away from us. We may, however, plausibly conjecture that the hoodlike accumulation of nebulae in Coma Berenices is of the same approximate remoteness with the main galactic stream, and may thus be said to constitute the polar cap of a sphere equatorially girdled by the Milky Way.

☆☆☆☆☆☆☆☆☆☆☆☆☆☆☆☆☆☆☆☆☆☆☆☆☆☆☆☆

---

[1] *Cape Results*, p. 387.

Introduction to Henrietta Leavitt's Paper

Henrietta Swan Leavitt

Because the following paper by Henrietta Swan Leavitt (1868–1921) of Harvard Observatory was crucially important for subsequent developments, it deserves very careful reading. Given here is a brief explication of its contents, which may be read either before or after Leavitt's paper, which appeared over the signature of E. C. Pickering, the director of Harvard Observatory.

An important distinction in astronomy is that between brightness or apparent magnitude and luminosity or absolute magnitude. The need for that distinction becomes evident from the following question: which is brighter, a hundred watt bulb or the moon? One can argue for the bulb by saying that you can read by it; or for the moon by noting that it will illuminate your path for a long journey. The brightness of a light source is relative; it depends on its distance. If the 100-watt bulb is one foot away, it is brighter than the moon; if the bulb is a mile away, the moon is brighter. To get around this confusion, astronomers distinguish luminosity (absolute magnitude) from brightness (apparent magnitude). Luminosity is not relative to distance; it is rather the measure of the quantity of light given off by a source. Clearly the moon is more luminous (has a greater absolute magnitude) than the light bulb. The precise relationship between luminosity and brightness is expressed in the following equation:

$$B = \frac{kL}{D^2},$$

where B = brightness, L = luminosity, D = distance from observer to source, and k is a proportionality constant. From that equation it is evident that removing a light source to twice the distance cuts its brightness to one-fourth. Doubling its luminosity, on the other hand, doubles its brightness. Most importantly, from knowing two of the three terms (B, L, and D) as well as the constant, we can calculate the third.

Sample Problem: Suppose you have two light bulbs with luminosities L_1 and L_2 and at distances D_1 and D_2. How are their brightnesses, B_1 and B_2, related?

Solution:

From knowing that $B_1 = \dfrac{kL_1}{D_1^2}$ and that $B_2 = \dfrac{kL_2}{D_2^2}$,

we get: $\dfrac{B_1}{B_2} = \dfrac{L_1 D_2^2}{L_2 D_1^2}$.

Thus if $L_1 = 2L_2$ and $D_1 = 6D_2$,

we infer that $\dfrac{B_1}{B_2} = \dfrac{2}{36}$ and $B_1 = \dfrac{B_2}{18}$.

The question arises as to how that equation can be applied to the stars. This could prove very interesting because of the importance of the distance of the stars. If it were possible to determine both the brightness and the luminosity for some star, this would permit a calculation of its absolute distance. The brightness is easily obtained; one needs only measure how bright the star appears. But alas there is no direct way of getting L or D or k.

Enter Miss Leavitt. Her achievement can be summarized in the following way: she found a method for comparing the luminosities (absolute magnitudes) of the members of a certain class of variable stars, which stars are known as *Cepheid variables.* For example, for two stars among this group of stars, she would be able to specify that the absolute magnitude of one is X times greater than that of the other. Her method worked only for this particular class of stars, and she had no way of determining the constant k. But even with these qualifications, the importance of her result was immense: she supplied a way to compare L values, which, when combined with known B values, provide the relative distance to all such stars (or objects in which such stars are located).

How did she do this? She discovered that variable stars of the Cepheid type have the property that the luminosity or absolute magnitude of each such star is proportional to the logarithm of its period (the time between two dimmings of the star). But how she was able to do this should not yet be clear, for it would appear that she could do this only *if* she knew the distances of some of those stars. And this was not knowledge she possessed.

Nonetheless, she found a way around that problem. Suppose that on a totally dark night you look out from a cliff and see five lights in the distance. You cannot, let us say, tell how far away they are or (equivalently) how luminous they are. Nor can you tell how luminous they are relative to each other. Suppose, however, someone tells you that the five lights are all on a large ship sailing in the distance. Can you now compare their luminosities? Yes; because they are all at the same distance, the brighter must be the more luminous.

In these terms, we can see what Leavitt did. Her ship was the smaller Magellanic Cloud, the stars of which are all at approximately the same distance. She found that for these stars, their brightnesses, and consequently their luminosities or absolute magnitudes, depend on their periods. The stars with longer periods of variability were always brighter. Thus if she were given two Cepheid variable stars, one in the Magellanic clouds, the other somewhere else, she could compare their observed periods, calculate their relative luminosities, and thence derive their relative distances. If she were to find by observing the period and brightness of the second star that its luminosity is equal to that of a certain star in the smaller Magellanic Cloud, but that it is four times brighter, she could infer that the star is positioned at a location half the distance of the smaller Magellanic Cloud. She could not, however, determine its true distance, only its distance relative to the smaller Magellanic Cloud; her achievement was, nonetheless, epoch making. She had found a method that would reach far, far beyond the limits of the previous method of measuring stellar distances, i.e., parallax.

Example: Suppose that through a parallax determination, it is known that a certain Cepheid variable star of period P is 10 light-years distant. Suppose that another star is found to have identical period, but that it is 100 times dimmer. Calculate the distance of this star.

Solution: Because the two stars have the same period, they must also be equal in luminosity. Consequently, distance must be what makes the second star appear dimmer. Knowing that brightness is inversely proportional to the distance squared, we have that the second star must be $\sqrt{100} = 10$ times more distant. Thus it is 100 light-years distant.

Important Note: This example is purely hypothetical. The reason for this is that no Cepheid variable is sufficiently near the earth that its parallax can be measured.

☆☆☆☆☆☆☆☆☆☆☆☆☆☆☆☆☆☆☆☆☆☆☆☆☆☆☆☆

E. C. Pickering [for Henrietta Leavitt], "Periods of 25 Variable Stars in the Small Magellanic Cloud," *Harvard College Observatory Circular 173* (March 3, 1912), 1–3.

The following statement regarding the periods of 25 variable stars in the Small Magellanic Cloud has been prepared by Miss Leavitt.

A Catalogue of 1777 variable stars in the two Magellanic Clouds is given in H.A. 60, No. 4.[1] The measurement and discussion of these objects present problems of unusual difficulty, on account of the large area covered by the two regions, the extremely crowded distribution of the stars contained in them, the faintness of the variables, and the shortness of their periods. As many of them never become brighter than the fifteenth magnitude, while very few exceed the thirteenth magnitude at maximum, long exposures are necessary, and the number of available photographs is small. The determination of absolute magnitudes for widely separated sequences of comparison stars of this degree of faintness may not be satisfactorily completed for some time to come. With the adoption of an absolute scale of magnitudes for stars in the North Polar Sequence, however, the way is open for such a determination.

---

[1][H. Leavitt, "1777 Variables in the Magellanic Clouds," *Harvard College Observatory Annals*, 60 (1908), 87–108.]

H.	Max.	Min.	Epoch	Period	Res. M	Res. m	H.	Max.	Min.	Epoch	Period	Res. M	Res. m
			d.	d.						d.	d.		
1505	14.8	16.1	0.02	1.25336	−0.6	−0.5	1400	14.1	14.8	4.0	6.650	+0.2	−0.3
1436	14.8	16.4	0.02	1.6637	−0.3	+0.1	1355	14.0	14.8	4.8	7.483	+0.2	−0.2
1446	14.8	16.4	1.38	1.7620	−0.3	+0.1	1374	13.9	15.2	6.0	8.397	+0.2	−0.3
1506	15.1	16.3	1.08	1.87502	+0.1	+0.1	818	13.6	14.7	4.0	10.336	0.0	0.0
1413	14.7	15.6	0.35	2.17352	−0.2	−0.5	1610	13.4	14.6	11.0	11.645	0.0	0.0
1460	14.4	15.7	0.00	2.913	−0.3	−0.1	1365	13.8	14.8	9.6	12.417	+0.4	+0.2
1422	14.7	15.9	0.6	3.501	+0.2	+0.2	1351	13.4	14.4	4.0	13.08	+0.1	−0.1
842	14.6	16.1	2.61	4.2897	+0.3	+0.6	827	13.4	14.3	11.6	13.47	+0.1	−0.2
1425	14.3	15.3	2.8	4.547	0.0	−0.1	822	13.0	14.6	13.0	16.75	−0.1	+0.3
1742	14.3	15.5	0.95	4.9866	+0.1	+0.2	823	12.2	14.1	2.9	31.94	−0.3	+0.4
1646	14.4	15.4	4.30	5.311	+0.3	+0.1	824	11.4	12.8	4.	65.8	−0.4	−0.2
1649	14.3	15.2	5.05	5.323	+0.2	−0.1	821	11.2	12.1	97.	127.0	−0.1	−0.4
1492	14.8	14.8	0.6	6.2926	−0.2	−0.4							

[Table I of Henrietta Leavitt's Paper]

Fifty-nine of the variables in the Small Magellanic Cloud were measured in 1904, using a provisional scale of magnitudes, and the periods of seventeen of them were published in H.A. 60, No. 4, Table VI. They resemble the variables found in globular clusters, diminishing slowly in brightness, remaining near minimum for the greater part of the time, and increasing very rapidly to a brief maximum. Table I gives all the periods which have been determined thus far, 25 in number, arranged in the order of their length. The first five columns contain the Harvard Number, the brightness at maximum and at minimum as read from the light curve, the epoch expressed in days following J.D. 2,410,000, and the length of the period expressed in days. The Harvard Numbers in the first column are placed in Italics, when the period has not been published hitherto. A remarkable relation between the brightness of these variables and the length of their periods will be noticed. In H.A. 60, No. 4, attention was called to the fact that the brighter variables have the longer periods, but at that time it was felt that the number was too small to warrant the drawing of general conclusions. The periods of 8 additional variables which have been determined since that time, however, conform to the same law.

The relation is shown graphically in Figure 1, in which the abscissas are equal to the periods, expressed in days, and the ordinates are equal to the corresponding magnitudes at maxima and at minima.

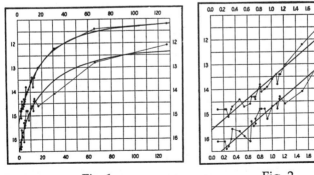

Fig. 1   Fig. 2

The two resulting curves, one for maxima and one for minima, are surprisingly smooth, and of remarkable form. In Figure 2, the abscissas are equal to the logarithms of the periods, and the ordinates to the corresponding magnitudes, as in Figure 1. A straight line can readily be drawn among each of the two series of points corresponding to maxima and minima, thus showing that there is a simple relation between the brightness of the variables and their periods. The logarithm of the period increases by about 0.48 for each increase of one magnitude in brightness. The residuals of the maximum and minimum of each star from the lines in Figure 2 are given in the sixth and seventh columns of Table I. It is possible that the deviations from a straight line may become smaller when an absolute scale of magnitudes is used, and they may even indicate the corrections that need to be applied to the provisional scale. It should be noticed that the average range, for bright and faint variables alike, is about 1.2 magnitudes. Since the variables are probably at nearly the same distance from the Earth, their periods are apparently associated with their actual emission of light, as determined by their mass, density, and surface brightness.

The faintness of the variables in the Magellanic Clouds seems to preclude the study of their spectra, with our present facilities. A number of brighter variables have similar light curves, as UY Cygni, and should repay careful study. The class of spectrum ought to be determined for as many such objects as possible. It is to be hoped, also, that the parallaxes of some variables of this type may be measured. Two fundamental questions upon which light may be thrown by such inquiries are whether there are definite limits to the mass of variable stars of the cluster type, and if the spectra of such variables having long periods differ from those of variables whose periods are short.

The facts known with regard to these 25 variables suggest many other questions with regard to distribution, relations to star clusters and nebulae, differences in the forms of the light curves, and the extreme range of the length of the periods. It is hoped that a systematic study of the light changes of all the variables, nearly two thousand in number, in the two Magellanic Clouds may soon be undertaken at this Observatory.

☆☆☆☆☆☆☆☆☆☆☆☆☆☆☆☆☆☆☆☆☆☆☆☆☆☆☆☆

Chapter Seven

Background to the Great Debate

Introduction

On 26 April 1920, the so-called Great Debate took place before the National Academy of Sciences. The debaters, Harlow Shapley of Mount Wilson Observatory and Heber Curtis of Lick Observatory, presented contrasting views on each of the three questions considered in these readings: (1) What is the size and structure of the Milky Way? (2) What and where are the nebulae? (3) Do island universes exist? The basic position advanced by Shapley was that the Milky Way is very large and that nebulae are relatively near it, if not actually part of it. Heber Curtis, on the other hand, championed a small Milky Way, but a large universe in which many nebulae are island universes. Their debate was a major development in the by-then nearly two-centuries-long quest by astronomers to answer these questions; moreover, in their presentations, Curtis and Shapley discussed a number of key developments and discoveries from the immediately preceding years. The present materials set the stage for consideration of that debate by outlining some of those key developments. The first item is a chronology that summarizes those advances; in reading it, you should ask yourself in what way and to what degree was each discovery or event significant for our set of three questions. Following the chronology, four short papers are presented; each contains information crucially relevant to the questions of the "Great Debate."

As background for understanding some aspects of these materials, it is helpful to have a basic knowledge of the scale used by astronomers to assign stellar magnitudes. For many years before the 1850s, the practice had been to describe the brightness of stars in terms of six classes of brightness. The brightest stars were designated first magnitude, whereas the dimmest naked eye stars were labeled sixth magnitude. This should not be taken to mean that a first magnitude star is 5 or 6 times brighter than a sixth magnitude star. In fact, it was known to be about 100 times brighter. In 1856, Norman R. Pogson successfully urged that a stellar magnitude scale be standardized. He proposed that a difference of five magnitude classes should, by definition, correspond exactly to a brightness difference of 100. This means that the difference in brightness between any star and a star in the next lowest class is equal to 2.512, which is the fifth root of 100. Example: given the fact that 5 − 2 = 3, a second magnitude star will appear ($2.512^3 =$) 15.85 times brighter than a fifth magnitude star.

**The 60"-Aperture Mount Wilson Reflector
Erected in 1908**

Chronology of Events Leading up to the Great Debate

1885 An extremely bright nova (new star) was observed near the center of Andromeda nebula. Before fading, this star, which was known as S Andromedae, rose to the 7th magnitude, that is, it became almost bright enough to be seen with the naked eye; in fact, it was about one tenth as bright as the entire nebula. Its brightness pointed to the conclusion either that Andromeda is quite near and not composed of stars or that S Andromedae surpassed all known stars in magnitude.

1887 Agnes Clerke in her *Popular History of Astronomy during the Nineteenth Century* declared: "There is no maintaining nebulae to be simply remote worlds of stars in the face of an agglomeration like the Nubecula Major [the larger Magellanic Cloud], containing in its (certainly capacious) bosom *both* stars and nebulae. Add the evidence of the spectroscope to the effect that a large proportion of these perplexing objects are gaseous, with the facts of their distribution telling of an intimate relation between the mode of their scattering and the lie of the Milky Way, and it becomes impossible to resist the conclusion that both nebular and stellar systems are parts of a single scheme."[1]

1888 The British astronomer Isaac Roberts, having made an especially fine photograph of the Andromeda nebula, remarked: "those who accept the nebular hypothesis [formation of planetary systems from a condensing nebulous mass] will be tempted to appeal to the constitution of this nebula for confirmation, if not for demonstration of the hypothesis." Roberts was suggesting that one seemingly can "see a new solar system in the process of condensation from the nebula—the central sun is now in the midst of nebulous

---

[1] Agnes Clerke, *Popular History of Astronomy during the Nineteenth Century*, 2nd ed. (Edinburgh, 1887), 456–7.

matter.... The two [smaller] nebulae [associated with the Andromeda Nebula] seem as though they were already undergoing their transformation into planets."[1]

1888   Lick Observatory opened on Mount Hamilton in California with a 36-inch-aperture refracting telescope, which was then the largest refractor in the world.

1896   Percival Lowell erected a 24-inch refractor at his Lowell Observatory in Flagstaff, Arizona.

1897   Yerkes Observatory of the University of Chicago opened in Williams Bay, Wisconsin, equipped with a 40-inch refractor, which is still the largest refractor in the world.

1898   Joseph Plassmann argued that some island universes exist, for example, Andromeda. Few of his contemporaries shared this view.

1898   Julius Scheiner, having detected dark lines in the spectrum from the Andromeda nebula, urged that it is "a cluster of sun-like stars." Some doubted Scheiner's observation.

1899   J. E. Keeler of Lick estimated that he could photographically detect as many as 120,000 nebulae. He also suggested that most nebulae are spirals.

1904   T. C. Chamberlin and F. R. Moulton, two American scientists, put forth their "planetismal" theory of the formation of solar systems. One aspect of their theory was their claim that some spiral nebulae are planetary systems in the process of formation. In fact,

---

[1] As quoted in Robert Smith, *The Expanding Universe: Astronomy's 'Great Debate' 1900–1931* (Cambridge: Cambridge University Press, 1982), p. 5.

Moulton at one point referred to the theory as the "spiral nebula theory."

1904 Jacobus Kapteyn of Groningen Observatory put forth carefully researched evidence that the nearest stars, after the solar motion is subtracted from them, show a definite tendency to move in two opposite directions or "streams."

1907 Max Wolf of Heidelberg Observatory stated that nebulae and star clusters are part of our "star-island" and added: "Distant, isolated Milky Ways have never been sighted by man."[1]

1907 Karl Bohlin, a Swedish astronomer, reported that he had measured a trigonometric parallax for the Andromeda nebula. On this basis, Bohlin set the distance of Andromeda at about 19 light-years.

1908 The 60-inch-aperture reflector (see photograph on p. 234) erected on Mount Wilson in California. This remained the largest reflector until 1917.

1911 Hugo von Seeliger, using statistical techniques, set the edge of the Milky Way at about 8,000 light-years from us.

1912 A. C. D. Crommelin of Greenwich Observatory published a paper opposing island universes, urging that the fact that the spirals appear off the plane of the Milky Way and toward the galactic poles indicates that they must be part of the structure of the Milky Way.

1912 Henrietta Leavitt published her study of the Smaller Magellanic Cloud, showing a period-luminosity relationship among the Cepheid variables in that

---

[1] As quoted in Stanley L. Jaki, *The Milky Way: An Elusive Road for Science* (New York: Science History, 1972), p. 278.

object. This in principle provided a method for calculating the relative distance of any Cepheid that has an observable period.

1912 Vesto Slipher of Lowell Observatory reported that the spectrum of the nebulosity surrounding the Pleiades is identical to that of the nearby stars; this indicated the existence of what have come to be known as reflection nebulae.

1913 Cornelius Easton, who in 1900 had argued for a Milky Way consisting of rings, suggested that the Milky Way has a spiral structure. Nonetheless, he did not conclude that all spirals are island universes; he remarked that "nobody will deny the existence of a whirlpool because he sees a number of small eddies in the convolution of the great one."[1]

1913 Ejnar Hertzsprung made the first attempt to determine a zero-point for the period-luminosity relationship of Leavitt, i.e., he attempted to affix absolute, not just relative values, to the scale that Leavitt had established. Because all the Cepheids were (and still are) beyond the reach of parallax measurements, he had to use approximation techniques. He used a statistical method based on the observed proper motions and radial motions of 13 Cepheids. On this basis, he assigned a distance of 30,000 light-years to the Smaller Magellanic Cloud.

1913 Henry Norris Russell of Princeton published a paper presenting his spectrum-luminosity diagram. Russell showed that when the luminosities of stars at known distances are plotted against their spectral types, they fall into a pattern resembling a reversed 7. This provided a rough method of determining the distances of stars located even far beyond the reach of

---

[1] As quoted in Richard Berendzen, Richard Hart, and Daniel Seeley, *Man Discovers the Galaxies* (New York: Science History, 1976), p. 58.

parallactic methods. Diagrams in which the spectral types of stars are plotted against their luminosities are known as Hertzsprung-Russell diagrams, Hertzsprung having also developed the notion of comparing spectral type with luminosity.

1914 Vesto Slipher and others reported the detection of a curvature in the spectral lines of some nebulae seen edge-on. This indicated that these nebulae are rotating.

1914 Jacobus Kapteyn noted that fainter stars, presumably at greater distances, show a reddening; this, he suggested, could be due to the presence of obscuring matter in space blocking blue rays.

1914 Arthur Stanley Eddington in his *Stellar Movements and the Structure of the Universe* argued for the Island Universe theory on heuristic grounds, stating: "If . . . it is assumed that [spiral nebulae] are external to the stellar system, that they are in fact systems co-equal with our own, we have at least an hypothesis which can be followed up, and may throw some light on the problems that have been before us. For this reason, the 'island universe' theory is much to be preferred as a working hypothesis; and its consequences are so helpful as to suggest a distinct probability of its truth."[1]

1915 Vesto Slipher published a study of the radial velocities as measured spectrographically of 15 spirals. He reported that these nebulae generally have high radial velocities, on the order of 300 to 1100 km/sec. He also reported that they are usually recessional velocities, that is, away from us.

---

[1] As quoted in Smith, *Expanding Universe*, p. 26.

1916	Adriaan van Maanen of Mount Wilson reported his observation of a rotation of 0.02" per year in M101, a prominent spiral that is seen face on. He later reported rotations in other spirals. His work was supported by observations of Schouten of Holland, Kostinsky of Russia, and others. James Jeans noted that these observations were in accord with some of his own theoretical studies.
1916	Harlow Shapley of Mount Wilson, who had been studying globular clusters, argued that they are quite distant objects. Finding blue stars in them, indicating that despite their distance, little if any reddening had taken place, he urged in a 1916 letter that Kapteyn's value for the effect of obscuring matter (about one magnitude per thousand parsecs) "must be from ten to a hundred times too large ... and the absorption in our immediate region of the stellar system must be entirely negligible."[1]
1917	G. W. Ritchey of Mount Wilson reported his detection in a spiral nebula of a nova that was visible at that time. Heber D. Curtis at Lick had already found three novae in earlier photographic plates of nebulae and published this fact after Ritchey's result was announced. Ritchey, Shapley of Mount Wilson, and others found additional novae. Curtis began to investigate these novae in spirals as a possible criterion of distance. In 1917, he noted that of 27 known novae in our Milky Way, the average magnitude is 5.5, whereas novae in spirals are about 10 magnitudes dimmer, meaning that they are about 10,000 times dimmer. This suggested that novae in spirals are about 100 times more remote than novae in the Milky Way. He stated: "The occurrence of these new stars in spirals must be regarded as having very

---

[1] As quoted in Smith, *Expanding Universe*, p. 59.

definite bearing on the 'island universe' theory...."[1] S Andromedae, however, remained a problem.

1917   100-inch reflector erected at Mount Wilson (see picture). It remained the largest reflector until 1948 when the Palomar reflector was completed.

The 100"-Aperture Mount Wilson Reflector Erected in 1917

1917   A. C. D. Crommelin, by then a convert to the Island Universe theory, advocated its acceptability on aesthetic grounds: "Whether true or false, the

---

[1] As quoted in Jaki, *Milky Way*, p. 292.

hypothesis of external galaxies is certainly a sublime and magnificent one. . . . Our conclusions in Science must be based on evidence, and not on sentiment. But we may express the hope that this sublime conception may stand the test of further examination."[1]

1917 Curtis, having observed dark lanes on the edges of spirals seen edge-on, suggested that were the Milky Way a spiral, it too would presumedly have dark lanes of obscuring matter. This obscuring matter would help explain why no spirals are seen in the plane of the Milky Way.

By 1918 Shapley had recalculated the zero point for the Cepheids, using a method similar to that used by Hertzsprung. On this basis, he calculated the distance of M13, the prominent globular cluster in Hercules, in which he had located Cepheids. Having determined the distance of M13 to his satisfaction and assuming that the other globular clusters are of comparable size, he estimated their distances by comparing their sizes and brightnesses to his figures for M13. He put the globular clusters at distances from 20,000 to 200,000 light-years and urged that they define the shape of the galaxy and that the majority are clustered around the center of the galaxy. He argued that the diameter of the Milky Way must be about 300,000 light-years with the sun located some distance from its center. Although he had earlier accepted island universes, his large Milky Way led him to view globular clusters and spirals as not comparable to it in size; this conclusion entailed rejecting them as island universes.

1919 Shapley argued that if spirals are Island Universes and if M101 is one fifth the Milky Way's size (using his value for the Milky Way's size), then the rotation reported by van Maanen for M101 entails that the edge

---

[1] As quoted in Smith, *Expanding Universe*, p. 16.

of M101 must be moving faster than the speed of light. This being contrary to the laws of physics, he concluded that M101 cannot be an Island Universe.

1919   E. E. Barnard of Yerkes Observatory published one of his many papers on the "Dark Markings" in the heavens. From more than a decade of observation of such regions, he had concluded that they are most probably due to obscuration by clouds of dark matter in interstellar space.

1920   J. H. Reynolds of England reported that the outer regions of spirals tend to be bluer than the central regions, whereas the Milky Way seems to be homogeneous. This seemed to indicate that the spirals are different in character from the Milky Way.

1920   Knut Lundmark of Sweden analyzed 22 novae found in Andromeda and set its distance on this basis at 600,000 light-years. Lundmark claimed that S Andromedae, the very, very bright nova that had appeared in 1885, belongs to a special class of novae different from all novae previously observed in Andromeda.

Commentary on H. N. Russell and the Spectrum-Luminosity Diagram

The following paper by Henry Norris Russell (1877–1957) of Princeton is of crucial importance in many ways. For our present concerns, the chief of these is that it provided another method of measuring distances to celestial objects. The following brief exposition suggests how his results made this possible.

As Russell stated early in his paper, the spectra of over 100,000 stars had become known. Over 99% of these stars were found to fall into six spectral types: B, A, F, G, K, M; that is, they range in a continuous fashion from whitish-blue to red stars. This continuity suggested that the difference in spectral type may be a function of a single physical variable, which Russell hypothesized to be their temperature.

Does a relationship exist between a star's spectral type and its luminosity? Were such a relationship to be found, one could infer the luminosity of a star from observation of its spectral type. This, in turn, would permit calculation of its distance. Russell set about plotting the spectral types of stars against their luminosities. He could do this because by 1913, the parallaxes of a substantial number of stars were known. From their parallaxes, their distances were determined and from observation of their brightness, their luminosities could be found. Russell's Figure 1 (see p. 249) is a plot of the spectral classes (horizontal coordinate) of many stars versus their luminosities (vertical component). It would have been ideal if a straight line resulted, but this did not occur.

Nonetheless, Russell found a striking result. With very few exceptions, the stars fell into a broad pattern, in particular, a backward seven. Further examination of his graph shows:
(1) Essentially all the white to bluish stars (the B and A types) fall on the top portion of the graph, specifically in the region of high luminosity or absolute magnitude. There was one exception: the star (o^2 Eridani) in the lower left corner. All the faint stars fall in the red group (K to M) at the lower right.
(2) Although essentially all faint stars are reddish, not all reddish stars are faint. Some reddish stars—upper right—possess high luminosities.
(3) Not only is the lower left section relatively empty, so is the middle right section. In other words, reddish stars have either high or low luminosities, not middle range values.

But how can Russell's result be related to distance? Consider a B-type star. Can any inference be drawn as to its distance from knowledge of its spectral type and apparent brightness? The graph shows that the range of luminosities of B-type stars is relatively restricted. Consequently, a reasonably good estimate can be made from the graph of its *range* of possible luminosities, and from this its distance can be estimated. This method is less effective for red stars; they may have high or low luminosities, but one cannot tell which. Consider now a cluster of stars; say a globular cluster. Because stars in a cluster are all at about the same distance, differences in brightness among them should be the result of differences in luminosity. Suppose one were to plot the brightness values of the stars in the cluster against their spectral types. A reversed seven should then become apparent. If this pattern emerges, then by superimposing this reversed seven on Russell's graph, one can read off the luminosities of the stars and consequently calculate the distance of the cluster.

Example: Take a star at the middle point of the vertical part of the reversed seven found for the globular cluster. Comparison with Russell's graph assigns the star a luminosity of about +6, and from this the distance of the star can be estimated.

Russell checked his method by using stars in various bright groups of stars (Figure 2; p. 253) at known distances. Again a reversed seven resulted.

Plots of spectral type versus luminosity are now known as Hertzsprung-Russell diagrams. Hertzsprung's name is included because he had anticipated Russell in some of his ideas. Diagrams of this form fill astronomy books; the evolution of stars, for example, is discussed in terms of such diagrams. As will become clear, Russell's analysis was very important for the great debate.

☆☆☆☆☆☆☆☆☆☆☆☆☆☆☆☆☆☆☆☆☆☆☆☆☆☆☆☆☆
Henry Norris Russell, Selection from his "Relations between the Spectra and Other Characteristics of Stars," *Popular Astronomy*, 22 (1914), 275–94.

Investigations into the nature of the stars must necessarily be very largely based upon the average characteristics of groups of stars selected in various ways,—as by brightness, proper motion, and the like. The publication within the last few years of a great wealth of accumulated observational material makes the compilation of such data an easy process; but some methods of grouping appear to bring out much more definite and interesting relations than others, and, of all the principles of division, that which separates the stars according to their spectral types has revealed the most remarkable differences, and those which most stimulate attempts at a theoretical explanation.

In the present discussion, I shall attempt to review very rapidly the principal results reached by other investigators, and shall then ask your indulgence for an account of certain researches in which I have been engaged during the past few years.

Thanks to the possibility of obtaining with the objective prism photographs of the spectra of hundreds of stars on a single plate, the number of stars whose spectra have been observed and classified now exceeds one hundred thousand, and probably as many more are within the reach of existing instruments. The vast majority of these spectra show only dark lines, indicating that absorption in the outer and least dense layers of the stellar atmospheres is the main cause of their production. Even if we could not identify a single line as arising from some known constituent of these atmospheres, we could nevertheless draw from a study of the spectra, considered merely as line-patterns, a conclusion of fundamental importance.

The spectra of the stars show remarkably few radical differences in type. More than ninety-nine per cent of them fall into one or other of the six great groups which, during the classic work of the Harvard College Observatory, were recognized as of fundamental importance, and received as designations, by the process of "survival of the fittest", the rather arbitrary series of letters B, A, F, G, K, and M. That there should be so few types is noteworthy, but much more remarkable is the fact that they form a continuous series. Every degree of gradation, for example, between the typical spectra denoted by B and A may be found in different stars, and the same is true to the end of the series, a fact recognized in the familiar decimal classification, in which B5, for example, denotes a spectrum half-way between the typical examples of B and A. This series is not merely continuous: it is *linear*. There exist indeed slight differences between the spectra of different stars of the same spectral class, such as A0; but these relate to minor details, which usually require a trained eye for their detection, while the difference between successive classes, such as A and F, are conspicuous to the novice. Almost all the stars of the small outstanding minority fall into three other classes, denoted by the letters O, N, and R. Of these O undoubtedly precedes B at the head of the series, while R and N, which grade into one another, come probably at its other end, though in this case the transition stages, if they exist, are not yet clearly worked out.

From these facts it may be concluded that the principal differences in stellar spectra, however they may originate, arise in the main from variations in a single physical condition in the stellar atmospheres. This follows at once from the linearity of the series. If the spectra depended, to a comparable degree, on two independently variable conditions, we should expect that we would be obliged to represent their relations, not by points on a line, but by points scattered over an area. The minor differences which are usually described as "peculiarities" may well represent the effects of other physical conditions than the controlling one.

The first great problem of stellar spectroscopy is the identification of this predominant cause of the spectral differences. The hypothesis which suggested itself immediately upon the first studies of stellar spectra was that the differences arose from variations in the chemical composition of the stars. Our knowledge of this composition is now very extensive. Almost every line in the spectra of all the principal classes can be produced in the laboratory, and the evidence so secured regarding the uniformity of nature is probably the most impressive in existence. The lines of certain elements are indeed characteristic of particular spectral classes; those of helium, for instance, appear only in Class B, and form its most distinctive characteristic. But negative conclusions are proverbially unsafe. The integrated spectrum of the Sun shows no evidence whatever of helium, but in that of the chromosphere it is exceedingly conspicuous. Were it not for the fact that we are near this one star of Class G, and can study it in detail, we might have erroneously concluded that helium was confined to the "helium stars." There are other cogent arguments against this hypothesis. For example, the members of a star-cluster, which are all moving together, and presumably have a common origin, and even the physically connected components of many double stars, may have spectra of very different types, and it is very hard to see how, in such a case, all the helium and most of the hydrogen could have collected in one star, and practically all the metals in the other. A further argument—and to the speaker a very convincing one—is that it is almost unbelievable that differences of chemical composition should reduce to a function of a single variable, and give rise to the observed linear series of spectral types.

I need not detain you with the recital of the steps by which astrophysicists have become generally convinced that the main cause of the differences of the spectral classes is difference of temperature of the stellar atmospheres. There is time only to review some of the most important evidence which, converging from several quarters, affords apparently a secure basis for this belief.[1]

Having thus made a rapid survey of the general field, I will now ask your attention in greater detail to certain relations which have been the more special objects of my study.

Let us begin with the relations between the spectra and the real brightness of the stars. These have been discussed by many investigators,—notably by Kapteyn and Hertzsprung,—and many of the facts which will be brought before you are not new: but the observational material here presented is, I believe, much more extensive than has hitherto been assembled. We can only determine the real brightness of a star when we know its distance; but the recent accumulation of direct measures of parallax, and the discovery of several moving clusters of stars whose distances can be determined, put at our disposal far more extensive data than were available a few years ago.

Figure 1 shows graphically the results derived from all the direct measures of parallax available in the spring of 1913 (when the diagram was constructed). The spectral class appears as the horizontal coordinate, while the vertical one is the absolute magnitude, according to Kapteyn's definition,— that is, the visual magnitude which each star would appear to have if it should be brought up to a standard distance, corresponding to a parallax of 0".1 (no account being taken of any possible absorption of light in space.)

---

[1][Seven pages of Russell's text, in which he presents the evidence that differences among the spectral classes of celestial objects correlates with differences in temperature among them, have been omitted at this point.]

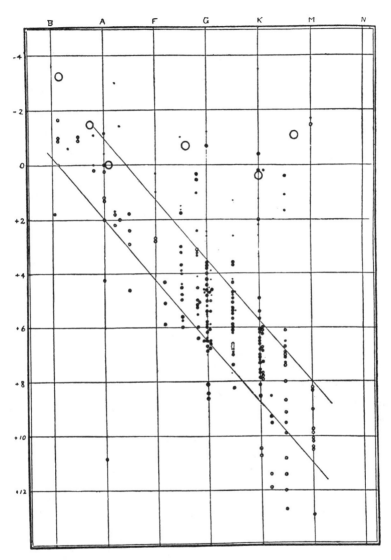

Figure 1.

The absolute magnitude –5. at the top of the diagram, corresponds to a luminosity 7500 times that of the Sun, whose absolute magnitude is 4.7. The absolute magnitude 14, at the bottom, corresponds to $\frac{1}{5000}$ of the Sun's luminosity. The larger dots denote the stars for which the computed probable error of the parallax is less than 42 per cent of the parallax itself, so that the probable error of the resulting absolute magnitude is less than ±1ʹ.0. This is a fairly tolerant criterion for a "good parallax", and the small dots, representing the results derived from the poor parallaxes, should hardly be used as a basis for any argument. The solid black dots represent stars whose parallaxes depend on the mean of two or more determinations; the open circles, those observed but once. In the latter case, only the results of those observers whose work appears to be nearly free from systematic error have been included, and in all cases the observed parallaxes have been corrected for the probable mean parallax of the comparison stars to which they were referred. The large open circles in the upper part of the diagram represent mean results for numerous bright stars of small proper motion (about 120 altogether) whose observed parallaxes hardly exceed their probable errors. In this case the best thing to do is to take means of the observed parallaxes and magnitudes for suitable groups of stars, and then calculate the absolute magnitudes of the typical stars thus defined. These will not exactly correspond to the mean of the individual absolute magnitudes, which we could obtain if we knew all the parallaxes exactly, but they are pretty certainly good enough for our purpose.

Upon studying Figure 1, several things can be observed.

1. All the white stars, of Classes B and A, are bright, far exceeding the Sun; and all the very faint stars,—for example, those less than $\frac{1}{50}$ as bright as the Sun,—are red, and of Classes K and M. We may make this statement more specific by saying, as Hertzsprung does,[1] that there is a certain limit of brightness for each spectral class, below which stars of this class are very rare, if they occur at all. Our diagram shows that this limit varies by rather more than two magnitudes from class to class. The single apparent exception is the faint double companion to o^2 Eridani, concerning whose parallax and brightness there can be no doubt, but whose spectrum, though apparently of Class A, is rendered very difficult of observation by the proximity of its far brighter primary.

---

[1] A.N. 4422, 1910.

2. On the other hand, there are many red stars of great brightness, such as Arcturus, Aldebaran and Antares, and these are as bright, on the average, as the stars of Class A, though probably fainter than those of Class B. Direct measures of parallax are unsuited to furnish even an estimate of the upper limit of brightness to which these stars attain, but it is clear that some stars of all the principal classes must be very bright. The range of actual brightness among the stars of each spectral class therefore increases steadily with increasing redness.

3. But it is further noteworthy that all the stars of Classes K5 and M which appear on our diagram are either very bright or very faint. There are none comparable with the Sun in brightness. We must be very careful here not to be misled by the results of the methods of selection employed by observers of stellar parallax. They have for the most part observed either the stars which appear brightest to the naked eye or stars of large proper motion. In the first case, the method of selection gives an enormous preference to stars of great luminosity, and, in the second, to the nearest and most rapidly moving stars, without much regard to their actual brightness. It is not surprising, therefore, that the stars picked out in the first way (and represented by the large circles in Figure 1) should be much brighter than those picked out by the second method, (and represented by the smaller dots). But if we consider the lower half of the diagram alone, in which all the stars have been picked out for proper-motion, we find that there are no very faint stars of Class G, and no relatively bright ones of Class M. As these stars were selected for observation entirely without consideration of their spectra, (most of which were then unknown) it seems clear that this difference, at least, is real, and that there is a real lack of red stars comparable in brightness to the Sun, relatively to the number of those 100 times fainter.

The appearance of Figure 1 therefore suggests the hypothesis that, if we could put on it some thousands of stars, instead of the 300 now available, and plot their absolute magnitudes without uncertainty arising from observational error, we would find the points representing them clustered principally close to two lines, one descending sharply along the diagonal, from B to M, the other starting also at B, but running almost horizontally. The individual points, though thickest near the diagonal line, would scatter above and below it to a vertical distance corresponding to at least two magnitudes, and similarly would be thickest near the horizontal line, but scatter above and below it to a distance which cannot so far be definitely specified, so that there would be two fairly broad bands in which most of the points lay. For Classes A and F, these two zones would overlap, while their outliers would still intermingle in Class G, and probably even in Class K. There would however be left a triangular space between the two zones, at the right-hand edge of the diagram, where very few, if any, points appeared; and the lower left-hand corner would be still more nearly vacant.

We may express this hypothesis in another form by saying that there are two great classes of stars,—the one of great brightness, (averaging perhaps a hundred times as bright as the Sun) and varying very little in brightness from one class of spectrum to another: the other of smaller brightness, which falls off very rapidly with increasing redness. These two classes of stars were first noticed by Hertzsprung,[1] who has applied to them the excellent names of *giant* and *dwarf* stars. The two groups, on account of the considerable internal differences in each, are only distinctly separated among the stars of Class K or redder. In Class F they are partially, and in Class A thoroughly intermingled, while the stars of Class B may be regarded equally well as belonging to either series.

In addition to the stars of directly measured parallax, represented in Figure 1, we know with high accuracy the distances and real brightness of about 150 stars which are members of the four moving clusters whose convergent points are known,—namely, the Hyades, the Ursa Major group, the 61 Cygni group, and the large group in Scorpius, discovered independently by Kapteyn, Eddington, and Benjamin Boss, whose motion appears to be almost entirely parallactic. The data for the stars of these four groups are plotted in Figure 2, on the same system as in Figure 1.

---

[1]*Zeitschrift für Wissenschaftliche Photographie*, vol. 3, p. 442, 1905.

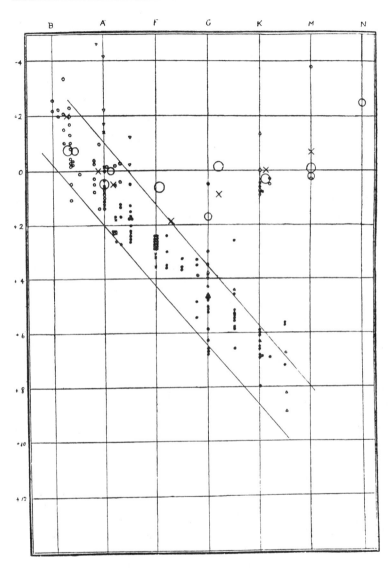

Figure 2.

The solid black dots denote the members of the Hyades; the open circles, those of the group in Scorpius, the crosses the Ursa Major group, and the triangles the 61 Cygni group. Our lists of the members of each group are probably very nearly complete down to a certain limiting (visual) magnitude, but fail at this point, owing to lack of knowledge regarding the proper motions of the fainter stars. The apparently abrupt termination of the Hyades near the absolute magnitude 7.0, and of the Scorpius group at 1.5 arises from this observational limitation.

The large circles and crosses in the upper part of Figure 2 represent the absolute magnitudes calculated from the mean parallaxes and magnitudes of the groups of stars investigated by Kapteyn, Campbell, and Boss [. . . .] The larger circles represent Boss's results, the smaller circles Kapteyn's, and the large crosses Campbell's.

It is evident that the conclusions previously drawn from Figure 1 are completely corroborated by these new and independent data. Most of the members of these clusters are dwarf stars, and it deserves particular notice that the stars of different clusters, which are presumably of different origin, are similar in absolute magnitude. But there are also a few giant stars, especially of Class K, (among which are the well known bright stars of this type in the Hyades); and most remarkable of all is Antares, which, though of Class M, shares the proper motion and radial velocity of the adjacent stars of Class B, and is the brightest star in the group, giving out about 2000 times the light of the Sun. It is also clear that the naked eye stars, studied by Boss, Campbell and Kapteyn, are for the most part giants.[1]

☆☆☆☆☆☆☆☆☆☆☆☆☆☆☆☆☆☆☆☆☆☆☆☆☆☆☆☆☆☆

Commentary on the Papers of
van Maanen, Slipher, and Eddington

The following papers by Adriaan van Maanen (1884–1946) and Vesto Slipher (1875–1969) respectively have at least three similarities: (1) their authors had access to superb telescopes, the former having worked at Mount Wilson in California, the latter at Lowell Observatory in Flagstaff, Arizona; (2) each determined motions in regard to spiral nebulae; (3) their determinations were extremely difficult, requiring that each press his instrumentation to its limits.

---

[1][The final four pages of Russell's paper, in which he discusses his results from a theoretical point of view, are not included in this selection.]

These similarities should not obscure a crucial difference: What van Maanen reported was that he had measured rotations of certain nebulae that are situated in such a way that their planes are approximately perpendicular to a line from the earth to their location, whereas what Slipher reported was that he had measured radial motions in certain nebulae, finding in most cases that they are moving rapidly away from the earth. The situation is somewhat more complicated than this, for as noted in the Eddington paper, Slipher had himself detected a rotation in at least one nebula. He had, however, detected it as a radial motion and did so while observing the nebula edge-on, not face-on, as in the case of van Maanen. Because these are important distinctions, it will prove useful to stress them by an analogy. A good way to think of it is that van Maanen observed a rotation as one usually observes the rotation of a Ferris wheel, whereas Slipher observed, as it were, the rotation of a merry-go-round. Moreover, Slipher reported that some of the nebulae that he observed edge-on were moving rapidly away from us.

An important consequence follows from van Maanen's results, in particular, his determination that M101 rotates with a period of 85,000 years. If assumptions are made about the distance and hence the size of M101, it is possible to calculate the speed of a point at its edge. The angular radius of M101 is about 8'. Suppose M101 is at an "island universe" distance, say, 10,000,000 light-years. On this assumption, we can calculate its radius, R.

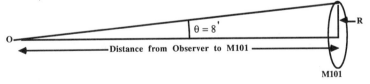

Let $\tan \theta = R/10^7$ light-years; then because $\tan 8' = .00233$, we have $R = 10^7 \cdot .00233$ light-years $= 23,300$ light-years. Such a radius would make M101 roughly comparable to the Milky Way in size, at least if one were to accept the maximum value (30,000 light-years) that Heber Curtis allowed for the diameter of the Milky Way.

Using van Maanen's 85,000-year figure for the rotational

period (T) of M101, it is easy to calculate the speed of a point on the edge of M101. The formula for the speed is: $v = \frac{2\pi R}{T}$. Making appropriate substitutions gives:

$$v = \frac{2\pi \cdot 2.33 \cdot 10^4 \cdot 365 \cdot 24 \cdot 60 \cdot 60 \cdot 186000 \text{miles}}{85000 \cdot 365 \cdot 24 \cdot 60 \cdot 60 \text{secs}} = 1.72 \cdot C,$$

where C is the speed of light. But this result contradicts Einstein's conclusion that no object can travel faster than the speed of light. Were we to set the distance of M101 at a significantly lower figure, say, 1,000,000 light-years, a point at its edge would still be moving at an incredible speed (32,000 miles per second). Thus it seems that if van Maanen's figure of 85,000 years for the rotation period of M101 is correct, then it is unlikely that M101 can be an island universe.

The final paper in this group of three was written by Arthur Stanley Eddington (1882–1944), a leading British theoretical astronomer. It summarizes the results reported by Slipher and van Maanen as well as commenting on their papers.

☆☆☆☆☆☆☆☆☆☆☆☆☆☆☆☆☆☆☆☆☆☆☆☆☆☆☆☆
Vesto M. Slipher, "Spectroscopic Observation of Nebulae," *Popular Astronomy*, 23 (1915), 21–3.

During the last two years the spectrographic work at Flagstaff has been devoted largely to nebulae. While the observations were chiefly concerned with the spiral nebulae they also include planetary and extended nebulae and globular star clusters.

Nebular spectra may be broadly divided into two general types (1) bright-line and (2) dark-line. The so-called gaseous nebulae are of the first type; the spiral nebulae of the second type.

Nebulae are faint and hence are generally difficult of spectrographic observation because of the extreme faintness of their dispersed light. In the bright-line spectrum the light is concentrated in a few points; in the dark line (continuous) spectrum it is spread out along its whole length. Hence linear dispersion does not affect directly the brightness of the one but vitally that of the other. Thus while the usual stellar spectrograph may serve in a limited way for the bright-line spectrum it is useless for the dark-line one. This suggests why, until recent years, observations of nebular spectra were devoted chiefly to objects having bright lines. The dark-line spectrum is faint in the extreme. It will not over-emphasize this matter to recall that Keeler in his classical observations of planetary (bright-line) nebulae was able to employ a linear dispersion equal to that given by twenty-four sixty-degree prisms, whereas Huggins was able to obtain only a faint photographic impression of the dark-line spectrum of the greatest of the spirals, the Andromeda nebula.

Unfortunately no choice of telescope—as regards aperture or focal-length or ratio of aperture to focus—will increase the brightness of the spectrum of an extended surface. But the spectrograph greatly influences such a spectrum; and of the spectrograph the camera is the determining factor for brightness. When the one-prism Flagstaff spectrograph used in stellar velocities has its 18.5-inch camera replaced by a 3 1/4-inch Voigtlander camera it is an efficient instrument for nebular work. This change of cameras increases the speed of the instrument fully 30-fold, while the linear scale of spectrum, in consequence of the powerful prism used, is still one-third that of some instruments now employed elsewhere in stellar velocity work. High angular dispersion is necessary, or at least a good means, for overcoming the photographic difficulty that an absorption line, no matter how dark, can not be recorded by the granular surface of a rapid plate if the line is too fine. In short, there is a limit beyond which it is no longer profitable to narrow the slit. This limit with the Flagstaff spectrograph is rather wide and I have profited by it.

The spectrograph has been attached to the 24-inch refractor and enclosed in a constant temperature case. Seed "30" plates were employed. The comparison spectrum was iron and vanadium.

When entering upon this work it seemed that the chief concern would be with the nebular spectra themselves, but the early discovery that the great Andromeda spiral had the quite exceptional velocity of –300 km showed the means then available, capable of investigating not only the spectra of the spirals but their velocities as well. I have given more attention to velocity since the study of the spectra had been undertaken with marked success by Fath at Lick and Mount Wilson, and by Wolf at Heidelberg.

Spectrograms were obtained of about 40 nebulae and star clusters. The spectrum shown by the spirals thus far observed is predominantly type II (G–K). The best observable nebula, that in Andromeda,[1] shows a pure stellar type of spectrum, with none of the composite features to be expected in the spectrum of the integrated light of stars of various types and such as are shown by the spectra of the globular star clusters which present a blend of the more salient features of type I and type II spectra.

In the table is a list of the spiral nebulae observed. As far as possible their velocities are given, although in many cases they are only rough provisional values.[2]

As far as the data go, the average velocity is 400 km. It is positive by about 325 km. It is 400 km on the north side and less than 200 km on the south side of the Milky Way. Before the observation of N.G.C. 1023, 1068, and 7331, which were among the last to be observed, the signs were all negative on one side and all positive on the other, and it then seemed as if the spirals might be drifting across the Milky Way.

N.G.C. 3115, 4565, 4594, and 5866 are spindle nebulae—doubtless spirals seen edge-on. Their average velocity is about 800 km, which is much greater than for the remaining objects and suggests that the spirals move edge forward.

---

[1]The bright lines Wolf thought to be present in this and other similar nebulae he has now come to believe were only contrast effects. He also writes that he gets from a recent plate a velocity of –350 km for this nebula.

[2][A number of the nebulae in Slipher's list are objects that you have encountered before although you may not recognize them because Slipher identified them in terms of the numbers assigned to them in the *New Galactic Catalogue*. For example, N.G.C. 221 = M32; N.G.C. 224 = M31, Andromeda; N.G.C. 598 = M33; N.G.C. 1068 = M77; N.G.C. 3031 = M81; N.G.C. 3627 = M66; N.G.C. 4594 = M104; N.G.C. 4826 = M64; N.G.C. 5194 = M51. MJC]

N.G.C.				
	221	Velocity	− 300 km[1]	
	224[2]		− 300	These are on the south side of the Milky Way.
	598		−	
	1023		+ 200 roughly	
	1068		+1100	
	7331		+ 300 roughly	
	3031		+ small	
	3115		+ 400 roughly	
	3627		+ 500	
	4565		+1000	These are on the north side of the Milky Way.
	4594		+1100	
	4736		+ 200 roughly	
	4826		+ small	
	5194		± small	
	5866		+ 600	

As well as may be inferred, the average velocity of the spirals is about 25 times the average stellar velocity.[3] This great velocity would place these nebulae a long way along the evolutionary chain if we undertook to apply the Campbell-Kapteyn discovery of the increase in stellar velocity with "advance" in stellar spectral type.

☆☆☆☆☆☆☆☆☆☆☆☆☆☆☆☆☆☆☆☆☆☆☆☆☆☆

☆☆☆☆☆☆☆☆☆☆☆☆☆☆☆☆☆☆☆☆☆☆☆☆☆☆
Adriaan van Maanen, "Preliminary Evidence of Internal Motions in the Spiral Nebula Messier 101," *Proceedings of the National Academy of Sciences*, 2 (1916), 386-90.

Inasmuch as data for the proper motions of stars, determined by photographic methods, have been rapidly

---

[1][That is, kilometers per second. MJC]
[2]Wright at Lick has kindly communicated to me by letter his observation of this object which results in a velocity of −304 km.
[3][The fact that the average speed of these nebulae was measured to be about twenty-five times greater than the average speed of a star was a very striking result, which required interpretation. MJC]

accumulating since the beginning of the century, it may seem strange that the first results for nebulae should have been published only in 1915. It must not be forgotten, however, that photographs of nebulae require much longer exposures, and that, even with the best plates, the measures are more difficult and less accurate, because of the unsymmetrical character of the points and condensations upon which settings must be made, than is the case with the round images of stars. A given point in a nebula may be bisected quite differently on different plates and the measures will therefore fail to reveal the proper motion with anything like the precision attainable in a star cluster, for instance. The difficulty of bisection may, however, be largely overcome by measuring corresponding points on different plates in immediate succession.

For such measures the monocular arrangement of the stereocomparator is an admirable instrument, and that it is capable of yielding very accurate results has already been pointed out.[1] When therefore Mr. Ritchey placed at my disposal for measurement two excellent plates of the spiral nebula Messier 101, taken in 1910 and 1915, there was no question but that the stereocomparator was the best instrument for the purpose.

Although the results showed striking evidence of internal motion, the necessity of additional plates was strongly felt. At my request Dr. Curtis kindly placed at my disposal three photographs made with the Crossley reflector of the Lick Observatory, one by Keeler in 1899, one by Perrine in 1908, and one by Curtis himself in 1914.

The pair taken by Mr. Ritchey was completely measured twice, the Lick pairs 1914–1899 and 1908–1899 once each; each pair was measured in four positions, with east, west, north and south, respectively, in the direction of increasing readings of the micrometer screw. On the Mount Wilson plates 87 nebulous points were measured; on the first Lick pair 46, and on the second 69 points, while on all the pairs the same 32 stars were used for comparison purposes. The measures and reductions, which will be published in full in the *Astrophysical Journal*,

---

[1] *Astronomical Journal*, 27, 140 (1912).

were made substantially in the manner described in my recent paper on the determination of stellar parallaxes.[1] The principal difference being that the quadratic terms could not here by neglected and were accordingly included in the reductions.

The results showed that to each pair of plates could be given the same weight, and the direct mean of the values found for the proper motions of each point is therefore used in the discussion. The resulting motions, which are those relative to the mean of the 32 comparison stars, are due partly to a motion of translation of the nebula as a whole and partly to a possible internal motion. The annual motion of translation, which was derived by three different methods of reduction, was found to be: $\mu_\alpha = +0''.005$, $\mu_\delta = -0''.043$. Subtracting these from the total motions, the results are what may be called the internal motions. The accompanying plate shows these mean internal motions for each of the 87 points of the nebula. The individual motions for the comparison stars, which are surrounded by circles are also shown. The scale for the annual motions is given at the bottom of the plate. The density of the center of the nebula has been reduced to show the motions more clearly.

If the results as illustrated on the plate could be taken at their face value, they would certainly indicate a motion of rotation, or possibly motion along the arms of the spiral. Without expressing a final opinion as to the character of the motion, which must be determined by future work, it may be of interest to examine the evidence afforded by the existing material. Comparing the motions with the directions of the branches of the spirals, we find from 52 points in which this direction can be specified with fair accuracy, that the mean divergence of the motions is $7° \pm 4°$ toward the concave side of the spirals.

---

[1]*Mt. Wilson Contr.*, No. 111, 4 seq. 1916.

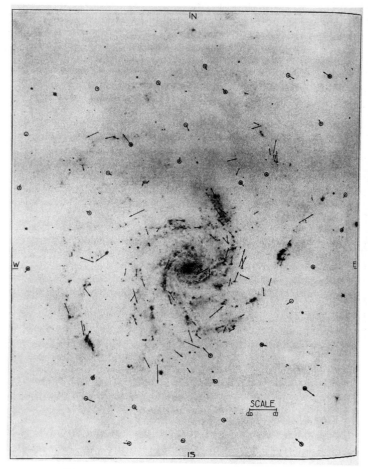

INTERNAL MOTIONS IN MESSIER 101

The arrows indicate the direction and magnitude of the mean annual motions. Their scale (0."1) is indicated on the plate. The scale of the nebula is 1 mm. = 10."5. The comparison stars are enclosed in circles.

To discuss the internal motions from the standpoint of rotation, they were analyzed into two components, along and perpendicular to the radius, the latter for convenience being spoken of as the rotational component.

The results are as follows:

78 points have a left-hand motion (N.W.S.E.), only 9 moving right-handedly; 58 points appear to be moving outward, while 28 show motion inward. The rotational motion

is the larger in the majority of cases, viz., 63 points. The mean rotational motion is 0".022 left-handed; the mean radial motion is 0".007 outward. The probable reality of the result is indicated by the satisfactory agreement of the pairs of plates, as shown by the following summary.

μ_{rot}	μ_{rot}				
+0 021	+0 004	Ritchey	1910	and	1915
+0.032	+0.012	Ritchey	1910	and	1915
+0.012	+0.007	Keeler and Curtis	1899	and	1914
+0.017	+0.006	Keeler and Perrine	1899	and	1908

The measures indicate a small but scarcely reliable decrease of rotational motion with increasing distance from the center, as shown by the following table.

Distance from Center	μ_{rot}	No. of Points
<3'.1	0".024	19
3.1 to 5.0	0.028	29
5.1 to 7.0	0.014	18
>7.0	0.019	21

The annual rotational component of 0.022 at the mean distance from the center of 5' corresponds to a rotational period of about 85,000 years. If we knew the parallax of the nebula, and if we could assume that the motions and the distances of the points from the center are mean values for elliptical orbits, the central mass could be calculated. Even though this assumption may be far from the truth it seems worth while to accept certain more or less hypothetical values of the parallax in order to get any idea of the order of the masses with which we are concerned.

Two estimates as to the parallax can be made, (1) by comparing the average motion of translation of 66 spiral nebulae, as given by Curtis[1] with those of the stars; (2) by comparing these cross-motions with the observed radial velocities of the few spirals for which such results are known. In the first case we derive a parallax of 0".005; in the second of

---

[1] *Publications Astronomical Society Pacific*, 27, 216 (1915).

0".0003. The corresponding central masses are in both cases very large, viz., 30,000 and 140,000,000 times that of the sun. Various objections to the acceptance of these results, even as rough guesses, can be made, but they are the best we have at the moment. The corresponding orbital motions would be 21 and 345 km./sec. These quantities do not seem absurd if we remember that Wolf by spectroscopic methods found a rotational component of ± 100 km./sec. in Messier 81.[1]

As the detailed results will soon be published I will only mention two more points which seem to confirm the reality of the motions. Mr. Nicholson very kindly spent much time in making check-measures on the plates taken by Mr. Ritchey, both with the stereocomparator and with another measuring machine in which two microscopes were mounted in such a way that they were directed toward corresponding points on the two plates, mounted on the same plate-carrier and moved by the same micrometer screw. His measures given satisfactory confirmation of my own results. Further, I have measured two plates of Messier 81, taken by Mr. Ritchey in 1910 and 1916, which show motion similar to that found above for Messier 101. It seems hardly necessary to suggest the importance of internal motions, such as are indicated here, in connection with the Chamberlin-Moulton hypothesis as to the origin of spiral nebulae, and it is to be hoped that material will soon be available for a fuller discussion.

☆☆☆☆☆☆☆☆☆☆☆☆☆☆☆☆☆☆☆☆☆☆☆☆☆☆☆☆☆

☆☆☆☆☆☆☆☆☆☆☆☆☆☆☆☆☆☆☆☆☆☆☆☆☆☆☆☆☆
A. S. E. [Arthur Stanley Eddington], "The Motions of Spiral Nebulae," *Royal Astronomical Society Monthly Notices*, 77 (1917), 375–7.

An accurate knowledge of the motions of spiral nebula, both radial and angular, would go far towards settling the status of these bodies in relation to the system of the stars. The spectroscopic linear velocities alone may indicate important dynamical relations (or absence of relation); and by comparing the radial and cross motions we could estimate the distance,

---

[1] *Vierteljahrsschrift Astronomischen Gesellschaft*, 49, 162 (1914).

and hence the size of these bodies. The relative motions of the different parts and particularly the rate of rotation, will also give evidence as to their nature, and perhaps determine the total mass involved. The interesting question has arisen whether the spiral nebulae are comparatively small bodies scattered among the stars and belonging to the stellar system, or whether they are great sub-systems on the fringe of the stellar system, or, thirdly, whether they are distant galaxies coequal with our own. For the last alternative the distances must be not merely great from the ordinary standpoint, but of the order of 100,000 parsecs at least; the proper motions would have to be absolutely insensible, since a motion of 0".01 per century at 100,000 parsecs corresponds to a linear velocity of 474 km. per sec.

Our object in this note is to survey the observational material that has been collected rather than to discuss theoretical consequences. It will be realised that the observations are of a very difficult character. The spectroscopic work involves very long exposures, during which the apparatus must be protected from changes, and the dispersion used is necessarily small. The measurements of position are made on diffuse knots which will not allow of high accuracy; and since the displacements are generally small, the unknown motions of the comparison stars may sometimes have an important influence.

The first determination of the radial velocity of a spiral nebula was made in 1913 by V. M. Slipher (*Lowell Obs. Bull.*, No. 58), who found for the Andromeda nebula a velocity of 300 km. per sec. approaching. Other investigators have found values in excellent agreement with this, as may be seen from the following:—

Wright	.	− 304 km. per sec.
Pease	.	− 329 " " "
Slipher	.	− 300 " " "

M. Wolf has also made several determinations, his latest result being −450 km. per sec. (*Vierteljahrs[s]chrift*, 51, p. 115), but his results have not been very consistent among themselves.

For fainter objects the values may be more uncertain. Thus

for N.G.C. 1068 the velocities + 1100, + 765, and + 910 km. per sec. have been given by Slipher, Pease, and Moore respectively (*Pub. Ast. Soc. Pac.*, 27, p. 192); but even here there is evidently some measure of agreement.

Slipher has given radial velocities for fifteen spirals, but except in these two cases there is no check on the results.[1] The average speed from his measures is between 300 and 400 km. per sec. In addition, R. E. Wilson (*Proc. N. A. Wash.*, 1, p. 183) has determined the velocity of a number of nebulae in the Magellanic Clouds. It is true that these were all gaseous nebulae but presumably the results give the velocities of the clouds as a whole, and these may perhaps be regarded as objects akin to the spirals. The values of the Greater and Lesser Clouds are + 275 and + 158 km. per sec. The whole material is evidently too scanty as yet to derive a value of the motion of the stellar system relative to a possible super-system of galaxies, and not much stress need be laid on the provisional determinations by Truman (*Pop. Ast.*, 24, p. 111), and by Young and Harper (*J.R.A.S. Canada*, 10, p. 134)—about 600 km. per sec. towards an apex at R.A. 20^h, Dec. $-20°$. It may, however, be recalled that Sir W. Herschel's solar apex derived from only seven stars turned out to be quite near the truth, and history may repeat itself in this case.

It seems to be very doubtful whether or not spiral nebulae show any detectable proper motions. H. D. Curtis (*Pub. Ast. Soc. Pac.*, 27, p. 216) has measured the motions of sixty-six spiral nebulae relative to comparison stars of the twelfth to fifteenth magnitude. The average was $3''.3$ per century, but this includes a number of large motions, to which the author attaches small weight. He considered, however, a motion of 8" per century for N.G.C. 253 to be worthy of confidence. The results have not yet been published in detail. A determination for a single nebula, Messier 101, by A. van Maanen (*Astrophysical Jour.*, 44 p. 210) deserves attention, since it depends on measures of a very large number of points. The motion is $1''.2$ per century relative to thirty-two comparison

---

[1] Since this was written another of Slipher's results, + 1100 km. per sec. for N.G.C. 4594, has been confirmed by Pease (*Pop. Ast.*, 25, p. 26).

stars, but since the parallactic motion of these stars would probably be of about this amount and in the appropriate direction, the result does not necessarily indicate a motion of the nebula. An attempt to determine the systematic proper motions of a large number of spirals has been made by C. Wirtz (*Astr. Nach.*, Nos. 4861, 4866), who compared recent measures with those made by Schultz at Upsala forty to fifty years earlier. He found that they were drifting towards R.A. 245°, Dec. − 3° (a direction almost opposite to the motion of the stars), whereas the gaseous nebulae were moving in a similar direction to the stars. These results have been criticised by Kobold and Seeliger, the latter showing that following another method of reduction the resulting apex was nearly the same as that of the stars.

Turning now to the internal motions, spectroscopic observations have shown in some spiral nebulae a curvature of the spectral lines, which points to a rapid velocity of rotation. Slipher detected a rotation in N.G.C. 4594 (Virgo), which is a spiral seen edgewise (*Lowell Obs. Bull.*, No. 62); Wolf found a rotation in M 81 (Ursa Major), which amounted to 100 km. per sec. not far from the nucleus; Pease found a difference between the velocities of the nucleus and a knot in M 33 (Triangulum) amounting to 200 km. per sec.; and in N.G.C. 4594 he found a linear velocity of rotation of 300 km. per sec. at 2' from the nucleus (*Pop. Ast.*, 25, p. 27).

Some very interesting results with regard to internal motions have recently been obtained by the comparison of early and recent photographs. A. van Maanen (loc. cit.) has studied the spiral M 101. From measures of a large number of points he has found that the flow is outwards along the spiral arms. Considering the rotation, it appears that the motion of the arms is concave side foremost—opposite to that of a Catherine-wheel; thus there would be a tendency to uncurl. The motion is surprisingly large, so that at 5' from the centre the rotation-period would be 85,000 years. For the nebula M 51 (Canes Venatici) S. Kostinsky has found indications of internal motion with the stereo-comparator (*Monthly Notices*, 77, No. 3). His conclusions appear to agree in the main with van Maanen's. It has been pointed out by J. H. Jeans (*Observatory*, 40, p. 60) that these results will fit in with a possible dynamical theory.

It would, we think, be premature to draw any general conclusions from these observations; but they show that a new field of astronomical investigation is rapidly being opened up, and we may hope to acquire a more definite knowledge in the near future.

☆ ☆

Chapter Eight

The Great Debate

Introduction

Having tracked through nearly two centuries the questions (1) of the size and structure of the Milky Way, and (2) of the existence of "island universes," we arrive at the "Great Debate." On 26 April 1920, Harlow Shapley and Heber Curtis, both of whom had by then established themselves as respected authorities on astronomy, presented their contrasting views on these questions before the National Academy of Sciences. Subsequently they redrafted their statements for publication. Their published papers, which follow this introduction, constitute the focus of this chapter.[1]

Harlow Shapley, who was thirteen years younger than Curtis, was born in 1885 on a farm near Carthage,

Harlow Shapley

---

[1] For information on what Shapley and Curtis said when they addressed the National Academy of Sciences, see Michael Hoskin, "The Great Debate: What Really Happened," *Journal for the History of Astronomy*, 7 (1976), 169–82. This paper contains a partial transcript of Shapley's statements as well as a reconstruction of Curtis's address based on the slides that he presented. It is worth noting that what "really happened" was there were two public debates, a spoken debate before the National Academy of Sciences, and the more fully developed published debate, the latter of which is presented in this chapter.

Missouri. By 1901, he had completed fifth grade and a business course as well as securing a position as a reporter for the *Daily Sun* of Chanute, Kansas. Shapley eventually applied to Carthage High School, but after having been denied admission, he enrolled at Carthage Collegiate Institute, from which he graduated a year later. In 1907, he entered the University of Missouri, planning to study journalism, only to find that the journalism school had not yet opened. Casting around for courses, he proceeded, according to his own account, alphabetically through the college catalogue, rejecting archaeology as unpronounceable, and settling on astronomy.[1] By 1910, Shapley had received a B.A. with high honors in mathematics and physics and by 1911 an M.A. He then entered Princeton, studying with H. N. Russell. At first interested in eclipsing binary stars (stars that periodically pass in front of each other), he went on to study cepheid variables. i.e., variable stars with periods of from about 1 to about 300 days and with a light curve such that they rise to their maximum brightness more rapidly than they dim (see diagram).

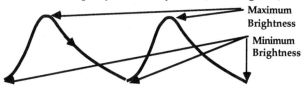

Light Curve for a Cepheid Variable Star

Completing his doctorate in 1913, Shapley toured Europe before taking in 1914 a position at Mount Wilson Observatory, where the 60-inch reflector was in use. He published very actively over the next few years, concentrating on cepheid variables and globular clusters. In 1921, already the author of a hundred papers, he became director of Harvard Observatory. He retired from that position in 1952 and died in 1972.

---

[1] Harlow Shapley, *Through Rugged Ways to the Stars* (New York: Charles Scribner's Sons, 1969), p. 17. Such at least is the account Shapley gave in this autobiographical work of how he decided to study astronomy. According to one plausible report, this story was meant as a jest.

Heber D. Curtis, born in Muskegon, Michigan in 1872, had by 1893 received both a B.A. and M.A. from the University of Michigan, where he had studied Greek, Latin, Assyrian, Hebrew, Sanskrit, and some science, but no astronomy. In 1894, he began teaching Greek and Latin at Napa College, which soon became part of the College of the Pacific. Having gained access in his first year to a telescope, Curtis was by 1896 teaching

Heber D. Curtis

mathematics and astronomy. In 1902, he finished a doctorate in astronomy at the University of Virginia, his dissertation being on comets. He then took a position at Lick Observatory, where by 1910 he was concentrating on nebulae. He departed from Lick in 1920 to direct the Allegheny Observatory in Pittsburgh. A decade later, he became the director of the observatory at the University of Michigan. Heber Curtis died in 1942.

Although the organizers of the "Great Debate" intended that Shapley and Curtis should focus on two issues—the size of the Milky Way and the question of island universes—Shapley concentrated on the former, Curtis on the latter. In his paper, Shapley stressed the use of blue stars as distance indicators; i.e., he relied heavily on Russell's correlation of spectral types with luminosities, in particular, on Russell's determination that the range in luminosities among blue stars (those on the far left portion of Russell's graph) is relatively small, whereas red stars, for example, were known to vary in luminosities over a far broader range. Shapley's emphasis on blue stars as distance indicators did not result primarily from the development of his own thought in which Cepheids had been centrally important for his globular cluster research. It resulted rather from the fact that in their pre-debate correspondence and in their actual verbal exchange, Curtis had attacked Shapley's reliance on

Cepheids, leading Shapley to shift his ground.

Two problems beset the use of blue stars as distance indicators. First, the range of their luminosities is rather broad, so that one can assign any one of a number of luminosities to an individual blue or B-type star. Second, Russell's first diagram was based on stars of known parallax, but this is not in every way a typical group of stars. Naturally, parallaxes had been sought in the brighter stars; consequently, stars of high luminosity were disproportionately present among the stars of known parallax. This effect made it doubtful whether that diagram gave an adequate indication of the relative proportions among the stars of the various spectral types. These problems combined to make it difficult to be sure what precise luminosity to attribute to any given B-type star. Shapley, for example, urged that the typical B-type is of giant magnitude and consequently that the B-type stars seen in globular clusters appear dim only because of their great distance. Curtis, on the other hand, claimed that the apparent dimness of the B-type stars in the clusters is probably due not to the remoteness of the clusters, but to the modest luminosities of those stars. This ambiguity made the argument over B-type stars less than decisive. Despite Shapley's emphasis on this argument, it was less central than the efforts to assign distances in terms of Cepheids (Shapley) or by means of novae (Curtis). In judging who had superior arguments in the debate, it is consequently unwise to give undue weight to the blue-star argument.

In analyzing the great debate, it is wise to be sensitive not only to substantive but also to methodological issues. For example, it may be useful to ask: whose position was more radical, whose more conservative? Which position was heuristically more promising? Which entailed abandoning the larger number of productive research techniques? And which position possessed greater internal consistency?

Although in 1920 astronomers had not yet reached a consensus on the issues in the great debate, by the late 1920s, a widely accepted position had emerged, partly because evidence unavailable in 1920 had been discovered during that decade, but also because new interpretations had been found for materials already discussed by Shapley and Curtis.

☆☆☆☆☆☆☆☆☆☆☆☆☆☆☆☆☆☆☆☆☆☆☆☆☆☆☆☆☆☆
Harlow Shapley, "The Scale of the Universe," *Bulletin of the National Research Council,* 3 (1921), 171–93.

CONTENTS

Evolution of the idea of galactic size
Surveying the solar neighborhood
On the distances of globular clusters
The dimensions and arrangement of the galactic system

EVOLUTION OF THE IDEA OF GALACTIC SIZE

The physical universe[1] was anthropocentric to primitive man. At a subsequent stage of intellectual progress it was centered in a restricted area on the surface of the earth. Still later, to Ptolemy and his school, the universe was geocentric; but since the time of Copernicus the sun, as the dominating body of the solar system, has been considered to be at or near the center of the stellar realm. With the origin of each of these successive conceptions, the system of stars has ever appeared larger than was thought before. Thus the significance of man and the earth in the sidereal scheme has dwindled with advancing knowledge of the physical world, and our conception of the dimensions of the discernible stellar universe has progressively changed. Is not further evolution of our ideas probable? In the face of great accumulations of new and relevant information can we firmly maintain our old cosmic conceptions?

As a consequence of the exceptional growth and activity of the great observatories, with their powerful methods of analyzing stars and of sounding space, we have reached an epoch, I believe, when another advance is necessary; our conception of the galactic system must be enlarged to keep in proper relationship the objects our telescopes are finding; the solar system can no longer maintain a central position. Recent studies of clusters and related subjects seem to me to leave no

---

[1]The word "universe" is used in this paper in the restricted sense, as applying to the total of sidereal systems now known to exist.

alternative to the belief that the galactic system is at least ten times greater in diameter—at least a thousand times greater in volume—than recently supposed.

Dr. Curtis,[1] on the other hand, maintains that the galactic system has the dimensions and arrangement formerly assigned it by students of sidereal structure—he supports the views held a decade or so ago by Newcomb, Charlier, Eddington, Hertzsprung, and other leaders in stellar astronomy. In contrast to my present estimate of a diameter of at least three hundred thousand light-years Curtis outlines his position as follows:[2]

> As to the dimensions of the galaxy indicated by our Milky Way, till recently there has been a fair degree of uniformity in the estimates of those who have investigated the subject. Practically all have deduced diameters of from 7,000 to 30,000 light-years. I shall assume a maximum galactic diameter of 30,000 light-years as representing sufficiently well this older view to which I subscribe though this is pretty certainly too large.

I think it should be pointed out that when Newcomb was writing on the subject some twenty years ago, knowledge of those special factors that bear directly on the size of the universe was extremely fragmentary compared with our information of today. In 1900, for instance, the radial motions of about 300 stars were known; now we know the radial velocities of thousands. Accurate distances were then on record for possibly 150 of the brightest stars, and now for more than ten times as many. Spectra were then available for less than one-tenth of the stars for which we have the types today. Practically nothing was known at that time of the photometric and spectroscopic methods of determining distance; nothing of the radial velocities of globular clusters or of spiral nebulae, or even of the phenomenon of star streaming.

As a further indication of the importance of examining anew the evidence on the size of stellar systems, let us consider the great globular cluster in Hercules—a vast sidereal organization concerning which we had until recently but vague

---

[1]See Part II of this article, by Heber D. Curtis.
[2]Quoted from a manuscript copy of his Washington address.

ideas. Due to extensive and varied researches, carried on during the last few years at Mount Wilson and elsewhere, we now know the positions, magnitudes, and colors of all its brightest stars, and many relations between color, magnitude, distance from the center, and star density. We know some of these important correlations with greater certainty in the Hercules cluster than in the solar neighborhood. We now have the spectra of many of the individual stars, and the spectral type and radial velocity of the cluster as a whole. We know the types and periods of light variation of its variable stars, the colors and spectral types of these variables, and something also of the absolute luminosity of the brightest stars of the cluster from the appearance of their spectra. Is it surprising, therefore, that we venture to determine the distance of Messier 13 and similar systems with more confidence than was possible ten years ago when none of these facts was known, or even seriously considered in cosmic speculations?

If he were writing now, with knowledge of these relevant developments, I believe that Newcomb would not maintain his former view on the probable dimensions of the galactic system.

For instance, Professor Kapteyn has found occasion, with the progress of his elaborate studies of laws of stellar luminosity and density, to indicate larger dimensions of the galaxy than formerly accepted. In a paper just appearing as *Mount Wilson Contribution*, No. 188,[1] he finds, as a result of the research extending over some 20 years, that the density of stars along the galactic plane is quite appreciable at a distance of 40,000 light-years—giving a diameter of the galactic system, exclusive of distant star clouds of the Milky Way, about three times the value Curtis admits as a maximum for the entire galaxy. Similarly Russell, Eddington, and, I believe, Hertzsprung, now subscribe to larger values of galactic dimensions; and Charlier, in a recent lecture before the Swedish Astronomical Association, has accepted the essential features of the larger galaxy, though formerly he identified the local system of B stars with the whole galactic system and obtained distances of the clusters and dimension of the galaxy

---

[1]The *Contribution* is published jointly with Dr. van Rhijn.

only a hundredth as large as I derive.

SURVEYING THE SOLAR NEIGHBORHOOD

Let us first recall that the stellar universe, as we know it, appears to be a very oblate spheroid or ellipsoid—a disk-shaped system composed mainly of stars and nebulae. The solar system is not far from the middle plane of this flattened organization which we call the galactic system. Looking away from the plane we see relatively few stars; looking along the plane, through a great depth of star-populated space, we see great numbers of sidereal objects constituting the band of light we call the Milky Way. The loosely organized star clusters, such as the Pleiades, the diffuse nebulae such as the great nebula of Orion, the planetary nebulae, of which the ring nebula in Lyra is a good example, the dark nebulosities—all these sidereal types appear to be a part of the great galactic system, and they lie almost exclusively along the plane of the Milky Way. The globular clusters, though not in the Milky Way, are also affiliated with the galactic system; the spiral nebulae appear to be distant objects mainly if not entirely outside the most populous parts of the galactic region.

This conception of the galactic system, as a flattened, watch-shaped organization of stars and nebulae, with globular clusters and spiral nebulae as external objects, is pretty generally agreed upon by students of the subject; but in the matter of the distances of the various sidereal objects—the size of the galactic system—there are, as suggested above, widely divergent opinions. We shall, therefore, first consider briefly the dimensions of that part of the stellar universe concerning which there is essential unanimity of opinion, and later discuss in more detail the larger field, where there appears to be a need for modification of the older conventional view.

Possibly the most convenient way of illustrating the scale of the sidereal universe is in terms of our measuring rods, going from terrestrial units to those of stellar systems. On the earth's surface we express distances in units such as inches, feet, or miles. On the moon, as seen in the accompanying photograph made with the 100-inch reflector,[1] the mile is still a usable measuring unit; a scale of 100 miles is indicated on the lunar scene.

Our measuring scale must be greatly increased, however, when we consider the dimensions of a star—distances on the surface of our sun, for example. The large sun-spots shown in the illustration cannot be measured conveniently in units appropriate to earthly distance—in fact, the whole earth itself is none too large. The unit for measuring the distances from the sun to its attendant planets, is, however, 12,000 times the diameter of the earth; it is the so-called astronomical unit, the average distance from earth to sun. This unit, 93,000,000 miles in length, is ample for the distances of planets and comets. It would probably suffice to measure the distances of whatever planets and comets there may be in the vicinity of other stars; but it, in turn, becomes cumbersome in expressing the distances from one star to another, for some of them are hundreds of millions, even a thousand million astronomical units away.

This leads us to abandon the astronomical unit and to introduce the light-year as a measure for sounding the depth of stellar space. The distance light travels in a year is something less than six million million miles. The distance from the earth to the sun is, in these units, eight light-minutes. The distance to the moon is 1.2 light-seconds. In some phases of our astronomical problems (studying photographs of stellar spectra) we make direct microscopic measures of a tenthousandth of an inch; and indirectly we measure changes in the wave-length of light a million times smaller than this; in discussing the arrangement of globular clusters in space, we must measure a hundred thousand light-years. Expressing these large and small measures with reference to the velocity of light, we have an illustration of the scale of the astronomer's universe—his measures range from the trillionth of a billionth

---

[1][Because Shapley's argument in this section is quite elementary and because the pictures (i.e., of areas of the moon and sun and of a nebula as seen with and without enhancement) that Shapley supplied to illustrate his argument are not in any way essential to his exposition, these illustrations have been omitted at this point.]

part of one light-second, to more than a thousand light-centuries. The ratio of the greatest measure to the smallest is as 10^{33} to 1.

It is to be noticed that light plays an all-important role in the study of the universe; we know the physics and chemistry of stars only through their light, and their distance from us we express by means of the velocity of light. The light-year, moreover, has a double value in sidereal exploration; it is geometrical, as we have seen, and it is historical. It tells us not only how far away an object is, but also how long ago the light we examine was started on its way. You do not see the sun where it is, but where it was eight minutes ago. You do not see faint stars of the Milky Way as they are now, but more probably as they were when the pyramids of Egypt were being built; and the ancient Egyptians saw them as they were at a time still more remote. We are, therefore, chronologically far behind events when we study conditions or dynamical behavior in remote stellar systems; the motions, light-emissions, and variations now investigated in the Hercules cluster are not contemporary, but, if my value of the distance is correct, they are the phenomena of 36,000 years ago. The great age of these incoming pulses of radiant energy is, however, no disadvantage; in fact, their antiquity has been turned to good purpose in testing the speed of stellar evolution, in indicating the enormous ages of stars, in suggesting the vast extent of the universe in time as well as in space.

Taking the light-year as a satisfactory unit for expressing the dimensions of sidereal systems, let us consider the distances of neighboring stars and clusters, and briefly mention the methods of deducing their space positions. For nearby stellar objects we can make direct trigonometric measures of distance (parallax), using the earth's orbit or the sun's path through space as a base line. For many of the more distant stars spectroscopic methods are available, using the appearance of the stellar spectra and the readily measurable apparent brightness of the stars. For certain types of stars, too distant for spectroscopic data, there is still a chance of obtaining the distance by means of the photometric method. This method is particularly suited to studies of globular clusters; it consists first in determining, by some means, the real luminosity of a star, that is, its so-called absolute magnitude, and second, in measuring its apparent magnitude. Obviously, if a star of known real brightness is moved away to greater and greater distances, its apparent brightness decreases; hence, for such stars of known absolute magnitude, it is possible, using a simple formula, to determine the distance by measuring the apparent magnitude.

It appears, therefore, that although space can be explored for a distance of only a few hundred light-years by direct trigonometric methods, we are not forced, by our inability to measure still smaller angles, to extrapolate uncertainly or to make vague guesses relative to farther regions of space, for the trigonometrically determined distances can be used to calibrate the tools of newer and less restricted methods. For example, the trigonometric methods of measuring the distance to moon, sun, and nearer stars are decidedly indirect, compared with the linear measurement of distance on the surface of the earth, but they are not for that reason inexact or questionable in principle. The spectroscopic and photometric methods of measuring great stellar distance are also indirect, compared with the trigonometric measurement of small stellar distance, but they, too, are not for that reason unreliable or of doubtful value. These great distances are not extrapolations. For instance, in the spectroscopic method, the absolute magnitudes derived from trigonometrically measured distances are used to derive the curves relating spectral characteristics to absolute magnitude; and the spectroscopic parallaxes for individual stars (whether near or remote) are, almost without exception, interpolations. Thus the data for nearer stars are used for purposes of calibration, not as a basis for extrapolation.

By one method or the other, the distances of nearly 3,000 individual stars in the solar neighborhood have now been determined; only a few are within ten light-years of the sun. At a distance of about 130 light-years we find the Hyades, the well known cluster of naked eye stars; at a distance of 600 light-years, according to Kapteyn's extensive investigations, we come to the group of blue stars in Orion—another physically-organized cluster composed of giants in luminosity. At distances comparable to the above values we also find the Scorpio-Centaurus group, the Pleiades, the Ursa Major system.

These nearby clusters are specifically referred to for two reasons.

In the first place I desire to point out the prevalence throughout all the galactic system of clusters of stars, variously organized as to stellar density and total stellar content. The gravitational organization of stars is a fundamental feature in the universe—a double star is one aspect of a stellar cluster, a galactic system is another. We may indeed, trace the clustering motive from the richest of isolated globular clusters such as the system in Hercules, to the loosely organized nearby groups typified in the bright stars of Ursa Major. At one hundred times its present distance, the Orion cluster would look much like Messier 37 or Messier 11; scores of telescopic clusters have the general form and star density of the Pleiades and the Hyades. The difference between bright and faint clusters of the galactic system naturally appears to be solely a matter of distance.

In the second place I desire to emphasize the fact that the nearby stars we use as standards of luminosity, particularly the blue stars of spectral type B, are members of stellar clusters. Therein lies a most important point in the application of photometric methods. We might, perhaps, question the validity of comparing the isolated stars in the neighborhood of the sun with stars in a compact cluster; but the comparison of nearby cluster stars with remote cluster stars is entirely reasonable, since we are now so far from primitive anthropocentric notions that it is foolish to postulate that distance from the earth has anything to do with the intrinsic brightness of stars.

ON THE DISTANCES OF GLOBULAR CLUSTERS

1. As stated above, astronomers agree on the distances to the nearby stars and stellar groups—the scale of the part of the universe that we may call the solar domain. But as yet there is lack of agreement relative to the distances of remote clusters, stars, and star clouds—the scale of the total galactic system.

The disagreement in this last particular is not a small difference of a few percent, an argument on minor detail; it is a matter of a thousand percent or more.

Curtis maintains that the dimensions I find for the galactic system should be divided by ten or more (see quotation [cited previously]); therefore, that galactic size does not stand in the way of interpreting spiral nebulae as comparable galaxies (a theory that he favors on other grounds but considers incompatible with the larger values of galactic dimensions). In his Washington address, however, he greatly simplified the present discussion by accepting the results of recent studies on the following significant points:

Proposition A.—The globular clusters form a part of our galaxy; therefore the size of the galactic system proper is most probably not less than the size of the subordinate system of globular clusters.

Proposition B.—The distances derived at Mount Wilson for globular clusters *relative to one another* are essentially correct. This implies among other things that (1) absorption of light in space has not appreciably affected the results, and (2) the globular clusters are much alike in structure and constitution, differing mainly in distance. (These relative values are based upon apparent diameters, integrated magnitudes, the magnitudes of individual giants or groups of giants, and Cepheid variables; Charlier has obtained much the same results from apparent diameters alone, and Lundmark from apparent diameters and integrated magnitudes.)

Proposition C.—Stars in clusters and in distant parts of the Milky Way are not peculiar—that is, uniformity of conditions and of stellar phenomena naturally prevails throughout the galactic system.

We also share the same opinion, I believe, on the following points:

a. The galactic system is an extremely flattened stellar organization, and the appearance of a Milky Way is partly due to the existence of distinct clouds of stars, and is partly the result of depth along the galactic plane.

b. The spiral nebulae are mostly very distant objects, probably not physical members of our galactic system.

c. If our galaxy approaches the larger order of dimensions, a serious difficulty at once arises for the theory that spirals are galaxies of stars comparable in size with our own: it would be necessary to ascribe impossibly great magnitudes to the new stars that have appeared in the spiral nebulae.

2. Through approximate agreement on the above points, the way is cleared so that the outstanding difference may be clearly stated: Curtis does not believe that the numerical value of the distance I derive for any globular cluster is of the right order of magnitude.

3. The present problem may be narrowly restricted therefore, and may be formulated as follows: Show that any globular cluster is approximately as distant as derived at Mount Wilson; then the distance of other clusters will be approximately right (see Proposition B), the system of clusters and the galactic system will have dimensions of the order assigned (see proposition A), and the "comparable galaxy" theory of spirals will have met with a serious, though perhaps not insuperable difficulty.

In other words, to maintain my position it will suffice to show that any one of the bright globular clusters has roughly the distance in light-years given below, rather than a distance one tenth of this value or less:[1]

Cluster	Distance in light-years	Mean photographic magnitude of brightest 25 stars	
		Apparent	Absolute
Messier 13	36,000	13.75	−1.5
Messier 3	45,000	14.23	−1.5
Messier 5	38,000	13.97	−1.4
Omega Centauri	21,000	12.3	−1.8

Similarly it should suffice to show that the bright objects in clusters are giants (cf. last column above), rather than stars of solar luminosity.

---

[1] In the final draft of the following paper Curtis has qualified his acceptance of the foregoing propositions in such a manner that in some numerical details the comparisons given below are no longer accurately applicable to his arguments; I believe, however, that the comparisons do correctly contrast the present view with that generally accepted a few years ago.

Some commentary should illuminate Shapley's argument. What is first needed is the distinction between the magnitude (or apparent magnitude) of an object and its absolute magnitude.

Magnitude: The magnitude of a celestial object is the measure of how bright it appears. The quantitative definition of magnitude was discussed previously. The following table giving magnitudes of various objects should provide a better sense of what this means.

Object	Magnitude
Sun	−26.5
Full Moon	−12.5
Venus (at brightest)	−4
Sirius (brightest star)	−1.4
Naked eye limit	+6.5
6-inch telescope limit	+13
Palomar (visual limit)	+20
Palomar (photographic limit)	+23.5

Absolute Magnitude: Absolute magnitude measures the quantity of light that an object emits. The absolute magnitude of an object is the brightness it would have if positioned 10 parsecs from the earth. Given below is a table of various absolute magnitudes. Note that some of these values were under dispute around 1920.

Object	Absolute Magnitude
Sun	+5
B-type Stars (acc. to Russell's spect.-luminosity diag.)	−3 to −0.7
M-type Stars (acc. to Russell's spect.-luminosity diag.)	−2 to +12
Ave. Absol. Magn. for Blues (acc. to Shapley)	0
Ave. Absol. Magn. for Blues (acc. to to Curtis)	+3
Ave. Absol. Magn. of Cepheids (acc. to Shapley)	−2.5
Ave. Absol. Magn. of Galactic Novae (acc. to Curtis)	−3

Shapley's claim can be summarized by stating that his positioning of the globular clusters at a far greater distance than Curtis did, entails that the brightest stars in the globular clusters must have higher absolute magnitudes than Curtis would assign them; otherwise they would not exhibit the apparent magnitudes that they do. This leads Shapley to develop a number of arguments directed at showing that the absolute magnitudes of the brightest stars in the clusters are close to those that he has attributed to them.

4. From observation we know that some or all of these four clusters contain:

    a. An interval of at least nine magnitudes (apparent and absolute) between the brightest and faintest stars.

    b. A range of color-index from –0.5 to +2.0, corresponding to the whole range of color commonly found among assemblages of stars.

> **Color Index:** The eye is primarily sensitive to light in the yellow to green region of the spectrum, whereas early photographic plates were primarily sensitive to light in the blue region. Consequently, the brightness of a star would usually be judged differently, depending on whether its brightness was measured by the eye or by its image on a blue-sensitive photographic plate. Eventually photographic plates were devised that had sensitivities comparable to that of the eye. The magnitude of a star determined from such plates is called its photovisual magnitude, whereas its magnitude as measured from blue-sensitive plates is called its photographic magnitude. Because for most stars these are quite different, the difference between them provides a measure of the color of the star. In particular, the color index of a star is equal to its photographic magnitude minus its photovisual magnitude, or, to put it in an equation: $CI = M_{pg} - M_{pv}$. As this equation suggests, a negative color index indicates a blue star, whereas a positive color index indicates a star richer in light from the red end of the spectrum. This is, in fact, how the color of stars is defined. As Shapley states, "For a negative color-index . . . the stars are called blue and the corresponding spectral type is B; for yellow stars, like the sun (type G), the color-index is about +0.8 mag.; for redder stars (types K, M) the color-index exceeds a magnitude."

    c. Stars of types B, A, F, G, K, M (from direct observations of spectra), and that these types are in sufficient agreement with the color classes to permit the use of the latter for ordinary statistical considerations where spectra are not yet known.

    d. Cepheid and cluster variables which are certainly analogous to galactic variables of the same types, in spectrum, color change, length of period, amount of light variation, and all characters of the light-curve.

> **Cluster Variables** are variable stars having a light curve more or less like that of Cepheid variables, but a period of less than one day. The contrast is with Cepheid variables, which have periods from one to about three hundred days. Cluster variables derive their name from the fact that they are usually found in globular clusters. In contemporary astronomy, such stars are characterized as RR Lyrae type stars.

    e. Irregular, red, small-range variables of the Alpha Orionis type, among the brightest stars of the cluster.

    f. Many red and yellow stars of approximately the same magnitude as the blue stars, in obvious agreement with the giant star phenomena of the galactic system, and clearly in disagreement with all we know of color and magnitude relations for dwarf stars.

    5. From these preliminary considerations we emphasize two special deductions:

First, a globular cluster is a pretty complete "universe" by itself, with typical and representative stellar phenomena, including several classes of stars that in the solar neighborhood are recognized as giants in luminosity.

Second, we are very fortunately situated for the study of distant clusters—outside rather than inside. Hence we obtain a comprehensive dimensional view, we can determine relative real luminosities in place of relative apparent luminosities, and we have the distinct advantage that the most luminous stars are easily isolated and the most easily studied. None of the brightest stars in a cluster escapes us. If giants or supergiants are there, they are necessarily the stars we study. We cannot deal legitimately with the average brightness of stars in globular clusters, because the faintest limits are apparently far beyond our present telescopic power. Our ordinary photographs record only the most powerful radiators—encompassing a range of but three or four magnitudes at the very top of the scale of absolute luminosity, whereas in the solar domain we have a known extreme range of 20 magnitudes in absolute brightness, and a generally studied interval of twelve magnitudes or more.

    6. Let us now examine some of the conditions that would

exist in the Hercules Cluster (Messier 13) on the basis of the two opposing values for its distance:

	36,000 light-years	3,600 light-years, or less
a. Mean absolute photographic magnitude of blue stars (C. I. <0.0)	0	+5, or fainter
b. Maximum absolute photographic magnitudes of cluster stars	Between −1.0 and −2.0	+3.2, or fainter
c. Median absolute photovisual magnitude of long-period Cepheids	−2	+3, or fainter
d. Hypothetical annual proper motion	0."004	0."04, or greater

a. *The blue stars.*—The colors of stars have long been recognized as characteristic of spectral types and as being of invaluable aid in the study of faint stars for which spectroscopic observations are difficult or impossible. The color-index, as used at Mount Wilson, is the difference between the so-called photographic (pg) and photovisual (pv) magnitudes—the difference between the brightness of objects in blue-violet and in yellow-green light. For a negative color-index (C. I. = pg. −pv. <0.0) the stars are called blue and the corresponding spectral type is B; for yellow stars, like the sun (type G), the color-index is about +0.8 mag.; for redder stars (types K, M) the color-index exceeds a magnitude.

An early result of the photographic study of Messier 13 at Mount Wilson was the discovery of large numbers of negative color-indices. Similar results were later obtained in other globular and open clusters, and among the stars of the galactic clouds. Naturally these negative color-indices in clusters have been taken without question to indicate B-type stars—a supposition that has later been verified spectroscopically with the Mount Wilson reflectors.[1]

---

[1] Adams and van Maanen published several years ago the radial velocities and spectral types of a number of B stars in the double cluster in Perseus, *Ast. Jour.*, Albany, N.Y., 27, 1913 (187-188).

The existence of 15th magnitude B-type stars in the Hercules cluster seems to answer decisively the question of its distance, because B stars in the solar neighborhood are invariably giants (more than a hundred times as bright as the sun, on the average), and such a giant star can appear to be of the fifteenth magnitude only if it is more than 30,000 light-years away.

We have an abundance of material on distances and absolute magnitudes of the hundreds of neighboring B's—there are direct measures of distance, as well as mean distances determined from parallactic motions, from observed luminosity curves, from stream motions, and from radial velocities combined with proper motion. Russell, Plummer, Charlier, Eddington, Kapteyn, and others have worked on these stars with the universal result of finding them giants.

Kapteyn's study of the B stars is one of the classics of modern stellar astronomy; his methods are mainly the well-tried methods generally used for studies of nearby stars. In his various lists of B's more than seventy percent are brighter than zero absolute photographic magnitude,[1] and only two out of 424 are fainter than +3. This result should be compared with the above-mentioned requirement that the absolute magnitudes of the blue stars in Messier 13 should be +5 or fainter in the mean, if the distance of the cluster is 3,600 light-years or less, and no star in the cluster should be brighter than +3.

A question might be raised as to the completeness of the material used by Kapteyn and others, for if only the apparently bright stars are studied, the mean absolute magnitudes may be too high. Kapteyn, however, entertains little doubt on this score, and an investigation[2] of the distribution of B-type stars, based on the *Henry Draper*

---

[1] Stars of types B8 and B9 are customarily treated with the A type in statistical discussion; even if they are included with the B's, 64 per cent of Kapteyn's absolute magnitudes are brighter than zero and only 4 per cent are fainter than +2. No stars of types B8 or B9 fainter than +3 are in Kapteyn's lists.

[2] Shapley, H., *Proceedings Nat. Acad. Sci.*, 5, 1900 (434-440); a further treatment of this problem is to appear in a forthcoming *Mount Wilson Contribution*.

Catalogue, shows that faint B's are not present in the Orion region studied by Kapteyn.

The census in local clusters appears to be practically complete without revealing any B stars as faint as +5. But if the Hercules cluster were not more distant than 3,600 light-years, its B stars would be about as faint as the sun, and the admitted uniformity throughout the galactic system (Proposition C) would be gain-said: for although near the earth, whether in clusters or not, the B stars are giants, away from the earth in all directions, whether in the Milky Way clouds or in clusters, they would be dwarfs—and the anthropocentric theory could take heart again.

Let us emphasize again that the near and the distant blue stars we are intercomparing are all cluster stars, and that there appears to be no marked break in the gradation of clusters, either in total content or in distance, from Orion through the faint open clusters to Messier 13.

b. *The maximum absolute magnitude of cluster stars.*—In various nearby groups and clusters the maximum absolute photographic brightness, determined from direct measures of parallax or stream motion or from both, is known to *exceed* the following values:

	M
Ursa Major system	−1.0
Moving cluster in Perseus	−0.5
Hyades	+1.0
Scorpio-Centaurus cluster	−2.5
Orion nebula cluster	−2.5
Pleiades	−1.0
61 Cygni group[1]	+1.0

No nearby physical group is known, with the possible exception of the 61 Cygni drift, in which the brightest stars are fainter than +1.0. The mean M of the above list of clusters is −0.8; yet for all distant physical groups it must be +3 or fainter

---

[1] The absolute visual magnitude of ε Virginis (spectrum G6) is 0.0 according to the Mount Wilson spectroscopic parallax kindly communicated by Mr. Adams.

(notwithstanding the certain existence within them of Cepheid variables and B-type stars), if the distance of Messier 13 is 3,600 light-years or less. Even if the distance is 8,000 light-years, as Curtis suggests in the following paper, the mean M would need to be +1.4 or fainter—a value still irreconcilable with observations on nearby clusters.

The requirement that the bright stars in a globular cluster should be in the maximum only two magnitudes brighter than our sun is equivalent to saying that in Messier 13 there is not one real giant among its thirty or more thousand stars. It is essentially equivalent, in view of Proposition B, to holding that of the two or three million stars in distant clusters (about half a million of these stars have been actually photographed) there is not one giant star brighter than absolute photographic magnitude +2. And we have just seen that direct measures show that all of our nearby clusters contain such giants; indeed some appear to be composed mainly of giants.

As a further test of the distances of globular clusters, a special device has been used with the Hooker reflector. With a thin prism placed in the converging beam shortly before the focus, we may photograph for a star (or for each of a group of stars) a small spectrum that extends not only through the blue region ordinarily photographed, but also throughout the yellow and red. By using specially prepared photographic plates, sensitive in the blue and red but relatively insensitive in the green-yellow, the small spectra are divided in the middle, and the relative intensity of the blue and red parts depends, as is well known, on spectral type and absolute magnitude; giants and dwarfs, of the same type in the Harvard system of spectral classification, show markedly different spectra. The spectral types of forty or fifty of the brighter stars in the Hercules cluster are known, classified as usual on the basis of spectral lines. Using the device described above, a number of these stars have been photographed side by side on the same plate with well known giants and dwarfs of the solar neighborhood for which distances and absolute magnitudes depend on direct measures of parallax. On the basis of the smaller distance for Messier 13, the spectra of these cluster stars (being then of absolute magnitude fainter than +4) should resemble the spectra of the dwarfs. But the plates clearly show that in absolute brightness the cluster stars equal, and in many cases even exceed, the giants—a result to be expected if the distance is of the order of 36,000 light-years.

The above procedure is a variation on the method used by Adams and his associates on brighter stars where sufficient dispersion can be obtained to permit photometric intercomparison of sensitive spectral lines. So far as it has been applied to clusters, the usual spectroscopic method supports the above conclusion that the bright red and yellow stars in clusters are giants.

An argument much insisted upon by Curtis is that the average absolute magnitude of stars around the sun is equal to or fainter than solar brightness, hence, that average stars we see in clusters are also dwarfs. Or, put in a different way, he argues that since the mean spectral class of a globular cluster is of solar type and the average solar-type star near the sun is of solar luminosity, the stars photographed in globular clusters must be of solar luminosity, hence not distant. This deduction, he holds, is in compliance with proposition C—uniformity throughout the universe. But in drawing the conclusions, Curtis apparently ignores, first, the very common existence of red and yellow giant stars in stellar systems, and second, the circumstance mentioned above in Section 5 that in treating a distant external system we naturally first observe its giant stars. If the material is not mutually extensive in the solar domain and in the remote cluster (and it certainly is not for stars of all types), then the comparison of averages means practically nothing because of the obvious and vital selection of brighter stars in the cluster. The comparison should be of nearby cluster with distant cluster, or of the luminosities of the same kinds of stars in the two places.

Suppose that an observer, confined to a small area in a valley, attempts to measure the distances of surrounding mountain peaks. Because of the short base line allowed him, his trigonometric parallaxes are valueless except for the nearby hills. On the remote peaks, however, his telescope shows green foliage. First he assumes approximate botanical uniformity throughout all visible territory. Then he finds that the average height of all plants immediately around him (conifers, palms, asters, clovers, etc.) is one foot. Correlating this average with the measured angular height of plants visible against the sky-line on the distant peaks he obtains values of the distances. If, however, he had compared the foliage on the nearby, trigonometrically-measured hills with that on the remote peaks, or had used some method of distinguishing various floral types he would not have mistaken pines for asters and obtained erroneous results for the distances of the surrounding mountains. All the principles involved in the botanical parallax of a mountain peak have their analogues in the photometric parallax of a globular cluster.

c. *Cepheid variables.*—Giant stars of another class, the Cepheid variables, have been used extensively in the exploration of globular clusters. After determining the period of a Cepheid, its absolute magnitude is easily found from an observationally derived period-luminosity curve, and the distance of any cluster containing such variables is determined as soon as the apparent magnitudes are measured. Galactic Cepheids and cluster Cepheids are strictly comparable by Proposition C—a deduction that is amply supported by observations at Mount Wilson and Harvard, of color, spectrum, light curves, and the brightness relative to other types of stars.

Curtis bases his strongest objections to the larger galaxy on the use I have made of the Cepheid variables, questioning the sufficiency of the data and the accuracy of the methods involved. But I believe that in the present issue there is little point in laboring over the details for Cepheids, for we are, if we choose, qualitatively quite independent of them in determining the scale of the galactic system, and it is only qualitative results that are now at issue. We could discard the Cepheids altogether, use instead either the red giant stars and spectroscopic methods, or the hundreds of B-type stars upon which the most capable stellar astronomers have worked for years, and derive much the same distance for the Hercules cluster, and for other clusters, and obtain consequently similar dimensions for the galactic system. In fact, the substantiating results from these other sources strongly fortify our belief in the assumptions and methods involved in the use of the Cepheid variables.

Since the distances of clusters as given by Cepheid variables are qualitatively in excellent agreement with the distances as given by blue stars and by yellow and red giants, discussed in the foregoing sub-sections *a* and *b*, I shall here refer only briefly to four points bearing on the Cepheid problem, first noting that if the distances of clusters are to be divided by 10 or 15, the same divisor should be also used for the distances derived for galactic Cepheids.

(1) The average absolute magnitude of typical Cepheids, according to my discussion of proper motions and magnitude correlations, is about −2.5. The material on proper motion has also been discussed independently by Russell, Hertzsprung, Kapteyn, Strömberg, and Seares; they all accept the validity of the method, and agree in making the mean absolute magnitude much the same as that which I derive. Seares finds, moreover, from a discussion of probable errors and of possible systematic errors, that the observed motions are irreconcilable with an absolute brightness five magnitudes fainter, because in that case the mean parallactic motion of the brighter Cepheids would be of the order of $0''.160$ instead of $0''.016 \pm 0.002$ as observed.

Both trigonometric and spectroscopic parallaxes of galactic Cepheids, as far as they have been determined, support the photometric values in demanding high luminosity; the spectroscopic and photometric methods are not wholly independent, however, since the zero point depends in both cases on parallactic motion.

(2) When parallactic motion is used to infer provisional absolute magnitudes for individual stars (a possible process only when peculiar motions are small and observations very good), the brighter galactic Cepheids indicate the correlation between luminosity and period.[1] The necessity, however, of neglecting individual peculiar motion and errors of observation

---

[1] Mr. Seares has called my attention to an error in plotting the provisional smoothed absolute magnitudes against log period for the Cepheids discussed in *Mount Wilson Contribution* No. 151. The preliminary curve for the galactic Cepheids is steeper than that for the Small Magellanic Cloud, Omega Centauri, and other clusters.

for this procedure makes the correlation appear much less clearly for galactic Cepheids than for those of external systems (where proper motions are not concerned), and little importance could be attached to the period-luminosity curve if it were based on local Cepheids alone. When the additional data mentioned below are also treated in this manner, the correlation is practically obscured for galactic Cepheids, because of the larger observational errors.

On account of the probably universal uniformity of Cepheid phenomena, however, we need to know only the *mean* parallactic motion of the galactic Cepheids to determine the zero point of the curve which is based on external Cepheids; and the *individual* motions do not enter the problem at all, except, as noted above, to indicate provisionally the existence of the period-luminosity relation. It is only this *mean* parallactic motion that other investigators have used to show the exceedingly high luminosity of Cepheids. My adopted absolute magnitudes and distances for all these stars have been based upon the final period-luminosity curve, and not upon individual motions.

(3) Through the kindness of Professor Boss and Mr. Roy of the Dudley Observatory, proper motions have been submitted for 21 Cepheids in addition to the 13 in the *Preliminary General Catalogue*. The new material is of relatively low weight, but the unpublished discussion by Strömberg of that portion referring to the northern stars introduces no material alteration of the earlier result for the mean absolute brightness of Cepheids.

It should be noted that the 18 pseudo-Cepheids discussed by Adams and Joy[1] are without exception extremely bright (absolute magnitudes ranging from −1 to −4); they are thoroughly comparable with the ordinary Cepheids in galactic distribution, spectral characteristics, and motion.

(4) From unpublished results kindly communicated by van Maanen and by Adams, we have the following verification of the great distance and high luminosity of the important, high-velocity, cluster-type Cepheid RR Lyrae:

---

[1] Adams, W. S., and A. H. Joy, *Publ. Ast. Soc. Pac.*, San Francisco, Calif., 31, 1919 (184-186).

Photometric parallax	0."003	(Shapley)
Trigonometric parallax	+0.006±0.006	(van Maanen)
Spectroscopic parallax	0.004	(Adams, Joy, and Burwell)

The large proper motion of this star, 0."25 annually, led Hertzsprung some years ago to suspect that the star is not distant, and that it and its numerous congeners in clusters are dwarfs. The large proper motion, however, indicates high real velocity rather than nearness, as the above results show. More recently Hertzsprung has reconsidered the problem and, using the cluster variables, has derived a distance of the globular cluster Messier 3 in essential agreement with my value.

d. *Hypothetical annual proper motion.*—The absence of observed proper motion for distant clusters must be an indication of their great distance because of the known high velocities in the line of sight. The average radial velocity of the globular clusters appears to be about 150 km/sec. By assuming, as usual, a random distribution of velocities, the transverse motion of Messier 13 and similar bright globular clusters should be greater than the quite appreciable value of 0."04 a year if the distance is less than 3,600 light-years. No proper motion has been found for distant clusters; Lundmark has looked into this matter particularly for five systems and concludes that the annual proper motion is less than 0."01.

7. Let us summarize a few of the results of accepting the restricted scale of the galactic system.

If the distances of globular clusters must be decreased to one-tenth, the light-emitting power of their stars can be only a hundredth that of local cluster stars of the same spectral and photometric types. As a consequence, I believe Russell's illuminative theory of spectral evolution would have to be largely abandoned, and Eddington's brilliant theory of gaseous giant stars would need to be greatly modified or given up entirely. Now both of these modern theories have their justification, first, in the fundamental nature of their concepts and postulates, and second, in their great success in fitting observational facts.

Similarly, the period-luminosity law of Cepheid variation would be meaningless; Kapteyn's researches on the structure of the local cluster would need new interpretation, because his luminosity laws could be applied locally but not generally; and a very serious loss to astronomy would be that of the generality of spectroscopic methods of determining star distances, for it would mean that identical spectral characteristics indicate stars differing in brightness by 100 to 1, depending only upon whether the star is in the solar neighborhood or in a distant cluster.

THE DIMENSIONS AND ARRANGEMENT OF THE GALACTIC SYSTEM

When we accept the view that the distance of the Hercules cluster is such that its stellar phenomena are harmonious with local stellar phenomena—its brightest stars typical giants, its Cepheids comparable with our own—then it follows that fainter, smaller, globular clusters are still more distant than 36,000 light-years. One-third of those now known are more distant than 100,000 light-years; the most distant is more than 200,000 light-years away, and the diameter of the whole system of globular clusters is about 300,000 light-years.

Since the affiliation of the globular clusters with the galaxy is shown by their concentration to the plane of the Milky Way and their symmetrical arrangement with respect to it, it also follows that the galactic system of stars is as large as this subordinate part. During the past year we have found Cepheid variables and other stars of high luminosity among the fifteenth magnitude stars of the galactic clouds; this can only mean that some parts of the clouds are more distant than the Hercules cluster. There seems to be good reason, therefore, to believe that the star-populated regions of the galactic system extend at least as far as the globular clusters.

One consequence of accepting the theory that clusters outline the form and extent of the galactic system, is that the sun is found to be very distant from the middle of the galaxy. It appears that we are not far from the center of a large local cluster or cloud, but that cloud is at least 50,000 light-years from the galactic center. Twenty years ago Newcomb remarked that the sun *appears* to be in the galactic plane because the Milky Way is a great circle—an encircling band of light—and that the sun also *appears* near the center of the universe because the star density falls off with distance in all directions. But he concludes as follows:

> "Ptolemy showed by evidence, which, from his standpoint, looked as sound as that which we have cited, that the earth was fixed in the center of the universe. May we not be the victim of some fallacy, as he was?"

Our present answer to Newcomb's question is that we have been victimized by restricted methods of measuring distance and by the chance position of the sun near the center of a subordinate system; we have been misled, by the consequent phenomena, into thinking that we are in the midst of things. In much the same way ancient man was misled by the rotation of the earth, with the consequent apparent daily motion of all heavenly bodies around the earth, into believing that even his little planet was the center of the universe, and that his earthly gods created and judged the whole.

If man had reached his present intellectual position in a later geological era, he might not have been led to these vain conceits concerning his position in the physical universe, for the solar system is rapidly receding from the galactic plane, and is moving away from the center of the local cluster. If that motion remains unaltered in direction and amount, in a hundred million years or so the Milky Way will be quite different from an encircling band of star clouds, the local cluster will be a distant object, and the star density will no longer decrease with distance from the sun in all directions.

Another consequence of the conclusion that the galactic system is of the order of 300,000 light-years in greatest diameter, is the previously mentioned difficulty it gives to the "comparable galaxy" theory of spiral nebulae. I shall not undertake a description and discussion of this debatable problem. Since the theory probably stands or falls with the hypothesis of a small galactic system, there is little point in discussing other material on the subject, especially in view of the recently measured rotations of spiral nebulae which appear fatal to such an interpretation.

> Comment: In other words, Shapley's point is that the great diameter (300,000 light-years) he attributes to the Milky Way almost certainly entails rejection of the comparable universe theory. The reason is that if the spirals were other universes, they would have to be, by definition, of a size comparable to the Milky Way. But if they were of such colossal size, their distances would have to be extremely great to accord with the fact that they appear so small. Moreover, Shapley noted that the rotations of spirals, i.e., the motions detected by Adriaan van Maanen, supply further evidence against island universes, because if these spirals were at island-universe distances, then points on their circumference, as explained earlier, would have to be moving at impossibly great speeds.

It seems to me that the evidence, other than the admittedly critical tests depending on the size of the galaxy, is opposed to the view that the spirals are galaxies of stars comparable with our own. In fact, there appears as yet no reason for modifying the tentative hypothesis that the spirals are not composed of typical stars at all, but are truly nebulous objects. Three very recent results are, I believe, distinctly serious for the theory that spiral nebulae are comparable galaxies—(1) Seares' deduction that none of the known spiral nebulae has a surface brightness as small as that of our galaxy; (2) Reynold's study of the distribution of light and color in typical spirals, from which he concludes they cannot be stellar systems; and (3) van Maanen's recent measures of rotation in the spiral M 33, corroborating his earlier work on Messier 101 and 81, and indicating that these bright spirals cannot reasonably be the excessively distant objects required by the theory.

But even if spirals fail as galactic systems, there may be elsewhere in space stellar systems equal to or greater than ours—as yet unrecognized and possibly quite beyond the power of existing optical devices and present measuring scales. The modern telescope, however, with such accessories as high-power spectroscopes and photographic intensifiers, is destined to extend the inquiries relative to the size of the universe much deeper into space, and contribute further to the problem of other galaxies.

☆☆☆☆☆☆☆☆☆☆☆☆☆☆☆☆☆☆☆☆☆☆☆☆☆☆☆☆☆☆☆☆

☆☆☆☆☆☆☆☆☆☆☆☆☆☆☆☆☆☆☆☆☆☆☆☆☆☆☆☆☆

Heber D. Curtis, "The Scale of the Universe," *Bulletin of the National Research Council*, 3 (1921), 194–217.

CONTENTS

Dimensions and structure of the galaxy
Evidence furnished by the magnitude of the stars
The spirals as external galaxies

DIMENSIONS AND STRUCTURE OF THE GALAXY

Definition of units employed.—The distance traversed by light in one year, 9.5×10^{12} km., or nearly six trillion miles, known as the light-year, has been in use for about two centuries as a means of visualizing stellar distances, and forms a convenient and easily comprehended unit. Throughout this paper the distances of the stars will be expressed in light-years.

The *absolute* magnitude of a star is frequently needed in order that we may compare the luminosities of different stars in terms of some common unit. It is that *apparent* magnitude which the star would have if viewed from the standard distance of 32.6 light-years (corresponding to a parallax of $0''.1$).

Knowing the parallax, or the distance, of a star, the absolute magnitude may be computed from one of the simple equations: Abs. Magn. = App. Magn. $+5 + 5 \times \log$ (parallax in seconds of arc) [or] Abs. Magn. = App. Magn. $+ 7.6 - 5 \times \log$ (distance in light-years).

Limitations in studies of galactic dimensions.—By direct methods the distances of individual stars can be determined with considerable accuracy out to a distance of about two hundred light-years.

At a distance of three hundred light-years (28×10^{14} km.) the radius of the earth's orbit (1.5×10^8 km.) subtends an angle slightly greater than $0''.01$, and the probable error of the best modern photographic parallax determinations has not yet been reduced materially below this value. The spectroscopic method of determining stellar distances through the absolute magnitude probably has, at present, the same limitations as the trigonometric method upon which the spectroscopic method depends for its absolute scale.

A number of indirect methods have been employed which extend our reach into space somewhat farther for the average distances of large groups or classes of stars, but give no information as to the individual distances of the stars of the group or class. Among such methods may be noted as most important the various correlations which have been made between the proper motions of the stars and the parallactic motion due to the speed of our sun in space, or between the proper motions and the radial velocities of the stars.

The limitations of such methods of correlation depend, at present, upon the fact that accurate proper motions are known, in general, for the brighter stars only. A motion of 20 km/sec. across our line of sight will produce the following annual proper motions:

Distance	100 l. y.	500 l. y.	1,000 l. y.
Annual p. m.	0".14	0".03	0".01

The average probable error of the proper motions of Boss is about 0".006. Such correlation methods are not, moreover, a simple matter of comparison of values, but are rendered difficult and to some extent uncertain by the puzzling complexities brought in by the variation of the space motions of the stars with spectral type, stellar mass (?), stellar luminosity (?), and still imperfectly known factors of community of star drift.

It will then be evident that the base-line available in studies of the more distant regions of our galaxy is woefully short, and that in such studies we must depend largely upon investigations of the distribution and of the frequency of occurrence of stars of the different apparent magnitudes and spectral types, on the assumption that the more distant stars, when taken in large numbers, will average about the same as known nearer stars. This assumption is a reasonable one, though not necessarily correct, as we have little certain knowledge of galactic regions as distant as five hundred light-years.

Were all the stars of approximately the same absolute magnitude, or if this were true even for the stars of any particular type or class, the problem of determining the general order of the dimensions of our galaxy would be comparatively easy.

But the problem is complicated by the fact that, taking the stars of all spectral types together, the dispersion in absolute luminosity is very great. Even with the exclusion of a small number of stars which are exceptionally bright or faint, this dispersion probably reaches ten absolute magnitudes, which would correspond to a hundred-fold uncertainty in distance for a given star. However, it will be seen later that we possess moderately definite information as to the *average* absolute magnitude of the stars of the different spectral types.

Dimensions of our galaxy.—Studies of the distribution of the stars and of the ratio between the numbers of stars of successive apparent magnitudes have led a number of investigators to the postulation of fairly accordant dimensions for the galaxy; a few may be quoted:

Wolf; about 14,000 light-years in diameter.
Eddington; about 15,000 light-years.
Shapley (1915); about 20,000 light-years.
Newcomb; not less than 7,000 light-years; later—perhaps 30,000 light-years in diameter and 5,000 light-years in thickness.
Kapteyn; about 60,000 light-years.[1]

General structure of the galaxy.—From the lines of investigation mentioned above there has been a similar general

---

[1] A complete bibliography of the subject would fill many pages. Accordingly references to authorities will in general be omitted. An excellent and nearly complete list of references may be found in Lundmark's paper,—"The Relations of the Globular Clusters and Spiral Nebulae to the Stellar Systems," in *K. Svenska Vet. Handlingar,* Bd. 60, No. 8, p. 71, 1920.

accord in the deduced results as to the shape and structure of the galaxy:

1. The stars are not infinite in number, nor uniform in distribution.

2. Our galaxy, delimited for us by the projected contours of the Milky Way, contains possibly a billion suns.

3. This galaxy is shaped much like a lens, or a thin watch, the thickness being probably less than one-sixth of the diameter.

4. Our Sun is located fairly close to the center of figure of the galaxy.

5. The stars are not distributed uniformly through the galaxy. A large proportion are probably actually within the ring structure suggested by the appearance of the Milky Way, or are arranged in large and irregular regions of greater star density. The writer believes that the Milky Way is at least as much a structural as a depth effect.

A spiral structure has been suggested for our galaxy; the evidence for such a spiral structure is not very strong, except as it may be supported by the analogy of the spirals as island universes, but such a structure is neither impossible nor improbable. The position of our Sun near the center of figure of the galaxy is not a favorable one for the precise determination of the actual galactic structure.

Relative paucity of galactic genera.—Mere size does not necessarily involve complexity; it is a remarkable fact that in a galaxy of a thousand million objects we observe, not ten thousand different types, but perhaps not more than five main classes, outside the minor phenomena of our own solar system.

1. *The stars.*—The first and most important class is formed by the stars. In accordance with the type of spectrum exhibited, we may divide the stars into some eight or ten main types; even when we include the consecutive internal gradations within these spectral classes it is doubtful whether present methods will permit us to distinguish as many as a hundred separate subdivisions in all. Average space velocities vary from 10 to 30 km/sec., there being a well-marked increase

in average space velocity as one proceeds from the blue to the redder stars.

2. The *globular star clusters* are greatly condensed aggregations of from ten thousand to one hundred thousand stars. Perhaps one hundred are known. Though quite irregular in grouping, they are generally regarded as definitely galactic in distribution. Space velocities are of the order of 300 km/sec.

3. *The diffuse nebulae* are enormous, tenuous, cloud-like masses; fairly numerous; always galactic in distribution. They frequently show a gaseous spectrum, though many agree approximately in spectrum with their involved stars. Space velocities are very low.

4. *The planetary nebulae* are small, round or oval, and almost always with a central star. Fewer than one hundred and fifty are known. They are galactic in distribution; spectrum is gaseous; space velocities are about 80 km/sec.

5. *The spirals.*—Perhaps a million are within reach of large reflectors; the spectrum is generally like that of a star cluster. They are emphatically non-galactic in distribution, grouped about the galactic poles, spiral in form. Space velocities are of the order of 1200 km/sec.

Distribution of celestial genera.—With one, and only one, exception, all known genera of celestial objects show such a distribution with respect to the plane of our Milky Way, that there can be no reasonable doubt that all classes, save this one, are integral members of our galaxy. We see that all the stars, whether typical, binary, variable, or temporary, even the rarer types, show this unmistakable concentration toward the galactic plane. So also for the diffuse and the planetary nebulae and, though somewhat less definitely, for the globular star clusters.

The one exception is formed by the spirals; grouped about the poles of our galaxy, they appear to abhor the regions of greatest star density. They seem clearly a class apart. *Never found in our Milky Way*, there is no other class of celestial objects with their distinctive characteristics of form, distribution, and velocity in space.

The evidence at present available points strongly to the

conclusion that the spirals are individual galaxies, or island universes, comparable with our own galaxy in dimensions and in number of component units. While the island universe theory of the spirals is not a vital postulate in a theory of galactic dimensions, nevertheless, because of its indirect bearing on the question, the arguments in favor of the island universe hypothesis will be included with those which touch more directly on the probable dimensions of our own galaxy.

Other theories of galactic dimensions.—From evidence to be referred to later Dr. Shapley has deduced very great distances for the globular star clusters, and holds that our galaxy has a diameter comparable with the distances which he has derived for the clusters, namely,—a galactic diameter of about 300,000 light-years, or at least ten times greater than formerly accepted. The postulates of the two theories may be outlined as follows:

Present Theory	Shapley's Theory
Our galaxy is probably not more than 30,000 light-years in diameter, and perhaps 5,000 light-years in thickness.	The galaxy is approximately 300,000 light-years in diameter, and 30,000, or more, light-years in thickness.
The clusters, and all other types of celestial objects except the spirals, are component parts of our own galactic system.	The globular clusters are remote objects, but a part of our own galaxy. The most distant cluster is placed about 220,000 light-years away.
The spirals are a class apart, and not intra-galactic objects. As island universes, of the same order of size as our galaxy, they are distant from us 500,000 to 10,000,000, or more, light-years.	The spirals are probably of nebulous constitution, and possibly not members of our own galaxy, driven away in some manner from the regions of greatest star density.

EVIDENCE FURNISHED BY THE MAGNITUDE OF THE STARS

The "average" star.—It will be of advantage to consider the two theories of galactic dimensions from the standpoint of the average star. What is the "average" or most frequent type of star of our galaxy or of a globular star cluster, and if we can with some probability postulate such an average star, what bearing will the characteristics of such a star have upon the question of its average distance from us?

No adequate evidence is available that the more distant stars of our galaxy are in any way essentially different from stars of known distance nearer to us. It would seem then that we may safely make such correlations between the nearer and the more distant stars, *en masse*. In such comparisons the limitations of spectral type must be observed as rigidly as possible, and results based upon small numbers of stars must be avoided, if possible.

Many investigations, notably Shapley's studies of the colors of stars in the globular clusters, and Fath's integrated spectra of these objects and of the Milky Way, indicate that the average star of a star cluster or of the Milky Way will, in the great majority of cases, be somewhat like our Sun in spectral type, *i.e.*, such an average star will be, in general, between spectral types F and K.

Characteristics of F–K type stars of known distance.—The distances of stars of type F–K in our own neighborhood have been determined in greater number, perhaps, than for the stars of any other spectral type, so that the average absolute magnitude of stars of this type seems fairly well determined. There is every reason to believe, however, that our selection of stars of these or other types for direct distance determinations has not been a representative one. Our parallax programs have a tendency to select stars either of great luminosity or of great space velocity.

Kapteyn's values for the average absolute magnitudes of the stars of the various spectral types are as follows:

Type	Average Abs. Magn.
B5	+1.6
A5	+3.4
F5	+7
G5	+10
K5	+13
M	+15

The same investigator's most recent luminosity-frequency curve places the maximum of frequency of the stars in general, taking all the spectral types together, at absolute magnitude +7.7.

A recent tabulation of about five hundred modern photographically determined parallaxes places the average absolute magnitude of stars of type F–K at about +4.5.

The average absolute magnitude of five hundred stars of spectral types F to M is close to +4, as determined spectroscopically by Adams.

It seems certain that the two last values of the average absolute magnitude are too low, that is,—indicate too high an average luminosity, due to the omission from our parallax programs of the intrinsically fainter stars. The absolute magnitudes of the dwarf stars are, in general, fairly accurately determined; the absolute magnitudes of many of the giant stars depend upon small and uncertain parallaxes. In view of these facts we may somewhat tentatively take the average absolute magnitude of F–K stars of known distance as not brighter than +6; some investigators would prefer a value of +7 or +8.

Comparison of Milky Way stars with the "average" stars.—We may take, without serious error, the distances of 10,000 and 100,000 light-years respectively, as representing the distance in the two theories from our point in space to the central line of the Milky Way structure. Then the following short table may be prepared:

Apparent magnitudes	Corresponding absolute magnitudes for distances of,—	
	10,000 light-years	100,000 light-years
10	−2.4	−7.4
12	−0.4	−5.4
14	+1.6	−3.4
16	+3.6	−1.4
18	+5.6	+0.6
20	+7.6	+2.6

It will be seen from the above table that the stars of apparent magnitudes 16 to 20, observed in our Milky Way structure in such great numbers, and, from their spectrum, believed to be predominantly F–K in type, are of essentially the same absolute luminosity as known nearer stars of these types, if assumed to be at the average distance of 10,000 light-years. The greater value postulated for the galactic dimensions requires, on the other hand, an enormous proportion of giant stars.

Proportion of giant stars among stars of known distance.— All existing evidence indicates that the proportion of giant stars in a given region of space is very small. As fairly representative of several investigations we may quote Schouten's results, in which he derives an average stellar density of 166,000 stars in a cube 500 light-years on a side, the distribution in absolute magnitude being as follows:

Absolute magnitudes	No. of stars	Relative percentages
−5 to −2	17	.01
−2 to +1	760	.5
+1 to +5	26800	16
+5 to +10	138600	83

Comparison of the stars of the globular clusters with the "average" star.—From a somewhat cursory study of the negatives of ten representative globular clusters I estimate the average apparent visual magnitude of all the stars in these clusters as in the neighborhood of the eighteenth. More powerful instruments may eventually indicate a somewhat fainter mean value, but it does not seem probably that this value is as much as two magnitudes in error. We then have:

Apparent magnitude of average cluster star	Corresponding absolute magnitudes if at distances of,—	
	10,000 light-years	100,000 light-years
18	+5.6	+0.6

Here again we see that the average F–K star of a cluster, if assumed to be at a distance of 10,000 light-years, has an average luminosity about the same as that found for known nearer stars of this type. The greater average distance of 100,000 light-years requires a proportion of giant stars enormously greater than is found in those regions of our galaxy of which we have fairly definite distance data.

While it is not impossible that the clusters are exceptional regions of space and that, with a tremendous spatial concentration of suns, there exists also a unique concentration of giant stars, the hypothesis that cluster stars are, on the whole, like those of known distance seems inherently the more probable.

It would appear, also, that galactic dimensions deduced from correlations between large numbers of what we may term average stars must take precedence over values found from small numbers of exceptional objects, and that, where deductions disagree, we have a right to demand that a theory of galactic dimensions based upon the exceptional object or class shall not fail to give an adequate explanation of the usual object or class.

The evidence for greater galactic dimensions.—The arguments for a much larger diameter for our galaxy than that hitherto held, and the objections which have been raised against the island universe theory of the spirals rest mainly upon the great distances which have been deduced for the globular star clusters.

I am unable to accept the thesis that the globular clusters are at distances of the order of 100,000 light-years, feeling that much more evidence is needed on this point before it will be justifiable to assume that the cluster stars are predominatingly giants rather than average stars. I am also influenced, perhaps unduly, by certain fundamental uncertainties in the data employed. The limitations of space available for the publication of this portion of the discussion unfortunately prevents a full treatment of the evidence. In calling attention to some of the uncertainties in the basal data, I must disclaim any spirit of captious criticism, and take this occasion to express my respect for Dr. Shapley's point of view, and my high appreciation of the extremely valuable work which he has done on the clusters. I am willing to accept correlations between large masses of stellar data, whether of magnitudes, radial velocities, or proper motions, but I feel that the dispersion in stellar characteristics is too large to permit the use of limited amounts of any sort of data, particularly when such data is of the same order as the probable errors of the methods of observation.

The deductions as to the very great distances of the globular clusters rest, in the final analysis, upon three lines of evidence:

1. Determination of the relative distances of the clusters on the assumption that they are objects of the same order of actual size.

2. Determination of the absolute distances of the clusters through correlations between Cepheid variable stars in the clusters and in our galaxy.

3. Determination of the absolute distances of the clusters through a comparison of their brightest stars with the intrinsically brightest stars of our galaxy.

Of these three methods, the second is given most weight by Shapley.

It seems reasonable to assume that the globular clusters are of the same order of actual size, and that from their apparent diameters the *relative* distances may be determined. The writer would not, however, place undue emphasis upon this relation. There would seem to be no good reason why there may not exist among these objects a reasonable amount of difference in actual size, say from three- to five-fold, differences which would not prevent them being regarded as of the same order of size, but which would introduce considerable uncertainty into the estimates of relative distance.

The evidence from the Cepheid variable stars.—This portion of Shapley's theory rests upon the following three hypotheses or lines of evidence:

A. That there is a close coordination between absolute magnitude and length of period for the Cepheid variables of our galaxy, similar to the relation discovered by Miss Leavitt among Cepheids of the Smaller Magellanic Cloud.

B. That, if of identical periods, Cepheids anywhere in the universe have identical absolute magnitudes.

C. This coordination of absolute magnitude and length of period for galactic Cepheids, the derivation of the absolute scale for their distances and the distances of the clusters, and, combined with A) and B), the deductions therefrom as to the much greater dimensions for our galaxy, depend almost entirely upon the sizes and the internal relationships of the proper motions of eleven Cepheid variables.

Under the first heading, it will be seen later that the actual evidence for such a coordination among galactic Cepheids is very weak. Provided that the Smaller Magellanic Cloud is not in some way a unique region of space, the behavior of the Cepheid variables in this Cloud is, through analogy, perhaps the strongest argument for postulating a similar phenomenon among the Cepheid variables of our galaxy.

Unfortunately there is a large dispersion in practically all the characteristics of the stars. That the Cepheids lack a reasonable amount of such dispersion is contrary to all experience for the stars in general. There are many who will regard the assumption made under B) above as a rather drastic one.

If we tabulate the proper motions of these eleven Cepheids, as given by Boss, and their probable errors as well, it will be seen that the average proper motion of these eleven stars is of the order of one second of arc per century in either coordinate;

that the average probable error is nearly half this amount, and that the probable errors of half of these twenty-two coordinates may well be described as of the same size as the corresponding proper motions.

Illustrations bearing on the uncertainty of proper motions of the order of 0."01 per year might be multiplied at great length. The fundamental and unavoidable errors in our star positions, the probable errors of meridian observations, the uncertainty in the adopted value of the constant of precession, the uncertainties introduced by the systematic corrections applied to different catalogues, all have comparatively little effect when use is made of proper motions as large as ten seconds of arc per century. Proper motions as small as one second of arc per century are, however, still highly uncertain quantities, entirely aside from the question of the possible existence of systematic errors. As an illustration of the differences in such minute proper motions as derived by various authorities, the proper motions of three of the best determined of this list of eleven Cepheids, as determined by Auwers, are in different quadrants from those derived by Boss.

There seems no good reason why the smaller coordinates of this list of twenty-two may not eventually prove to be different by once or twice their present magnitude, with occasional changes of sign. So small an amount of presumably uncertain data is insufficient to determine the scale of our galaxy and many will prefer to wait for additional material before accepting such evidence as conclusive.

In view of:

1. The known uncertainties of small proper motions, and,

2. The known magnitude of the purely random motions of the stars, the determination of *individual* parallaxes from *individual* proper motions can never give results of value, though the average distances secured by such methods of correlation from large numbers of stars are apparently trustworthy. The method can not be regarded as a valid one, and this applies whether the proper motions are very small or are of appreciable size.

As far as the galactic Cepheids are concerned, Shapley's curve of coordination between absolute magnitude and length of period, though found through the mean absolute magnitude of the group of eleven, rests in reality upon individual parallaxes determined from individual proper motions, as may be verified by comparing his values for the parallax of these eleven stars with[1] the values found directly from the upsilon component of the proper motion (namely,—that component which is parallel to the Sun's motion) and the solar motion. The differences in the two sets of values, 0.0002 in the mean, arise from the rather elaborate system of weighting employed.

The final test of a functional relation is the agreement obtained when applied to similar data not originally employed in deducing the relation. We must be ready to allow some measure of deviation in such a test, but when a considerable proportion of other available data fails to agree within a reasonable amount, we shall be justified in withholding our decision.

If the curve of correlation deduced by Shapley for galactic Cepheids is correct in both its absolute and relative scale, and if it is possible to determine individual distances from individual proper motions, the curve of correlation, using the same method as far as the proper motions are concerned (the validity of which I do not admit), should fit fairly well with other available proper motion and parallax data. The directly determined parallaxes are known for five of this group of eleven, and for five other Cepheids. There are, in addition, twenty-six other Cepheids for which proper motions have been determined. One of these was omitted by Shapley because of irregularity of period, one for irregularity of the light curve, two because the proper motions were deemed of insufficient accuracy, two because the proper motions are anomalously large; the proper motions of the others have been recently investigated at the Dudley Observatory, but have less weight than those of the eleven Cepheids used by Shapley.

---

[1] *Mt. Wilson Contr.* No. 151, Table V.

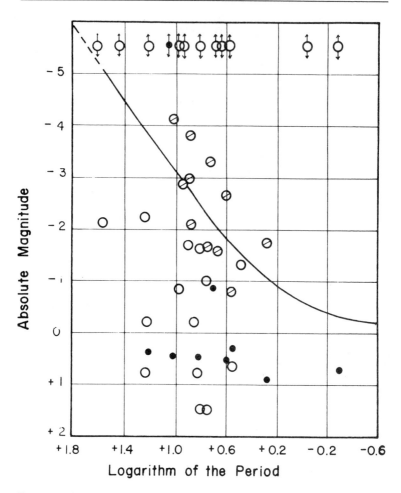

Fig. 1.—Agreement of other data with the luminosity-period correlation curve. Absolute magnitudes calculated from the upsilon component of the proper motion are indicated by circles; the eleven employed by Shapley are marked with a bar. Black dots represent directly determined parallaxes. The arrows attached to the circles at the upper edge of the diagram indicate that either the parallax or the upsilon component of the proper motion is negative, and the absolute magnitude indeterminate in consequence.

In Figure 1 the absolute magnitudes are plotted against the logarithm of the period; the curve is taken from *Mt. Wilson Contr.* No. 151, and is that finally adopted by Shapley after the introduction of about twelve Cepheids of long periods in clusters, twenty-five from the Smaller Magellanic Cloud, and a large number of short period cluster-type variables in clusters with periods less than a day, which have little effect on the general shape of the curve. The barred circles represent the eleven galactic Cepheids employed by Shapley, the black dots those Cepheids for which parallaxes have been determined, while the open circles indicate variables for which proper motions have since become available, or not employed originally by Shapley. For the stars at the upper edge of the diagram, the attached arrows indicate that either the parallax, or the upsilon component of the proper motion is negative, so that the absolute magnitude is indeterminate, and may be anything from infinity down.

From the above it would seem that available observational data lend little support to the fact of a period-luminosity relation among galactic Cepheids. In view of the large discrepancies shown by other members of the group when plotted on this curve, it would seem wiser to wait for additional evidence as to proper motion, radial velocity, and, if possible, parallax, before entire confidence can be placed in the hypothesis that the Cepheids and cluster-type variables are invariably super-giants in absolute luminosity.

Argument from the intrinsically brightest stars.—If the luminosity-frequency law is the same for the stars of the globular clusters as for our galaxy, it should be possible to correlate the intrinsically brightest stars of both regions and thus determine cluster distances. It would seem, *a priori*, that the brighter stars of the clusters must be giants, or at least approach that type, if the stars of the clusters are like the general run of stars. Through the application of a spectroscopic method Shapley has found that the spectra of the brighter stars in clusters resemble the spectra of galactic giant stars, a method which should be exceedingly useful after sufficient tests have been made to make sure that in this phenomenon, as is unfortunately the case in practically all stellar characteristics, there is not a large dispersion, and also whether slight differences in spectral type may at all materially affect the deductions.

The average "giant" star.—Determining the distance of Messier 3 from the variable stars which it contains, Shapley then derives absolute magnitude –1.5 as the mean luminosity of the twenty-five brightest stars in this cluster. From this mean value, –1.5, he then determines the distances of other clusters. Instead, however, of determining cluster distances of the order of 100,000 light-years by means of correlations on a limited number of Cepheid variables, a small and possibly exceptional class, and from the distances thus derived deducing that the absolute magnitudes of many of the brighter stars in the clusters are as great as –3, while a large proportion are greater than –1, it would seem preferable to begin the line of reasoning with the attributes of known stars in our neighborhood, and to proceed from them to the clusters.

What is the average absolute magnitude of a galactic giant star? On this point there is room for honest difference of opinions, and there will doubtless be many who will regard the conclusions of this paper as ultra conservative. Confining ourselves to existing observational data, there is no evidence that a group of galactic giants, of average spectral type about G5, will have a mean absolute magnitude as great as –1.5; it is more probably in the neighborhood of +1.5, or three absolute magnitudes fainter, making Shapley's distances four times too large.

Russell's suggestion is worth quoting in this connection, written in 1913, when parallax data were far more limited and less reliable than at present:

> The giant stars of all the spectral classes appear to be of about the same mean brightness,—averaging a little above absolute magnitude zero, that is, about a hundred times as bright as the Sun. Since the stars of this series ... have been selected by apparent brightness, which gives a strong preference to those of greatest luminosity, the average brightness of all the giant stars in a given region of space must be less than this, perhaps considerably so.

Some reference has already been made to the doubtful value of parallaxes of the order of $0''.010$, and it is upon such small or negative parallaxes that most of the very great absolute luminosities in present lists depend. It seems clear that parallax work should aim at using as faint comparison stars as possible, and that the corrections applied to reduce relative parallaxes to absolute parallaxes should be increased very considerably over what was thought acceptable ten years ago.

From a study of the plotted absolute magnitudes by spectral type of about five hundred modern direct parallaxes, with due regard to the uncertainties of minute parallaxes, and keeping in mind that most of the giants will be of types F to M, there seems little reason for placing the average absolute magnitude of such giant stars as brighter than +2.

The average absolute magnitude for the giants in Adams's list of five hundred spectroscopic parallaxes is +1.1. The two methods differ most in the stars of type G, where the spectroscopic method shows a maximum at +0.6, which is not very evident in the trigonometric parallaxes.

In such moving star clusters as the Hyades group, we have thus far evidently observed only the giant stars of such groups.

The mean absolute magnitude of forty-four stars believed to belong to the Hyades moving cluster is +2.3. The mean absolute magnitude of the thirteen stars of types F, G, and K, is +2.4. The mean absolute magnitude of the six brightest stars is +0.8 (two A5, one G, and three of K type).

The Pleiades can not fittingly be compared with such clusters or the globular clusters; its composition appears entirely different as the brightest stars average about B5, and only among the faintest stars of the cluster are there any as late as F in type. The parallax of this group is still highly uncertain. With Schouten's value of $0\overset{\prime\prime}{.}037$ the mean absolute magnitude of the six brightest stars is +1.6.

With due allowance for the redness of the giants in clusters, Shapley's mean visual magnitude of the twenty-five brightest stars in twenty-eight globular clusters is about 14.5. Then, from the equation given in the first section of this paper we have,—

$+2 = 14.5 + 7.6 - 5 \times \log \text{distance}$,

or, log distance = 4.02 = 10,500 light-years as the average distance.

If we adopt instead the mean value of Adams +1.1, the distance becomes 17,800 light-years.

Either value for the average distance of the clusters may be regarded as satisfactorily close to those postulated for a galaxy of the smaller dimensions held in this paper, in view of the many uncertainties in the data. Either value, also, will give on the same assumptions a distance of the order of 30,000 light-years for a few of the faintest and apparently most distant clusters. I consider it very doubtful whether any cluster is really so distant as this, but find no difficulty in provisionally accepting it as a possibility, without thereby necessarily extending the main structure of the galaxy to such dimensions. While the clusters seem concentrated toward our galactic plane, their distribution in longitude is a most irregular one, nearly all lying in the quadrant between 270° and 0°. If the spirals are galaxies of stars, their analogy would explain the existence of frequent nodules of condensation (globular clusters?) lying well outside of and distinct from the main structure of a galaxy.

It must be admitted that the B-type stars furnish something of a dilemma in any attempt to utilize them in determining cluster distances.

From the minuteness of their proper motions, most investigators have deduced very great luminosities for such stars in our galaxy. Examining Kapteyn's values for stars of this type, it will be seen that he finds a range in absolute magnitude from +3.25 to −5.47. Dividing the 433 stars of his lists into two magnitude groups, we have:

Mean abs. magn. 249 B stars, brighter than abs. magn. 0	= −1.32
Mean abs. magn. 184 B stars, fainter than abs. magn. 0	= +0.99
Mean abs. magn. all	= −0.36

Either the value for the brighter stars, −1.32, or the mean of all, −0.36, is over a magnitude brighter than the average absolute magnitude of the giants of the other spectral types among nearer galactic stars.

Now this galactic relation is apparently *reversed* in such clusters as M. 3 or M. 13, where the B-type stars are about three magnitudes fainter than the brighter K and M stars and about a magnitude fainter than those of G type. Supposing that the present very high values for the galactic B-type stars are correct, if we assume similar luminosity for those in the clusters we must assign absolute magnitudes of −3 to −6 to the F to M stars of the clusters, for which we have no certain galactic parallel, with a distance of perhaps 100,000 light-years. On the other hand, if the F to M stars of the cluster are like the brighter stars of these types in the galaxy, the average absolute magnitude of the B-type stars will be only about +3, and too low to agree with present values for galactic B stars. I prefer to accept the latter alternative in this dilemma, and to believe that there may exist B-type stars of only two to five times the brightness of the Sun.

While I hold to a theory of galactic dimensions approximately one-tenth of that supported by Shapley, it does not follow that I maintain this ratio for any particular cluster distance. All that I have tried to do is to show that 10,000 light-years is a reasonable *average* cluster distance.

There are so many assumptions and uncertainties involved that I am most hesitant in attempting to assign a given distance to a given cluster, a hesitancy which is not diminished by a consideration of the following estimates of the distance of M. 13 (The Great Cluster in Hercules).

Shapley, 1915, provisional	100,000 light-years
Charlier, 1916	170 light-years
Shapley, 1917	36,000 light-years
Schouten, 1918	4,300 light-years
Lundmark, 1920	21,700 light-years

It should be stated here that Shapley's earlier estimate was merely a provisional assumption for computational illustration, but all are based on modern material, and illustrate the fact that good evidence may frequently be interpreted in different ways.

My own estimate, based on the general considerations outlined earlier in this paper, would be about 8,000 light-years, and it would appear to me, at present, that this estimate is perhaps within fifty per cent of the truth.

THE SPIRALS AS EXTERNAL GALAXIES

The spirals.—If the spirals are island universes it would seem reasonable and most probable to assign to them dimensions of the same order as our galaxy. If, however, their dimensions are as great as 300,000 light-years, the island universes must be placed at such enormous distances that it would be necessary to assign what seem impossibly great absolute magnitudes to the novae which have appeared in these objects. For this reason the island universe theory has an indirect bearing on the general subject of galactic dimensions, though it is, of course, entirely possible to hold both to the island universe theory and to the belief in the greater dimensions for our galaxy by making the not improbable assumption that our own island universe, by chance, happens to be several fold larger than the average.

Some of the arguments against the island universe theory of the spirals have been cogently put by Shapley, and will be quoted here for reference. It is only fair to state that these earlier statements do not adequately represent Shapley's present point of view, which coincides somewhat more closely with that held by the writer.

> With the plan of the sidereal system here outlined, it appears unlikely that the spiral nebulae can be considered separate galaxies of stars. In addition to the evidence heretofore existing, the following points seem opposed to the "island universe" theory: (a) the dynamical character of the region of avoidance; (b) the size of the galaxy; (c) the maximum luminosity attainable by a star; (d) the increasing commonness of high velocities among other sidereal objects, particularly those outside the region of avoidance ... the cluster work strongly suggests the hypothesis that spiral nebulae ... are, however, members of the galactic organization ... the novae in spirals may be considered as the engulfing of a star by the rapidly moving nebulosity. (*Publ. Astron. Soc. of the Pacific,* Feb. 1918, p. 53.)

The recent work on star clusters, in so far as it throws some light on the probable extent and structure of the galactic system, justifies a brief reconsideration of the question of external galaxies, and apparently leads to the rejection of the hypothesis that spiral nebulae should be interpreted as separate stellar systems.

Let us abandon the comparison with the galaxy and assume an average distance for the brighter spirals that will give a reasonable maximum absolute magnitude for the novae (*and in a footnote;*—provisionally, let us say, of the order of 20,000 light-years). Further, it is possible to explain the peculiar distribution of the spirals and their systematic recession by supposing them repelled in some manner from the galactic system, which appears to move as a whole through a nebular field of indefinite extent. But the possibility of these hypotheses is of course not proposed as competent evidence against the "island universe" theory.... Observation and discussion of the radial velocities, internal motions, and distribution of the spiral nebulae, of the real and apparent brightness of novae, of the maximum luminosity of galactic and cluster stars, and finally of the dimensions of our own galactic system, all seem definitely to oppose the "island universe" hypothesis of the spiral nebulae.... (*Publ. Astron. Soc. of the Pacific,* Oct. 1919, pp. 261 ff.)

The dilemma of the apparent dimensions of the spirals.—In apparent size the spirals range from a diameter of 2° (Andromeda), to minute flecks 5", or less, in diameter.

They may possibly vary in actual size, roughly in the progression exhibited by their apparent dimensions.

The general principle of approximate equality of size for celestial objects of the same class seems, however, inherently the more probable, and has been used in numerous modern investigations, *e.g.*, by Shapley in determining the relative distances of the clusters.

On this principle of approximate equality of actual size:

As Island Universes	As Galactic Phenomena
Their probable distances range from about 500,000 light-years (Andromeda), to distances of the order of 100,000,000 light-years.	If the Nebula of Andromeda is but 20,000 light-years distant, the minute spirals would need to be at distances of the order of 10,000,000 light-years, or far outside the greater dimensions postulated for the galaxy.
At 500,000 light-years the Nebula of Andromeda would be 17,000 light-years in diameter, or of the same order of size as our galaxy.	If all are galactic objects, equality of size must be abandoned, and the minute spirals assumed to be about a thousand-fold smaller than the largest.

The spectrum of the spirals. —

As Island Universes	As Galactic Phenomena
The spectrum of the average spiral is indistinguishable from that given by a star cluster.	If the spirals are intragalactic, we must assume that they are some sort of finely divided matter, or of gaseous constitution.
It is approximately F-G in type, and in general character resembles closely the integrated spectrum of our Milky Way. It is just such a spectrum as would be expected from a vast congeries of stars.	In either case we have no adequate and actually existing evidence by which we may explain their spectrum. Many diffuse nebulosities of our galaxy show a bright-line gaseous spectrum. Others, associated with bright stars, agree with their involved stars in spectrum, and are well explained as a reflection or resonance effect.
The spectrum of the spirals offers no difficulties on the island universe theory.	Such an explanation seems untenable for most of the spirals.

The distribution of the spirals.—The spirals are found in greatest numbers just where the stars are fewest (at the galactic poles), and not at all where the stars are most numerous (in the galactic plane). This fact makes it difficult, if not impossible, to fit the spirals into any coherent scheme of stellar evolution, either as a point of origin, or as a final evolutionary product. No spiral has as yet been found actually within the structure of the Milky Way. This peculiar distribution is admittedly difficult to explain on any theory. This factor of distribution in the two theories may be contrasted as follows:

As Island Universes

It is most improbable that our galaxy should, by mere chance, be placed about half way between two great groups of island universes.

So many of the edgewise spirals show peripheral rings of occulting matter that this dark ring may well be the rule rather than the exception.

If our galaxy, itself a spiral on the island universe theory, possesses such a peripheral ring of occulting matter, this would obliterate the distant spirals in our galactic plane, and would explain the peculiar apparent distribution of the spirals.

There is some evidence for such occulting matter in our galaxy.

With regard to the observed excess of velocities of recession, additional observations may remove this. Part of the excess may well be due to the motion of our own galaxy in space. The Nebula of Andromeda is approaching us.

As Galactic Phenomena

If the spirals are galactic objects, they must be a class apart from all other known types: why none in our neighborhood?

Their abhorrence of the regions of greatest star density can only be explained on the hypothesis that they are, in some unknown manner, repelled by the stars.

We know of no force adequate to produce such a repulsion, except perhaps light-pressure.

Why should this repulsion have invariably acted essentially at right angles to our galactic plane?

Why have not some been repelled in the direction of our galactic plane?

The repulsion theory, it is true, is given some support by the fact that most of the spirals observed to date are receding from us.

The space velocities of the spirals.—

As Island Universes

The spirals observed to date have the enormous average space velocity of 1200 km/sec.

In this velocity factor they stand apart from all galactic objects.

Their space velocity is one hundred times that of the galactic diffuse nebulosities, about thirty times the average velocity of the stars, ten times that of the planetary nebulae, and five times that of the clusters.

Such high speeds seem possible for individual galaxies.

Our own galaxy probably has a space velocity, relatively to the system of the spirals, of several hundred km/sec. Attempts have been made to derive this from the velocities of the spirals, but are uncertain as yet, as we have the radial velocities of but thirty spirals.

As Galactic Phenomena

Space velocities of several hundred km/sec. have been found for a few of the fainter stars.

It has been argued that an extension of radial velocity surveys to the fainter stars would possibly remove the discrepancy between the velocities of the stars and those of the spirals.

This is possible, but does not seem probable. The faint stars thus far selected for investigation have been stars of known large proper motions. They are exceptional objects through this method of selection, not representative objects.

High space velocities are the rule, not the exception, for the spirals.

High space velocities are still the exception, not the rule, for the stars of our galaxy.

Proper motions of the spirals.—Should the results of the next quarter-century show *close agreement among different observers* to the effect that the annual motions of translation or rotation of the spirals equal or exceed $0''.01$ in average value, it would seem that the island universe theory must be definitely abandoned.

A motion of 700 km/sec. across our line of sight will produce the following annual proper motions:

Distance in light years	1,000	10,000	100,000	1,000,000
Annual proper motion	$''.48$	$''.048$	$''.005$	$''.0005$

The older visual observations of the spirals have so large a probable error as to be useless for the determination of proper motions, if small; the available time interval for photographic determinations is less than twenty-five years.

The first proper motion given above should inevitably have been detected by either visual or photographic methods, from which it seems clear that the spirals can not be relatively close to us at the poles of our flattened galactic disk. In view of the hazy character of the condensations measured, I consider the trustworthy determination of the second proper motion given above impossible by present methods without a much longer time interval than is at present available; for the third and the fourth, we should need centuries.

New stars in the spirals.—Within the past few years some twenty-seven new stars have appeared in spirals, sixteen of these in the Nebula of Andromeda, as against about thirty-five which have appeared in our galaxy in the last three centuries. So far as can be judged from such faint objects, the novae in spirals have a life history similar to that of the galactic novae, suddenly flashing out, and more slowly, but still relatively rapidly, sinking again to a luminosity ten thousandfold less intense. Such novae form a strong argument for the island universe theory and furnish, in addition, a method of determining the approximate distances of the spirals.

With all its elements of simplicity and continuity, our universe is too haphazard in its details to warrant deductions from small numbers of exceptional objects. Where no other correlation is available such deductions must be made with caution, and with a full appreciation of the uncertainties involved.

It seems certain, for instance, that the dispersion of the novae in the spirals, and probably also in our galaxy, may reach at least ten absolute magnitudes, as is evidenced by a comparison of *S Andromedae* with the faint novae found recently in this spiral. A division into two magnitude classes is not impossible.

Tycho's Nova, to be comparable in absolute magnitude with some recent galactic novae, could not have been much more than ten light-years distant. If as close to us as one hundred light-years it must have been of absolute magnitude −8 at maximum; if only one thousand light-years away, it would have been of absolute magnitude −13 at maximum.

The distances and absolute magnitudes of but four galactic novae have been thus far determined; the mean absolute magnitude is −3 at maximum, and +7 at minimum.

These mean values, though admittedly resting upon a very limited amount of data, may be compared with the fainter novae which have appeared in the Nebula of Andromeda somewhat as follows: where 500,000 light-years is assumed for this spiral on the island universe hypothesis and, for comparison, the smaller distance of 20,000 light-years.

| | Apparent magnitudes ||
	Thirty galactic novae	Sixteen novae in Neb. Andromedae
At maximum	+5	+17
At minimum	+15±	+27 (?; conjectured from the analogy of the galactic novae)

| | Absolute magnitudes |||
| | Four galactic novae of known distances | Sixteen novae in Andr. if at distances of,— ||
		500,000 l.y.	20,000 l.y.
At maximum	−3	−4	+3
At minimum	+7	+6?	+13?

It will be seen from the above that, at the greater distance of the island universe theory, the agreement in absolute magnitude is quite good for the galactic and the spiral novae. If as close as 20,000 light-years, however, these novae must be unlike similar galactic objects, and of unusually low absolute magnitude at minimum. Very few stars have thus far been found as low in luminosity as absolute magnitude +13, corresponding, at this distance, to apparent magnitude 27.

The simple hypothesis that the novae in spirals represent the running down of ordinary galactic stars by the rapidly moving nebulosity becomes a possibility on this basis of distance (i. e., 20,000 light-years) for the brighter spirals are within the edges of the galactic system (Shapley).

This hypothesis of the origin of the novae in spirals is open to grave objections. It involves:

1. That the stars thus overtaken are of smaller absolute luminosity than the faintest thus far observed, with very few exceptions.

2. That these faint stars are extraordinarily numerous, a conclusion which is at variance with the results of star counts, which seem to indicate that there is a marked falling off in the number of stars below apparent magnitude 19 or 20.

As an illustration of the difficulties which would attend such a hypothesis, I have made a count of the stars in a number of areas about the Nebula of Andromeda, including, it is believed, stars at least as faint as magnitude 19.5, and find a star density, including all magnitudes, of about 6,000 stars per square degree.

If no more than 20,000 light-years distant this spiral will lie 7,000 light-years from the plane of the Milky Way, and if moving at the rate of 300 km/sec., it will sweep through 385 cubic light-years per year.

To make the case as favorable as possible for the hypothesis suggested, assume that none of the 6,000 stars per square degree are as close as 15,000 light-years, but that all are arranged in a stratum extending 5,000 light-years each way from the spiral.

Then the Nebula of Andromeda should encounter one of these stars every 520 years. Hence the actual rate at which novae have been found in this spiral would indicate a star density about two thousand times as great as that shown by the count; each star would occupy about one square second of arc on the photographic plate.

The spirals as island universes: summary.—

1. On this theory we avoid the almost insuperable difficulties involved in an attempt to fit the spirals in any coherent scheme of stellar evolution, either as a point of origin, or as an evolutionary product.

2. On this theory it is unnecessary to attempt to coordinate the tremendous space velocities of the spirals with those of the average star.

3. The spectrum of the spirals is such as would be expected from a galaxy of stars.

4. A spiral structure has been suggested for our own galaxy, and is not improbable.

5. If island universes, the new stars observed in the spirals seem a natural consequence of their nature as galaxies. Correlations between the novae in the spirals and those in our galaxy indicate distances ranging from perhaps 500,000 light-years in the case of the Nebula of Andromeda, to 10,000,000 or more light-years for the more remote spirals.

6. At such distances, these island universes would be of the same order of size as our own galaxy.

7. Very many spirals show evidence of peripheral rings of occulting matter in their equatorial planes. Such a phenomenon in our galaxy, regarded as a spiral, would serve to obliterate the distant spirals in our galactic plane, and would furnish an adequate explanation of the otherwise inexplicable distribution of the spirals.

There is a unity and an internal agreement in the features of the island universe theory which appeals very strongly to me. The evidence with regard to the dimensions of the galaxy, on both sides, is too uncertain as yet to permit of any dogmatic pronouncements. There are many points of difficulty in either theory of galactic dimensions, and it is doubtless true that many will prefer to suspend judgment until much additional evidence is forthcoming. Until more definite evidence to the contrary is available, however, I feel that the evidence for the smaller and commonly accepted galactic dimensions is still the stronger; and that the postulated diameter of 300,000 light-years must quite certainly be divided by five, and perhaps by ten.

I hold, therefore, to the belief that the galaxy is probably not more than 30,000 light-years in diameter; that the spirals are not intra-galactic objects but island universes, like our own galaxy, and that the spirals, as external galaxies, indicate to us a greater universe into which we may penetrate to distances of ten million to a hundred million light-years.

☆ ☆

Chapter Nine

The Resolution of the Issues in the Great Debate

Edwin Powell Hubble: Introduction

This chapter presents a number of the most important developments involved in the resolution of the issues central to the Great Debate and the establishment of our present view of the structure of the universe at large. The central figure in this process was Edwin Powell Hubble (1889–1953), about whom N. K. Mayall made the remark that "it is tempting to think that Hubble may have been to the observable region of the universe what the Herschels were to the Milky Way system, and what Galileo was to the solar system."[1]

Edwin Powell Hubble

Hubble, the son of a lawyer, was born in Marshfield, Missouri, but spent his early years in Kentucky, followed by high school in Wheaton, Illinois. He received his undergraduate education in astronomy at the University of Chicago, where he won letters in track, basketball, and boxing, in the last of which he was sufficiently skilled that he was asked to turn professional. After graduating in 1910, he studied jurisprudence at Oxford University as a Rhodes Scholar. Upon leaving Oxford, he taught high school for a year, but in 1914

---

[1] N. U. Mayall, "Edwin Hubble: Observational Cosmologist," *Sky and Telescope*, 13 (Jan., 1954), 85.

decided to return to the University of Chicago to pursue a doctorate in astronomy. This he received in 1917 with a thesis entitled "Photographic Investigations of Faint Nebulae," in which he concluded that spirals are extragalactic. Although offered a position at Mount Wilson Observatory, he joined the army, leaving it two years later as a major to take a position on the Mount Wilson staff. He remained at Mount Wilson until 1942, when he began four years work with the military on ballistic problems. Between 1946 and his death in 1953, he served as director of research at Mount Wilson, which in 1948 acquired the 200-inch-aperture Palomar reflector.

Hubble's Paper of 1925

The first paper by Hubble is something of a sleeper; its title, "Cepheids in Spiral Nebulae," does, however, suggest its significance for the key questions treated in this book. The events directly leading up to that paper began in 1922, when John C. Duncan, who periodically worked at Mount Wilson, discovered three variable stars in M33 (the spiral nebula in Triangulum). In 1923, Hubble succeeded in resolving the outer regions of M31 and M33 into "dense swarms of images." On 6 October 1923, Hubble detected a variable in M31. Before long, he had located a substantial number of Cepheids in M31, M33, and N.G.C. 6822. Finding the period and brightness values for these variables to be in agreement with Leavitt's period-luminosity law, he used Shapley's 1916 calculation of the zero point to estimate the distances of M31 and M33. Word of these results spread in 1924 to Curtis, Shapley, Russell, and others, but Hubble hesitated to publish his findings because they contradicted the studies of rotations by his colleague Adriaan van Maanen. Nonetheless, on 1 January 1925, Hubble's results were presented before the Washington meeting of the American Astronomical Society. Hubble was not himself in attendance, although Russell had urged him to submit his results.

A letter to Hubble from Joel Stebbins of the University of Wisconsin provides some interesting background regarding Hubble's presentation. Stebbins wrote: "On the first evening of the meeting, I happened to take dinner with Russell . . . one of the first things that he enquired about was whether you had sent in any contribution. On my answering no, he then said: 'Well, he is an ass. With a perfectly good thousand dollars available, he refuses to take it.'"[1] This was in reference to the $1,000 prize for the best paper at the meeting (half the prize eventually went to Hubble).

Stebbins and Russell, the letter goes on to state, decided to telegraph Hubble, urging him to send his main results by night letter, so that they could be made up into a paper to be presented before the conference ended. Leaving the hotel to go to a telegraph office, Russell and Stebbins stopped to pick up a telegram form at the main desk, where they noticed an envelope addressed to Russell that had been sent to him by Hubble. It contained the paper. Stebbins added in his letter: "we walked back to the group in the lobby saying that we had got quick service.... At the time, that coincidence seemed a miracle." Thus on 1 January 1925, Hubble's paper was presented by Russell.[2] As Hubble later wrote to Russell: "The real reason for my reluctance in hurrying to press was, as you may have guessed, the flat contradiction to van Maanen's rotations."[3] In his paper, Hubble used data from 22 Cepheids in M33 and 12 Cepheids in M31 to place them at a distance of about 285,000 parsecs or about 930,000 light-years. Because these are clearly "island universe" distances, Hubble's paper can be seen as a verification of the claims of Curtis in regard to the existence of island universes. Hubble's method was basically correct, although, due mainly to an error in Shapley's determination of the zero point for the period-luminosity law, Hubble's figures were too low, Andromeda being actually about 2.2 million light-years away and M33 being 2.3 million light-years distant. Later we shall see how this error was corrected.

---

[1] As quoted in Richard Berendzen and Michael Hoskin, "Hubble's Announcement of Cepheids in Spiral Nebulae," *Astronomical Society of the Pacific Leaflet No. 504* (June, 1971), p. 6.
[2] Word of Hubble's results had spread to some extent earlier. For example, he presented them at the Royal Astronomical Society in April, 1924 and they were reported in the *New York Times* in November, 1924. (I am indebted to Robert Smith for this information.)
[3] As quoted in Berendzen and Hoskin, "Hubble's Announcement," p. 11.

☆☆☆☆☆☆☆☆☆☆☆☆☆☆☆☆☆☆☆☆☆☆☆☆☆☆☆☆
Edwin Hubble, "Cepheids in Spiral Nebulae," *Popular Astronomy*, 33 (1925), 252–5.

Messier 31 and 33, the only spirals that can be seen with the naked eye, have recently been made the subject of detailed investigations with the 100-inch and 60-inch reflectors of the Mount Wilson Observatory. Novae are a common phenomenon in M 31, and Duncan has reported three variables within the area covered by M 33.[1] With these exceptions there seems to have been no definite evidence of actual stars involved in spirals. Under good observing conditions, however, the outer regions of both spirals are resolved into dense swarms of images in no way differing from those of ordinary stars. A survey of the plates made with the blink-comparator has revealed many variables among the stars, a large proportion of which show the characteristic light-curve of the Cepheids.

Up to the present time some 47 variables, including Duncan's three, and one true nova have been found in M 33. For M 31, the numbers are 36 variables and 46 novae, including the 22 novae previously discovered by Mount Wilson observers. Periods and photographic magnitudes have been determined for 22 Cepheids in M 33 and 12 in M 31. Others of the variables are probably Cepheids, judging from their sharp rise and slow decline, but some are definitely not of this type. One in particular, Duncan's No. 2 in M 33, has been brightening fairly steadily with only minor fluctuations since about 1906. It has now reached the 15th magnitude and has a spectrum of the bright line B type.

For the determinations of periods and normal curves of the Cepheids, 65 plates are available for M 33, and 130 for M 31. The latter object is too large for the area of good definition on one plate, so attention has been concentrated on three regions: around BD +41°151, BD +40°145, and a region some 45' along the major axis south preceding the nucleus.

Photographic magnitudes have been determined from twelve comparisons with selected areas No. 21 and 45, made

---

[1] *Publications of the Astronomical Society of the Pacific*, 35, 290, 1922.

with the 100-inch using exposures from 30 to 40 minutes. This procedure seemed preferable to the much longer exposures required for direct polar comparisons with the 60-inch. It involves, however, a considerable extrapolation based on scales determined from the faintest magnitudes available for the selected areas.

TABLE I.
CEPHEIDS IN M 33.

Var. No.	Period in Days	Log. P	Photographic Magnitudes	
			Max.	Min.
30	46.0	1.66	18.35	19.25
3	41.6	1.62	18.45	19.4
36	38.2	1.58	18.45	19.1
31	37.3	1.57	18.30	19.2
29	37.2	1.57	18.55	19.15
20	35.95	1.56	18.50	19.2
18	35.5	1.55	18.45	19.15
35	31.5	1.50	18.55	19.35
42	31.1	1.49	18.65	19.35
44	30.2	1.48	18.70	
40	26.0	1.41	19.00	
17	23.6	1.37	18.80	
11	23.4	1.37	18.85	
22	21.75	1.34	19.00	
12	21.2	1.33	18.80	
27	21.05	1.32	18.85	
43	20.8	1.32	18.95	
33	20.8	1.32	18.75	
10	19.6	1.29	18.80	
41	19.15	1.28	18.75	
37	18.05	1.26	18.95	
15	17.65	1.25	19.05	

TABLE II.
CEPHEIDS IN M 31.

Var. No.	Period in Days	Log. P	Photographic Magnitude Max.
5	50.17	1.70	18.4
7	45.04	1.65	18.15
16	41.14	1.61	18.6
9	38	1.58	18.3
1	31.41	1.50	18.2
12	22.03	1.34	19.0
13	22	1.34	19.0
10	21.5	1.33	18.75
2	20.10	1.30	18.5
17	18.77	1.28	18.55
18	18.54	1.27	18.9
14	18	1.26	19.1

Tables I and II give the data for the Cepheids in M 33 and M 31 respectively. No magnitudes fainter than 19.5 are recorded, because of the uncertainty involved in their precise determinations. The now familiar period-luminosity relation is conspicuously present.

For more detailed investigation of the relation, the magnitudes at maxima have been plotted against the logarithm of the period in days. This procedure is necessary, not only because of the uncertainties in the fainter magnitudes, but also because most of the fainter variables at minimum are below the limiting magnitude of the plates. It assumes that there is no relation between period and range, for otherwise a systematic error in the slope of the period-luminosity curve is introduced. Among the brighter Cepheids of M 33 the assumption appears to be allowable, for the ranges show a very small dispersion about the mean value of 0.8 magnitude. The average range and the dispersion are somewhat larger in M 31, but the data are too limited for a complete investigation.

The curve for M 33 appears to be very definite. The average deviation is about 0.1 magnitude, although a considerable systematic error is allowable in the slope. For M 31 the slope is very closely the same but the dispersion is much greater, averaging about 0.2 magnitude. This is probably greater than the accidental errors of measurement.

Shapley's period-luminosity curve[1] for Cepheids, as given in his study of globular clusters, is constructed on a basis of visual magnitudes. It can be reduced to photographic magnitudes by means of his relation between period and color-index, given in the same paper, and the result represents his original data. The slope is of the order of that for the spirals, but is not precisely the same. In comparing the two, greater weight must be given the brighter portion of the curve for the spirals, because of the greater reliability of the magnitude determinations. When this is done, the resulting values of $M - m$ are -21.8 and -21.9 for M 31 and M 33, respectively. These must be corrected by half the average ranges of the Cepheids in the two spirals, and the final values are then on the order of -22.3 for both nebulae. The corresponding distance is about 285,000 parsecs. The greatest uncertainty is probably in the zero point of Shapley's curve.

The results rest on three major assumptions: (1) The variables are actually connected with the spirals. (2) There is no serious amount of absorption due to amorphous nebulosity in the spirals. (3) The nature of Cepheid variation is uniform throughout the observable portion of the universe. As for the first, besides the weighty arguments based on analogy and probability, it may be mentioned that no Cepheids have been found on the several plates of the neighboring selected areas No. 21 and 45, on a special series of plates centered on BD + 35°207, just midway between the two spirals, nor in ten other fields well distributed in galactic latitude, for which six or more long exposures are available. The second assumption is very strongly supported by the small dispersion in the period-luminosity curve for M 33. In M 31, in spite of the somewhat larger dispersion, there is no evidence of an absorption effect to be measured in magnitudes.

These two spirals are not unique. Variables have also been found in M 81, M 101 and N.G.C. 2403, although as yet sufficient

---

[1] *Mt. Wilson Contribution* No. 151. *Astrophysical Journal*, 48, 89, 1918.

plates have not been accumulated to determine the nature of their variation.

☆ ☆

Hubble's 1926 Classification of Nebulae

After a number of years working on a classification of extra-galactic nebulae (or galaxies, to use the modern term, as Hubble did not do), Hubble published in 1926 a paper[1] in which he proposed a classification of nebulae into four groups: (1) ellipticals; (2) spirals; (3) barred spirals; and (4) irregular forms, such as the Magellanic Clouds. Only about 3% of galaxies belong to the fourth type. These types were arranged in a diagram, the so-called tuning-fork diagram, which is given below in the slightly revised form in which Hubble published it in 1936 in his *Realm of the Nebulae*.

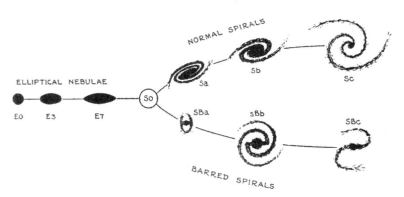

Hubble's Diagram of His Four Types of Nebulae

This classification has stood the test of time very well; it will be found in most current day astronomy texts. There is, however, one difference: Hubble seems to have believed that the direction of galactic evolution is from the base of the tuning fork to the ends of its prongs, so that the spirals are opening up. The trend today is to believe that the direction of evolution is just the opposite.

---

[1]Edwin Hubble, "Extra-galactic Nebulae," *Astrophysical Journal*, 64 (1926), 321–69.

Hubble's 1929 Paper and the Velocity-Distance Relationship

In 1929, Hubble made a discovery that Otto Struve and Velta Zebergs, writing in 1962, described as "the most spectacular discovery in astronomy that has been made in the past 60 years."[1] G. J. Whitrow has attributed a comparable significance to it; in 1972, he declared that Hubble's discovery "has come to be generally regarded as the outstanding discovery in twentieth-century astronomy."[2]

As noted previously, Slipher by 1914 had measured radial velocities for 13 spiral nebulae, finding that most show rather large recessional velocities. By 1925, Slipher had determined radial velocities for 39 galaxies. Hubble set to work with the 100-inch Mount Wilson reflector and by 1929 had radial velocities for 46 galaxies as well as approximate distances for 24 of them, having used Cepheids, novae, and blue stars to determine their distances. He combined these figures to announce the velocity-distance relationship or law and to give the "Hubble Constant." Hubble had found that when the velocities of the galaxies are corrected by eliminating the component due to the sun's motion in relation to the Milky Way, the velocities of the galaxies are all recessional and, moreover, are of magnitudes directly proportional to the distances of the galaxies. Hubble's law is simply $V = HD$, where V is the velocity of a galaxy, D its distance, and H is a constant (the Hubble constant), which Hubble originally set as 500 kilometers/second for every million parsecs in distance, i.e., for every 3.26×10^6 light-years. That is to say that for every million parsecs that a galaxy is distant from the Milky Way, it moves 500 km/sec faster. This value was eventually reduced to about 75 km/sec per million parsecs. Hubble's paper thus provided the empirical basis for the idea that we live in an expanding universe. Various theories, e.g., the big bang theory, were devised to account for this empirical result. Hubble's relationship can also be used to find both the size and age of

---

[1] Otto Struve and Velta Zebergs, *Astronomy of the 20th Century* (New York: Macmillan, 1962), p. 469.
[2] G. J. Whitrow, "Edwin Powell Hubble," *Dictionary of Scientific Biography*, vol. 6 (1972), 531.

the universe. It is worth noting that some astronomers, e.g., C. A. Wirtz, had investigated, even in the years immediately following Slipher's first detection of the radial velocities of the spirals, the possibility of detecting a pattern in these velocities, but until the late 1920s, the empirical basis for such a determination had remained unsatisfactory. Moreover, Hubble's 1929 paper does not contain a statement of the expanding universe theory; in fact, in his paper, he suggested that the effect he determined could possibly be explained in terms of a cosmology put forward by Willem de Sitter, which was a static rather than an expanding model of the cosmos.

To use Hubble's law to determine the distance of a galaxy, one secures the galaxy's radial velocity, which is done by measuring the shift in position of its spectral lines. Then divide this quantity by the value of H. The results obtained by this method are approximate, partly because the velocity of a galaxy will be influenced by the presence of other galaxies near it. And clearly this method depends heavily on the accuracy of the value of H.

Hubble's law can also be used to determine the age of the universe. Consider the relative motion of two objects from the time of the big bang until the present. The time T required for this distance D to open up between the objects will be the age of the universe. We can calculate the value of T from the Hubble law: By substituting the value for D into Hubble's law $V = HD$, we have that $V = HVT$. Dividing by V gives: $HT = 1$, which entails that $T = \frac{1}{H}$. The magnitude of T (Age of Universe) thus depends crucially on the Hubble Constant. If H = 500km/s/million parsecs, as Hubble thought in 1929, then Age = $\frac{1}{500\text{km/s/million parsecs}}$ = 2 billion yrs. If, however, we use the modern estimate of the value of H, i. e., 75km/s/million parsecs, then the age of universe must be about 13 billion years.

☆ ☆
Edwin Hubble, "A Relation between Distance and Radial Velocity among Extra-Galactic Nebulae," *National Academy of Sciences Proceedings,* 15 (1929), 168–73.

Determinations of the motion of the sun with respect to the extragalactic nebulae have involved a K term of several hundred kilometers which appears to be variable. Explanations of this paradox have been sought in a correlation between apparent radial velocities and distances, but so far the results have not been convincing. The present paper is a re-examination of the question, based on only those nebular distances which are believed to be fairly reliable.

Distances of extra-galactic nebulae depend ultimately upon the application of absolute-luminosity criteria to involved stars whose types can be recognized. These include, among others, Cepheid variables, novae, and blue stars involved in emission nebulosity. Numerical values depend upon the zero point of the period-luminosity relation among Cepheids, the other criteria merely check the order of the distances. This method is restricted to the few nebulae which are well resolved by existing instruments. A study of these nebulae, together with those in which any stars at all can be recognized, indicates the probability of an approximately uniform upper limit to the absolute luminosity of stars, in the late-type spirals and irregular nebulae at least, of the order of M (photographic) $= -6.3$[1] The apparent luminosities of the brightest stars in such nebulae are thus criteria which, although rough and to be applied with caution, furnish reasonable estimates of the distances of all extra-galactic systems in which even a few stars can be detected.

---

[1]*Mt. Wilson Contr.,* No. 324; *Astroph. J., Chicago, Ill.,* 64, 1926 (321).

TABLE I.
NEBULAE WHOSE DISTANCES HAVE BEEN ESTIMATED FROM STARS INVOLVED OR FROM MEAN LUMINOSITIES IN A CLUSTER

Object[1]		m_s	r	v	m_t	M_t
S. Mag.			0.032	+170	1.5	−16.0
L. Mag.		..	0.034	+290	0.5	17.2
N.G.C.	6822	..	0.214	−130	9.0	12.7
	598	..	0.263	−70	7.0	15.1
	221	..	0.275	−185	8.8	13.4
	224	..	0.275	−220	5.0	17.2
	5457	17.0	0.45	+200	9.9	13.3
	4736	17.3	0.5	+290	8.4	15.1
	5194	17.3	0.5	+270	7.4	16.1
	4449	17.8	0.63	+200	9.5	14.5
	4214	18.3	0.8	+300	11.3	13.2
	3031	18.5	0.9	−30	8.3	16.4
	3627	18.5	0.9	+650	9.1	15.7
	4826	18.5	0.9	+150	9.0	15.7
	5236	18.5	0.9	+500	10.4	14.4
	1068	18.7	1.0	+920	9.1	15.9
	5055	19.0	1.1	+450	9.6	15.6
	7331	19.0	1.1	+500	10.4	14.8
	4258	19.5	1.4	+500	8.7	17.0
	4151	20.0	1.7	+960	12.0	14.2
	4382	..	2.0	+500	10.0	16.5
	4472	..	2.0	+850	8.8	17.7
	4486	..	2.0	+800	9.7	16.8
	4649	..	2.0	+1090	9.5	<u>17.0</u>
Mean						−15.5

m_s = photographic magnitude of brightest stars involved.
r = distance in units of 10^6 parsecs. The first two are Shapley's values.
v = measured velocities in km./sec. N.G.C. 6822, 221, 224 and 5457 are recent determinations by Humason.
m_t = Holetschek's visual magnitude as corrected by Hopmann. The first three objects were not measured by Holetschek, and the values of m_t represent estimates by the author based upon such data as are available.
M_t = total visual absolute magnitude computed from m_t and r.

---

[1][A number of these galaxies are ones that have repeatedly appeared in these materials, e.g., N.G.C. 224 = M31 (Andromeda), N.G.C. 5194 = M51 (Whirlpool). MJC]

Finally, the nebulae themselves appear to be of a definite order of absolute luminosity, exhibiting a range of four or five magnitudes about an average value M (visual) = -15.2.[1] The application of this statistical average to individual cases can rarely be used to advantage, but where considerable numbers are involved, and especially in the various clusters of nebulae, mean apparent luminosities of the nebulae themselves offer reliable estimates of the mean distances.

Radial velocities of 46 extra-galactic nebulae are now available, but individual distances are estimated for only 24. For one other, N.G.C. 3521, an estimate could probably be made, but no photographs are available at Mount Wilson. The data are given in table 1. The first seven distances are the most reliable, depending, except for M 32 the companion of M 31, upon extensive investigations of many stars involved. The next thirteen distances, depending upon the criterion of a uniform upper limit of stellar luminosity, are subject to considerable probable errors but are believed to be the most reasonable values at present available. The last four objects appear to be in the Virgo Cluster. The distance assigned to the cluster, 2×10^6 parsecs, is derived from the distribution of nebular luminosities, together with luminosities of stars in some of the later-type spirals, and differs somewhat from the Harvard estimate of ten million light years.[2]

The data in the table indicate a linear correlation between distances and velocities, whether the latter are used directly or corrected for solar motion, according to the older solutions. This suggests a new solution for the solar motion in which the distances are introduced as coefficients of the K term, i.e., the velocities are assumed to vary directly with the distances, and hence K represents the velocity at unit distance due to this effect. The equations of condition then take the form

$$rK + X \cos \alpha \cos \delta + Y \sin \alpha \cos \delta + Z \sin \delta = v$$

Two solutions have been made, one using the 24 nebulae

---

[1] *Mt. Wilson Contr.*, No. 324; *Astroph. J., Chicago, Ill.*, 64, 1926 (321).
[2] *Harvard Coll. Obs. Circ.*, 294, 1926.

Resolution of the Great Debate 341

individually, the other combining them into 9 groups according to proximity in direction and in distance. The results are

	24 Objects	9 Groups	
X	-65 ± 50	$+3 \pm 70$	
Y	$+226 \pm 95$	$+230 \pm 120$	
Z	-195 ± 40	-133 ± 70	
K	$+465 \pm 50$	$+513 \pm 60$ km./sec. per 10^6 parsecs.
A	$286°$	$269°$	
D	$+40°$	$+33°$	
V_O	306 km./sec.	247 km./sec.	

For such scanty material, so poorly distributed, the results are fairly definite. Differences between the two solutions are due largely to the four Virgo nebulae, which, being the most distant objects and all sharing the peculiar motion of the cluster, unduly influence the value of K and hence of V_O. New data on more distant objects will be required to reduce the effect of such peculiar motion. Meanwhile round numbers, intermediate between the two solutions, will represent the probable order of the values. For instance, let $A = 277°$, $D = +36°$ (Gal. long. = $32°$, lat. = $+18°$), $V_O = 280$ km./sec., $K = +500$ km./sec. per million parsecs. Mr. Strömberg has very kindly checked the general order of these values by independent solutions for different groupings of the data.

A constant term, introduced into the equations, was found to be small and negative. This seems to dispose of the necessity for the old constant K term. Solutions of this sort have been published by Lundmark,[1] who replaced the old K by $k + lr + mr^2$. His favored solution gave $k = 513$, as against the former value of the order of 700, and hence offered little advantage.

---

[1] *Mon. Not. R. Astr. Soc.*, 85, 1925 (865-894).

TABLE 2
NEBULAE WHOSE DISTANCES ARE ESTIMATED FROM RADIAL VELOCITIES

OBJECT		v	v_s	r	m_t	M_t
N.G.C.	278	+650	−110	1.52	12.0	−13.9
	404	−25	−65	..	11.1	..
	584	+1800	+75	3.45	10.9	16.8
	936	+1300	+115	2.37	11.1	15.7
	1023	+300	−10	0.62	10.2	13.8
	1700	+800	+220	1.16	12.5	12.8
	2681	+700	−10	1.42	10.7	15.0
	2683	+400	+65	0.67	9.9	14.3
	2841	+600	−20	1.24	9.4	16.1
	3034	+290	−105	0.79	9.0	15.5
	3115	+600	+105	1.00	9.5	15.5
	3368	+940	+70	1.74	10.0	16.2
	3379	+810	+65	1.49	9.4	16.4
	3489	+600	+50	1.10	11.2	14.0
	3521	+730	+95	1.27	10.1	15.4
	3623	+800	+35	1.53	9.9	16.0
	4111	+800	−95	1.79	10.1	16.1
	4526	+580	−20	1.20	11.1	14.3
	4565	+1100	−75	2.35	11.0	15.9
	4594	+1140	+25	2.23	9.1	17.6
	5005	+900	−130	2.06	11.1	15.5
	5866	+650	−215	1.73	<u>11.7</u>	<u>−14.5</u>
Mean					10.5	−15.3

The residuals for the two solutions given above average 150 and 110 km./sec. and should represent the average peculiar motions of the individual nebulae and of the groups, respectively. In order to exhibit the results in a graphical form, the solar motion has been eliminated from the observed velocities and the remainders, the distance terms plus the residuals, have been plotted against the distances. The run of the residuals is about as smooth as can be expected, and in general the form of the solutions appears to be adequate.

The 22 nebulae for which distances are not available can be treated in two ways. First, the mean distance of the group derived from the mean apparent magnitudes can be compared with the mean of the velocities corrected for solar motion. The result, 745 km./sec. for a distance of 1.4×10^6 parsecs, falls between the two previous solutions and indicates a value for K of 530 as against the proposed value, 500 km./sec.

Secondly, the scatter of the individual nebulae can be examined by assuming the relation between distances and velocities as previously determined. Distances can then be calculated from the velocities corrected for solar motion, and absolute magnitudes can be derived from the apparent magnitudes. The results are given in table 2 and may be compared with the distribution of absolute magnitudes among the nebulae in table 1, whose distances are derived from other criteria.

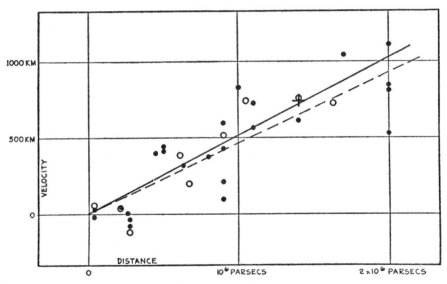

Velocity-Distance Relation among Extra-Galactic Nebulae.
Radial velocities, corrected for solar motion, are plotted against distances estimated from involved stars and mean luminosities of nebulae in a cluster. The black discs and full line represent the solution for solar motion using the nebulae individually; the circles and broken line represent the solution combining the nebulae into groups; the cross represents the mean velocity corresponding to the mean distance of 22 nebulae whose distances could not be estimated individually.

N. G. C. 404 can be excluded, since the observed velocity is so small that the peculiar motion must be large in comparison with the distance effect. The object is not necessarily an exception, however, since a distance can be assigned for which the peculiar motion and the absolute magnitude are both within the range previously determined. The two mean magnitudes, −15.3 and −15.5, the ranges, 4.9 and 5.0 mag., and the frequency distributions are closely similar for these two entirely independent sets of data; and even the slight difference in mean magnitudes can be attributed to the selected, very bright, nebulae in the Virgo Cluster. This entirely unforced agreement supports the validity of the velocity-distance relation in a very evident matter. Finally, it is worth recording that the frequency distribution of absolute magnitudes in the two tables combined is comparable with those found in the various clusters of nebulae.

The results establish a roughly linear relation between velocities and distances among nebulae for which velocities have been previously published, and the relation appears to dominate the distribution of velocities. In order to investigate the matter on a much larger scale, Mr. Humason at Mount Wilson has initiated a program of determining velocities of the most distant nebulae that can be observed with confidence. These, naturally, are the brightest nebulae in clusters of nebulae. The first definite result,[1] $v = +3779$ km./sec. for N. G. C. 7619, is thoroughly consistent with the present conclusions. Corrected for the solar motion, this velocity is +3910, which, with $K = 500$, corresponds to a distance of 7.8×10^6 parsecs. Since the apparent magnitude is 11.8, the absolute magnitude at such a distance is −17.65, which is of the right order for the brightest nebulae in a cluster. A preliminary distance, derived independently from the cluster of which this nebula appears to be a member, is of the order of 7×10^6 parsecs.

New data to be expected in the near future may modify the significance of the present investigation or, if confirmatory, will lead to a solution having many times the weight. For this reason it is thought premature to discuss in detail the obvious

---

[1] These PROCEEDINGS, 15, 1929 (167).

consequences of the present results. For example, if the solar motion with respect to the clusters represents the rotation of the galactic system, this motion could be subtracted from the results for the nebulae and the remainder would represent the motion of the galactic system with respect to the extra-galactic nebulae.

The outstanding feature, however, is the possibility that the velocity-distance relation may represent the de Sitter effect, and hence that numerical data may be introduced into discussions of the general curvature of space. In the de Sitter cosmology, displacements of the spectra arise from two sources, an apparent slowing down of atomic vibrations and a general tendency of material particles to scatter. The latter involves an acceleration and hence introduces the element of time. The relative importance of these two effects should determine the form of the relation between distances and observed velocities; and in this connection it may be emphasized that the linear relation found in the present discussion is a first approximation representing a restricted range in distance.

☆☆☆☆☆☆☆☆☆☆☆☆☆☆☆☆☆☆☆☆☆☆☆☆☆☆☆☆☆☆

The Tension between Hubble's and van Maanen's Researches

Earlier you read Adriaan van Maanen's 1916 report of his observation that M101 rotates at a rate of one rotation per 85,000 years. This conclusion, as you also saw, created problems for the island universe theory. The tensions between van Maanen's observations and the island universe theory further increased as van Maanen reported subsequent detections of rotations in other nebulae. By 1923, he had reported rotations in no less than seven spirals and was able, moreover, to cite supportive observations by a few other observers as well as theoretical support from James Jeans of England.

Hubble, who was a colleague of van Maanen at Mount Wilson Observatory, was acutely aware of the degree to which van Maanen's results went against his evidence that spirals are other universes. In fact, we know from one of Hubble's letters that he hesitated to publish his 1924–1925 paper on the distances of M31 and M33 as determined by Cepheids because his distances were far beyond what could be reconciled with van Maanen's rotations. Shapley was also influenced by van Maanen's results; in fact, in his 1969 autobiography, he explained his rejection of island universes in the great debate as due to van Maanen's result. Shapley wrote: "They wonder why Shapley made this blunder. The reason he made it was that Van [sic] Maanen was his friend and he believed in friends!"[1]

After Hubble's 1925 determination of the distances of M31 and M33 and subsequently of other galaxies, doubts about the reliability of van Maanen's observations became widespread. Nonetheless, van Maanen stood by his results and even measured more rotations. Hubble, whose exasperation was mounting, decided in the early 1930s to examine some of van Maanen's observations for himself. Upon doing so, he found no evidence of rotations. He proceeded to prepare a strongly worded attack on van Maanen's results, but at the urging of the director at Mount Wilson, he agreed to a compromise that consisted of the publication of two papers, which appeared together in the 1935 issue of *The Astrophysical Journal*. Hubble began his paper by stating: "The outstanding discrepancy in the current conception of nebulae as extra-galactic systems lies in the large angular rotations announced more than a decade ago by Dr. van Maanen. The extreme significance of the phenomena, if real, has led the writer to remeasure four of the principal nebulae, M 81, M 51, M 33, and M 101."[2] After recording his determinations, he stated that the "results in all cases are comparable with the uncertainties of the measures, estimated as of the order of 0.0002mm," i.e., any shifts detected were of a magnitude such as would result from the imprecision of the instrumentation and were far smaller than the shifts van Maanen had reported. Hubble also reported that as a check, he had asked two colleagues, Walter Baade and Seth Nicholson, to make measurements as well. Their results, Hubble reported, agreed with his own, rather than van Maanen's. In response,

---

[1] Harlow Shapley, *Through Rugged Ways to the Stars* (New York: Charles Scribner's Sons, 1969), p. 80.
[2] Edwin Hubble, "Angular Rotations of Spiral Nebulae," *Astrophysical Journal*, 81 (1935), 334–5:334.

van Maanen reported on a number of recent determinations he had made, noting that in these cases the amount of rotation was substantially smaller than he had detected earlier. Then van Maanen stated: "In consideration of the difficulty of avoiding systematic errors in this special problem, these results, together with the measures of Hubble, Baade, and Nicholson which are given in the preceding article, make it desirable to view the motions with reserve. Although my own measures of recent plates ... show considerably smaller values of the apparent rotational component than those first obtained, the persistence of the positive sign is very marked and will require the most searching investigation in the future."[1]

Historians of astronomy have made a number of investigations as to the possible sources of van Maanen's errors. The conclusion reached has been, in the words of Robert Smith, "van Maanen was 'seeing' what he expected to see."[2] The case of van Maanen indicates that although the giant Mount Wilson 60- and 100-inch-aperture telescopes contributed importantly to the achievements of Shapley, Hubble, and others, they worked no magic for van Maanen. Great telescopes may be a necessary, but they are scarcely a sufficient condition for truly creative contributions to astronomy.

Interstellar Obscuring Matter

A question that had bothered astronomers for about a century or more was settled in 1930; this was the question of the existence of interstellar obscuring matter, which would act to dim or to block out the light of stars or to alter their color.

By around 1920, studies carried out by E. E. Barnard of "dark nebulae" made it seem probable that there are large clouds of dark matter in space, e.g., the Coal Sack. But is there matter besides this, perhaps in a tenuous state? Kapteyn in the first decade of the twentieth century had uncovered some

---

[1] Adriaan van Maanen, "Internal Motions in Spiral Nebulae," *Astrophysical Journal*, 81 (1935), 334–5:334–5.
[2] Robert Smith, *The Expanding Universe: Astronomy's "Great Debate" 1900–1931* (Cambridge: Cambridge University Press, 1982), p. 129.

evidence of this; he had found that there are fewer dim stars than would be expected if one assumes that stars are homogeneously distributed throughout the Milky Way. Moreover, he reported that the more distant stars seem to be reddened, which effect would result were interstellar matter acting to dim the blue end of the spectrum, as the laws of light scattering lead one to expect. These laws indicate that light of short wavelength, say blue, tends to be scattered more than that of longer wavelength, for example, red. Also it was found that the spectra of certain stars contain dark spectral lines that are shifted due to radial velocities—and other dark lines showing no shift, indicating the possible existence of stationary interstellar obscuring matter in the intermediate regions of space. Around 1916, however, Shapley investigated whether there is a reddening of the stars of the globular clusters, which he argued are very distant. Finding no reddening in those stars, he argued that space must be transparent. Most astronomers seem to have been inclined to accept this conclusion.

In the "Great Debate" of 1920, Curtis had suggested the existence of interstellar obscuring matter. His argument was that dark lanes are seen in the edges of some spirals and if the Milky Way is also a spiral, obscuring matter should lie in its plane. This would explain why no spirals are seen in the plane of the Milky Way. When Hubble showed the correctness of the island universe theory, this supported the claims made by Curtis for obscuring matter. As late as 1929 and 1930, however, Shapley could still muster arguments for the transparency of space, except for such dark clouds as the Coal Sack.

In 1930, very strong evidence for the presence of obscuring matter was presented by Robert J. Trumpler of Lick Observatory.[1] Trumpler's work was based on a study of open clusters, which are also called galactic clusters. Open clusters are clusters of stars in which the stars are widely separated from each other, e.g., the Pleiades. Such clusters generally lie in the plane of the Milky Way. Trumpler determined the distances of about one hundred open clusters, using the

---

[1] Robert J. Trumpler, "Preliminary Results on the Distances, Dimensions, and Space Distribution of Open Star Clusters," *Lick Observatory Bulletin*, 14, no. 420 (1930), 154–88.

brightness and spectral types of their stars as distance criteria because Cepheids could not generally be found in them. He pointed out that if we assume that the apparent brightness of open clusters is a function of their distance and consequently use their apparent brightness to determine their distances, then it follows that the nearer open clusters are only half the size of the more remote ones. To put it differently, this assumption seemed to entail the result that the farther away an open cluster is, the larger it is—which, of course, made no sense. Consequently, he argued that interstellar obscuring matter is dimming the more distant open clusters. And this is correct.

An analogy will help in understanding this. Imagine looking down a street some night at a row of street lights. Were you able to measure accurately, you would expect to find that both the apparent areas and apparent brightnesses of the street lights would decrease in proportion to $1/\text{distance}^2$. But suppose you find that the lights get smaller less rapidly than they grow dimmer; that even the dimmest ones are of relatively large area. What could you conclude? Try to recall when you have seen such an effect. This is the effect seen in a fog where lights grow dimmer more rapidly than they decrease in area. Trumpler had found an analogous effect; in fact, he had found a sort of interstellar fog or obscuring matter. Trumpler's result explained why we see almost no spirals in the plane of the Milky Way and also played a part in the scaling down of the size of the Milky Way from Shapley's diameter of 300,000 light-years to about 100,000 light-years, which is the current value. This interstellar obscuring matter has created all sorts of problems for astronomers by limiting the extent of the universe they are able to see.

Two Important Discoveries Made by Walter Baade

In the early 1940s and again in the early 1950s, Walter Baade made major discoveries relating to the matters we have been discussing, and also to the question of the uniformity of nature among celestial objects.

As of the early 1940s, although the spiral arms of Andromeda had been resolved, no one had succeeded in resolving its central region. Walter Baade, working with the telescopes on Mount Wilson and taking advantage of the improved observational possibilities created by the war-time "black outs," was able to resolve the central region of Andromeda. When he did this, he found that the nucleus contains mainly bright red stars, whereas the stars seen in the spiral arms are dominantly bluish. This led him to urge that there are two different classes of stars:

Population I Stars, which are characteristically in the spiral arms and are younger stars than Population II stars; and
Population II Stars, which are older stars characteristically found in the galactic nucleus, in globular clusters, and in elliptical galaxies.

Baade also asserted that these two types of stars have somewhat different Hertzsprung-Russell diagrams.[1] Thus the assumption of the uniformity of nature as applied to the stars suffered a set-back. The differences between these two types of stars also played a role in misleading Shapley and Curtis, who assumed uniformity of nature in their debate.

Baade made an even more important discovery in the early 1950s. It also related to the uniformity of nature and specifically to the Cepheid variables and to the zero point in the period-luminosity law. Even in the early 1930s, some evidence indicated that the period-luminosity zero point might be somewhat off. For example, Hubble noted that the globular clusters around Andromeda tend to be about one and one half magnitudes dimmer than those around the Milky Way, at least if the then current values for the distance of Andromeda are assumed. Moreover, although Andromeda appeared to be structurally similar to the Milky Way, it was calculated to be substantially smaller. Nonetheless, in his *Realm of the Nebulae* (1936), Hubble stated concerning Shapley's zero point for the period-luminosity law that "Further revision is expected to be of minor importance." (p.16)

---

[1] Walter Baade, "The Resolution of Messier 32, NGC 205, and the Central Portion of the Andromeda Nebula," *Astrophysical Journal*, 100 (1944), 137–46.

In 1952, Baade found that the necessary revision was not so minor.[1] Using Shapley's formulation of the period-luminosity law, Baade urged that the RR Lyrae variable stars in Andromeda ought to be visible in the Palomar 200-inch telescope. The RR Lyrae stars are short-period variables—periods of about 1/2 day—that seemed to fit the period-luminosity law. These variables are usually found in globular clusters. According to Shapley's formulation of the period-luminosity law, the RR Lyrae stars should have been visible with the Palomar telescope, which at 30-minute exposure times could photograph stars as dim as the 22.4 magnitude.

From the failure of the Palomar telescope to detect such RR Lyrae stars in Andromeda, Baade postulated that just as there are Population I and II type stars, so also there are two types of Cepheids:

Type I Cepheids are characteristically found in the spiral arms, whereas

Type II Cepheids are generally located in globular clusters and in galactic nuclei. Type I Cepheids are about four to five times brighter than Type II Cepheids. One of the problems with the period-luminosity law as discussed by Shapley is that it was based on Type II Cepheids in globular clusters, whereas the distance of Andromeda had been calculated from the Cepheids in its spiral arms. Baade realized that, because of this, the luminosities of the Andromeda spiral arm Cepheids must be about four times greater than had been thought and that consequently Andromeda must be about twice as distant as previously believed. A similar effect applied to all spirals. Thus Baade's result led to a doubling of the size of the universe and also to a doubling of its age. This revision in the period-luminosity law and its zero point was hardly minor. In light of Baade's work, the period-luminosity law for Cepheids now takes the form indicated in the next diagram:

---

[1] Walter Baade, "A Revision of the Extra-Galactic Distance Scale," *Transactions of the International Astronomical Union*, 8 (1952), 397–8.

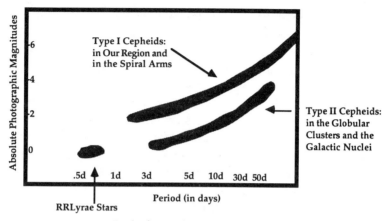

Period-Luminosity Diagram
for Cepheids and RR Lyrae Variables

Harlow Shapley's Reaction to the Resolution of the Issues in the Great Debate

It is interesting to examine Shapley's reaction to the discoveries announced by Hubble in his 1924 paper. In 1930, Shapley published his *Star Clusters*, in which he suggested that recent researches on the universe "seem to leave no doubt that our galactic system is extremely large compared with typical external systems." He went on to add: "Among the hundreds of thousands, possibly millions, ... of island universes ... our own system tends to be continental in dimensions."[1]

In 1969, Shapley published his autobiographical *Through Rugged Ways to the Stars*, in one chapter of which he reminisced about the "Great Debate":

> During the Mount Wilson years my series of papers on "Colors and Magnitudes in Stellar Clusters," continued to develop. ... Heber D. Curtis ... , who was one of the best observers and talkers about the universe, was skeptical concerning the Mount Wilson work. He started out as a classicist, which appealed to me but didn't help much when he began talking about where the globular clusters are. This was not jealousy on his part; he was just doubtful as to whether the spirals were going down the right groove. His

---

[1] Harlow Shapley, *Star Clusters* (New York: McGraw-Hill, 1930), p. 179.

skepticism included doubt about Van [sic] Maanen's work to some extent, and especially about mine on clusters which had led rather naturally and cleanly to my new method of measuring big distances and therefore to my new views on the size of the universe and how it is put together. If one looks at the papers Curtis wrote at the time, one wonders how he got that way, because the evidence is so much against his conservative conclusion.[1]

Later in the chapter, Shapley added:

> As for the actual "debate," I must point out that I had forgotten about the whole thing long ago, and nobody had mentioned it to me for many years. Then, beginning about eight or ten years ago, it was talked about again. To have it come up suddenly as an issue, and as something historic, was a surprise, for at the time I had just taken it for granted. Curtis was right partly, and I was right partly. Then this Great Debate was featured as parts of two chapters of *Astronomy of the 20th Century* (1962) by Otto Struve and Velta Zebergs and is described in any number of other publications. I don't know who first dug it up or when, and I'm also puzzled as to why. I don't think the word "debate" was used at the time (in 1920). Actually it was a sort of symposium, a paper by Curtis and a paper by me, and a rebuttal apiece. Now I would know how to dodge things a little better. Curtis did a moderately good job. Some of his science was wrong, but his delivery was all right. As I have said, he was a classicist.
>
> There were two or three hundred people present; the Academy was not as big as it is now. Not many astronomers were there. . . .
>
> Anyway, it was a pleasant meeting, and our subject matter was the scale of the universe. That was what I was prepared to talk about and did talk about, and I think I won the "debate" from the standpoint of the assigned subject matter. I was right and Curtis was wrong on the main point—the scale, the size. It is a big universe, and he viewed it as a small one. From the beginning Curtis picked on another matter: are the spiral galaxies inside our system or outside? He said that they are outside systems. I said, "I don't know what they are, but according to certain evidences they are not outside."
>
> But that was not the assigned subject. Curtis, having set

---

[1] Harlow Shapley, *Through Rugged Ways to the Stars* (New York: Charles Scribner's Sons, 1969), p. 75.

up this straw man, won on that. I was wrong because I was banking on Van Maanen's measures of large proper motions in spirals. If you have large proper motions you are dealing with things near at hand. I consider this as a blunder of mine because I faithfully went along with my friend Van Maanen and *he* was wrong on the proper motions of galaxies—that is, their cross motions. Although Curtis and Hubble and some others discredited Van Maanen's measures and questioned his conclusions, I stood by Van Maanen. I was wrong, not so much in any statement that I made, but in the inference that the spirals must be inside . . . our system.

So it was a double win and a double loss. In a sense we were talking about two different things. The most nearly correct description of the affair, which has been much misrepresented, is by Struve and Zebergs. They have it practically the way I have told it here, but a good many others have not got it straight. They wonder why Shapley made this blunder. The reason he made it was that Van Maanen was his friend and he believed in friends!

As I remember it, I read my paper and Curtis presented his paper. . . . Then I replied, and then he replied. . . .

Later I too recognized that Van Maanen's measurements were wrong. Knut Lundmark came over from Sweden to go through the measuring of the rotations of the galaxies, and he could not verify Van Maanen's figures. But he was a gentleman; he did not want to criticize them publicly. When I found it was a false trail I switched immediately to the truth; I have been complimented more than once on that since. Curtis also yielded pretty soon on the scale of the universe.

Our papers were published by the National Research Council, which was probably the hand of Hale again. . . . When the papers finally appeared, they were somewhat different from the actual speeches. There was some yielding on both sides.

The strongest comment I can make about the affair now is how little I remember about it, although it was important in my career. At the time the Harvard job was hovering. . . . (pp. 79–81)

At about the same time that Shapley's autobiography appeared, the historian of astronomy Richard Berendzen was engaged in research on the history of the great debate and on

Shapley's early career. Berendzen's review[1] pointed out such an array of errors in Shapley's remarks on the great debate that the review can be seen as confirming Shapley's comment on "how little I remember about it."

Edwin Hubble's *The Realm of the Nebulae*

In 1936, Edwin Hubble published his *Realm of the Nebulae*, in which he summarized the state of stellar astronomy at that time. Based on his 1935 Silliman Lectures at Yale, this book presented a widely accessible report on the many advances that had enriched this area. Hubble's expository skills as well as his central role in a number of those developments led many persons to recognize his book as a classic of stellar astronomy. For its 1958 reprinting, the distinguished astronomer Allan Sandage wrote a new preface in which, after noting some changes in stellar astronomy in the intervening years, he concluded by stressing: "But these are changes in numerical detail and not in fundamental philosophy or direction of attack. Hubble's original approach to observational cosmology remains." As a sidelight, it is interesting to note one change not mentioned by Sandage. This is that by 1958 the term "nebulae" itself had taken on yet another new meaning. Whereas Hubble distinguished in his book between galactic nebulae and extragalactic nebulae, specifying that he would use the term in the latter sense (pp. 16-17), standard usage for "nebula" has come to be restricted to nebulous patches located within the Milky Way, the term "galaxy" being used for what Hubble called "extragalactic nebulae." Consequently, were Hubble's book renamed so as to conform to recent usage, it would be called *The Realm of the Galaxies*.

The first section of Chapter One of *Modern Theories of the Universe from Herschel to Hubble* consists of a presentation of some of the views Hubble championed regarding the proper methodology for astronomy. This presentation was derived from a number of Hubble's publications, including his *Realm of the Nebulae*. We shall now look at this subject again,

---

[1] Richard Berendzen, "[Review of] Harlow Shapley, *Through Rugged Ways to the Stars*," *Journal for the History of Astronomy*, 1 (1970), 85-7.

specifically in reference to this book so as to highlight some of Hubble's themes. You should ask yourself whether his claims are supported by the historical materials that you have studied.

In his preface to *The Realm of the Nebulae,* Hubble stressed the importance of powerful telescopes:

> The conquest of the Realm of the Nebulae is an achievement of great telescopes. It began with the identification of nebulae as independent stellar systems, comparable to our own system of the Milky Way. Once the nature of the nebulae was known, methods of estimating distances were readily developed, and the new field was open to investigation. (p. ix)

In his introduction, he commented more specifically on the nature of astronomical method, remarking in its opening sentence that "Science is the one human activity that is truly progressive." One of the reasons he cited for such progress is the following:

> This remarkable attribute of science is bought at a price—the strict limitation of the subject matter. "Science," as Campbell remarks, "deals with judgments concerning which it is possible to obtain universal agreement." These data are not individual events, but the invariable associations of events or properties which are known as laws of science. Agreement is secured by means of observation and experiment. The tests represent external authorities which all men must acknowledge, by their actions if not by their words, in order to survive. (p. 1)

Hubble went on to state that science "is necessarily barred from the world of values. There no external authority is known. Each man appeals to his private god and recognizes no superior court of appeal." (p. 1) In support for this position, Hubble cited the following passage from George Sarton:

> The saints of today are not necessarily more saintly than those of a thousand years ago; our artists are not necessarily greater than those of early Greece; they are likely to be inferior; and, of course, our men of science are not necessarily more intelligent than those of old; yet one thing is certain, their knowledge is at once more extensive and more

accurate. The acquisition and systemization of positive knowledge is the only human activity that is truly cumulative and progressive. (As quoted on p. 2)

In a subsequent section of the introduction to his *Realm of the Nebulae*, Hubble distinguished on methodological grounds between two types of astronomers: "One emphasizes the observational approach, the other, the theoretical point of view. The observer commonly starts by accumulating an isolated group of data, together with their estimated uncertainties." (pp. 2-3) Having secured satisfactory observations, the observer "selects from the possible relations, the simplest one that is consistent with the body of general knowledge." (p. 3) After illustrating this procedure in regard to his own work on the velocity-distance relationship, Hubble stressed the permanent character of observational results: "The observations and the laws which express their relations are permanent contributions to the body of knowledge [whereas] theories change with the spreading background." (p. 4) He contrasted this feature of observational results with those obtained by theoreticians: "Many theories are formulated but relatively few endure the tests. ... less competent minds are embarrassed by the custom of testing predictions." (p. 5) After admitting that his sharp distinction between observationalists and theoreticians is sometimes less clear in actual practice, he stressed that "One of the few universal characteristics [of scientific researchers] is healthy skepticism toward unverified speculations. These are regarded as topics for conversation until tests can be devised." (p. 6) At the conclusion of his introduction, Hubble specified another reason for the progressive character of astronomy: "Astronomy, as the other [scientific] disciplines, has its own technical vocabulary of words and phrases with precise definitions. The terms always carry the same significance and substitutes are not employed. Variety is sacrificed for precision." (p. 7) Hubble then proceeded to present some basic concepts of stellar astronomy, including the standards of measurement used, how stellar magnitudes are defined, and Shapley's determination of the scale for the period-luminosity relationship, concerning which Hubble predicted: "Further revision is expected to be of minor importance." (p. 16)

Much of Hubble's first chapter consists of a historical survey of the development of nebular studies, in which he repeated his emphasis on the contribution made by the great telescopes: "But now, thanks to great telescopes, we know something of [the nature of the nebulae], something of their real size and brightness...." (p. 20) In this context, he stressed the crucial role of distance determinations in the history of stellar astronomy; in understanding the nebulae: "The distances were the essential data. Until they were found, no progress was possible." (p. 22) As part of his historical account, Hubble remarked that

> ... the conception of a stellar system, isolated in space, was formulated as early as 1750. The author was Thomas Wright (1711-86)[,] an English instrument maker and private tutor. ... Wright's speculations went beyond the Milky Way. A single stellar system, isolated in the universe, did not satisfy his philosophical mind. He imagined other, similar systems and, as visible evidence of their existence, referred to the mysterious clouds called "nebulae." (p. 23)

Turning to Kant, Hubble recounted that "Five years later, Immanuel Kant (1724-1804) developed Wright's conception in a form that endured, essentially unchanged, for the following century and a half." (p. 23) After detailing Kant's conceptions, Hubble added the historical remark: "The theory, which came to be called the theory of island universes, found a permanent place in the body of philosophical speculation. The astronomers themselves took little part in the discussions: they studied the nebulae." (p. 25) Hubble then praised the observations and catalogues of William and John Herschel as well as Huggins's spectroscopic results, suggesting that two crucial observational results in unraveling the question of island universes were the detection of S Andromedae, the bright nova sighted in the Andromeda nebula in 1885, and Vesto Slipher's measurements around 1914 of the radial velocities of various spirals. And Hubble added: "The solution came ten years later, largely with the help of a great telescope, the 100-inch [Mount Wilson] reflector...." (p.28)

The body of Hubble's *Realm of the Nebulae* consists of a

clear and interesting presentation of the remarkable developments that had occurred in the previous two decades and to which he had contributed so much. The final two sentences of Hubble's book illustrate again both Hubble's high esteem for astronomical observation as well as his confidence in the progressive character of astronomy: "The search will continue. Not until the empirical resources are exhausted, need we pass on to the dreamy realms of speculation." (p. 202)

Epilogue

Humanity and the Universe: Some Quotations

Scriptural Passages

Genesis 1:14–18
And God said, "Let there be lights in the firmament of the heavens to separate the day from the night; and let them be for signs and for seasons and for days and years, and let them be lights in the firmament of the heavens to give light upon the earth." And it was so. And God made the two great lights, the greater to rule the day, and the lesser light to rule the night; he made the stars also. And God set them in the firmament of the heavens to give light upon the earth, to rule over the day and over the night, and to separate the light from the darkness. And God saw that it was good.[1]

Psalms 8:1–4
O Lord, our Lord,
how majestic is thy name in all the earth!
Thou whose glory above the heavens is chanted
    by the mouth of babes and infants,
thou hast founded a bulwark because of thy foes,
    to still the enemy and the avenger.
When I look at thy heavens, the work of thy fingers,
    the moon and the stars which thou hast established;
what is man that thou art mindful of him,
    and the son of man that thou dost care for him?

Psalms 19:1–4
The heavens are telling the glory of God:
and the firmament proclaims his handiwork.
Day to day pours forth speech,
    and night to night declares knowledge.

---

[1] This and the following passages from Scripture are taken from *The New Oxford Annotated Bible with the Apocrypha*, ed. Herbert G. May and Bruce M. Metzger (New York: Oxford University Press, 1962).

There is no speech, nor are there words;
   their voice is not heard;
yet their voice goes out through all the earth,
   and their words to the end of the world.

Colossians I:15–20
He [Christ] is the image of the invisible God, the first-born of all creation; for in him all things were created, in heaven and on earth, visible and invisible, whether thrones or dominions or principalities or authorities—all things were created through him and for him. He is before all things, and in him all things hold together. He is the head of the body, the church; he is the beginning, the first-born from the dead, that in everything he might be preeminent. For in him all the fulness of God was pleased to dwell, and through him to reconcile to himself all things, whether on earth or in heaven, making peace by the blood of his cross.

Ancient and Medieval Periods

Plato, *Timaeus* 47b–c
God invented and gave us sight to the end that we might behold the courses of intelligence in the heaven, and apply them to the courses of our own intelligence which are akin to them, . . . and that we, learning them and partaking of the natural truth of reason, might imitate the absolutely unerring courses of God and regulate our own vagaries.[1]

Aristotle, *Parts of Animals*, Bk I, ch. 5
Both departments, however, have their special charm. The scanty conceptions to which we can attain of celestial things give us, from their excellence, more pleasure than all our knowledge of the world in which we live; just as a half glimpse of persons that we love is more delightful than a leisurely view of other things, whatever their number and dimensions. On the

---

[1]Translated by Benjamin Jowett and published in *The Collected Dialogues of Plato, Including the* Letters, ed. Edith Hamilton and Huntington Cairns (New York: Random House, 1963), p. 1175.

other hand, in certitude and in completeness our knowledge of terrestrial things has the advantage.[1]

Lucretius (c. 99–55 B.C.), *On the Nature of the Universe*, Bk. II
Take first the pure and undimmed lustre of the sky and all that it enshrines; the stars that roam across its surface, the moon and the surpassing splendour of the sunlight. If all these sights were now displayed to mortal view for the first time by a swift unforeseen revelation, what miracle could be recounted greater than this? What would men before the revelation have been less prone to conceive as possible? Nothing, surely. ... For the mind wants to discover by reasoning what exists in the infinity of space that lies out there, beyond the ramparts of this world—that region into which the intellect longs to peer and into which the free projection of the mind does actually extend its flight.[2]

Cicero (106–43 B.C.)
In the heavens there is nothing fortuitous, unadvised, inconstant, or variable; all there is order, reason, and constancy.[3]

Plutarch (c. 45–125), *On the Tranquility of the Mind*
Alexander wept when he heard from Anaxarchus that there was an infinite number of worlds; and his friends asking him if any accident had befallen him, he returns this answer; "Do you not think it a matter worthy of lamentation that, when there is such a vast multitude of them, we have not yet conquered one?"[4]

---

[1] Translated by William Ogle and published in *The Basic Works of Aristotle*, ed. Richard McKeon (New York: Random House, 1941), p. 656.
[2] Lucretius, *The Nature of the Universe*, trans. Ronald Latham (Baltimore: Penguin, 1963), pp. 90–1.
[3] As quoted in Henry C. King, *The Background of Astronomy* (London: Watts, 1957), p. 90.
[4] John Bartlett, *Familiar Quotations*, 12th ed., ed. Christopher Morley (Boston: Little, Brown and Company, 1951), p. 1116.

Saint Francis of Assisi, from his "Canticle of the Sun"
Be praised, my Lord, with all thy works whate'er they be,
    Our noble Brother Sun especially,
    Whose brightness makes the light by which we see,
    And he is fair and radiant, splendid and free,
    A likeness and a type, Most High, of Thee.
Be praised, my Lord, for Sister Moon and every Star
    That thou hast formed to shine so clear from heaven afar.[1]

Sixteenth and Seventeenth Century

William Shakespeare, *Hamlet*, I, V, 66
There are more things in heaven and earth, Horatio,
Than are dreamt of in your philosophy.[2]

John Donne, "Anatomie of the World, The First Anniversary," lines 279–80.
Man hath weave'd out a net, and this net throwne
Upon the Heavens, and now they are his owne.[3]

Blaise Pascal (1623–1662)

From Pascal's *Pensées*, #72 (Brunschvicg numeration) = #199 (Lafuma numeration)
Let man then contemplate the whole of nature in her full and grand majesty, and turn his vision from the low objects which surround him. Let him gaze on that brilliant light, set like an eternal lamp to illumine the universe; let the earth appear to him a point in comparison with the vast circle described by the sun; and let him wonder at the fact that this vast circle is itself but a very fine point in comparison with that described by the stars in their revolution round the firmament. But if our view be arrested there, let our imagination pass beyond; it will

---

[1] *The Little Flowers of St. Francis*, trans. Raphael Brown (New York: Image, 1958), p. 317.
[2] *The Oxford Shakespeare*, ed. G. R. Hibbard (Oxford: Oxford University Press, 1987), p. 194.
[3] John Donne, *Poetical Works*, ed. Herbert J. C. Grierson (London: Oxford University Press, 1966), pp. 215–16.

sooner exhaust the power of conception than nature that of supplying material for conception. The whole visible world is only an imperceptible atom in the ample bosom of nature. No idea approaches it. We may enlarge our conceptions beyond all imaginable space; we only produce atoms in comparison with the reality of things. It is an infinite sphere, the centre of which is everywhere, the circumference nowhere. In short it is the greatest sensible mark of the almighty power of God, that imagination loses itself in that thought.

Returning to himself, let man consider what he is in comparison with all existence; let him regard himself as lost in this remote corner of nature; and from the little cell in which he finds himself lodged, I mean the universe, let him estimate at their true value the earth, kingdoms, cities, and himself. What is a man in the Infinite?[1]

From Pascal's *Pensées*, #205 (Brunschvicg) = #68 (Lafuma)

When I consider the short duration of my life, swallowed up in the eternity before and after, the little space which I fill, and even can see, engulfed in the infinite immensity of spaces of which I am ignorant, and which know me not, I am frightened, and am astonished at being here rather than there; for there is no reason why here rather than there, why now rather than then. Who has put me here? By whose order and direction have this place and time been allotted to me? ....

From Pascal's *Pensées*, #206 (Brunschvicg) = #201 (Lafuma)
The eternal silence of these infinite spaces frightens me.

From Pascal's *Pensées*, # 242 (Brunschvicg) = #781 (Lafuma)

I admire the boldness with which these persons undertake to speak of God. In addressing their argument to infidels, their first chapter is to prove Divinity from the works of nature. I should not be astonished at their enterprise, if they were addressing their argument to the faithful; for it is certain that

---

[1]This and the following quotations from Pascal are from his *Pensées* as trans. W. F. Trotter, which translation cites each item by the number in the ordering of Léon Brunschvicg. The numberings of the *pensées* assigned in the ordering of Louis Lafuma are also provided.

those who have the living faith in their heart see at once that all existence is none other than the work of the God whom they adore. But for those in whom this light is extinguished, and in whom we purpose to rekindle it, persons destitute of faith and grace, who, seeking with all their light whatever they see in nature that can bring them to this knowledge, find only obscurity and darkness; to tell them that they have only to look at the smallest things which surround them, and they will see God openly, to give them, as a complete proof of this great and important matter, the course of the moon and planets, and to claim to have concluded the proof with such an argument, is to give them ground for believing that the proofs of our religion are very weak. And I see by reason and experience that nothing is more calculated to arouse their contempt.

From Pascal's *Pensées*, #346 (Brunschvicg) = #759 (Lafuma)
Thought constitutes the greatness of man.

From Pascal's *Pensées*, #347 (Brunschvicg) = #200 (Lafuma)
Man is but a reed, the most feeble thing in nature; but he is a thinking reed. The entire universe need not arm itself to crush him. A vapour, a drop of water suffices to kill him. But, if the universe were to crush him, man would still be more noble than that which killed him, because he knows that he dies and the advantage which the universe has over him; the universe knows nothing of this.

All our dignity consists, then, in thought. By it we must elevate ourselves, and not by space and time which we cannot fill. Let us endeavour then to think well; this is the principle of morality.

From Pascal's *Pensées*, #348 (Brunschvicg) = #113 (Lafuma)
A thinking reed.—It is not from space that I must seek my dignity, but from the government of my thought. I shall have no more if I possess worlds. By space the universe encompasses and swallows me up like an atom; by thought I comprehend the world.

Bernard de Fontenelle, *Plurality of Worlds*, Conversation of the Fifth Day

You have made the Universe so large, *says she,* that I know not where I am, or what will become of me; what is it all to be divided into heaps confusedly, one among another? Is every Star the centre of a Vortex, as big as ours? Is that vast space which comprehends our Sun and Planets, but an inconsiderable part of the Universe? and are there as many such spaces, as there are fix'd Stars? I protest it is dreadful, Dreadful, Madam, *said I;* I think it very pleasant, when the Heavens were a little blue Arch, stuck with Stars, methought the Universe was too strait and close, I was almost stifled for want of Air; but now it is enlarg'd in height and breadth, and a thousand & a thousand Vortex's taken in; I begin to breathe with more freedom, and think the Universe to be incomparably more magnificent than it was before. Nature hath spar'd no cost, even to profuseness, and nothing can be so glorious, as to see such a prodigious number of Vortex's, whose several centres are possess'd by a particular Sun, which makes the Planets turn round it.[1]

Eighteenth Century[2]

Joseph Addison, "Ode" from "Spectator #465" (1712), written with reference to Psalm 19

I

The Spacious Firmament on high,
With all the blue Etherial Sky
And spangled Heav'ns, a Shining Frame,
Their great Original proclaim:
Th'unwearied Sun, from day to day,
Does his Creator's Pow'r display,

---

[1]Bernard de Fontenelle, *Plurality of Worlds,* trans. by John Glanvill (London: Nonesuch), pp. 114–15.

[2]For background information and commentary on a number of the quotations from the eighteenth and nineteenth centuries, see M. J. Crowe, *The Extraterrestrial Life Debate 1750–1900: The Idea of a Plurality of Worlds from Kant to Lowell* (Cambridge: Cambridge University Press, 1986).

And publishes to every Land
The Work of an Almighty Hand.
            II
Soon as the Evening Shades prevail,
The Moon takes up the wondrous Tale,
And nightly to the listning Earth
Repeats the Story of her Birth:
Whilst all the Stars that round her burn,
And all the Planets, in their turn,
Confirm the Tidings as they rowl,
And spread the Truth from Pole to Pole.
            III
What though, in solemn Silence, all
Move round the dark terrestrial Ball?
What tho' nor real Voice nor Sound
Amid their radiant Orbs be found?
In Reason's Ear they all rejoice,
And utter forth a glorious Voice,
For ever singing, as they shine,
'The Hand that made us is Divine.'[1]

Isaac Watts, from his hymn "The Creator and Creatures"
Thy voice produc'd the seas and spheres;
Bid the waves roar and the planets shine;
But nothing like thy Self appears,
Through all these spacious works of thine.[2]

J. G. Gottsched in an inscription below a drawing of the universe in his *Erste Gründe der gesammten Weltweisheit* (1731)
At this stare mind and wit, the soul doth fade away,
Amidst these wondrous worlds, amisdst this grand display;
Oh what in this is man? Nothing would he be named,
Should he not see the grandeur in what God hath framed.[3]

---

[1] *The Spectator*, vol. 4, ed. Donald F. Bond (Oxford: Clarendon, 1965), pp. 144–5.
[2] *Minor English Poets 1660–1780*, ed. David P. French, vol. 3 (New York: Benjamin Blom, 1963), p. 635.
[3] From Gottsched's *Erste Gründe der gesammten Weltweisheit* (Frankfurt, 1965 reprinting of the 1731 original), frontispiece, as

Alexander Pope, *Essay on Man* (1733–4), from Epistle I, lines 23–28
He who thro' vast immensity can pierce,
See worlds on worlds compose one universe,
Observe how system into system runs,
What other planets circle other suns,
What varied being peoples every star,
May tell why Heav'n has made us as we are.[1]

Alexander Pope, *Essay on Man*, from Epistle II, lines 1–7
    Know then thyself, presume not God to scan;
The proper study of mankind is Man.
Placed on this isthmus of a middle state,
A being darkly wise and rudely great:
With too much knowledge for the Sceptic side,
With too much weakness for the Stoic's pride,
He hangs between, ...[2]

Edward Young, *Night Thoughts* (1742–5), "Night the Ninth," lines 742–50, 772–3, 1018–21
This gorgeous apparatus! this display!
This ostentation of creative power!
This theatre!—what eye can take it in?
By what divine enchantment was it raised,
For minds of the first magnitude to launch
In endless speculation, and adore?
One sun by day, by night ten thousand shine;
And light us deep into the Deity;
How boundless in magnificence and might!
    ....
    Devotion! daughter of Astronomy!
An undevout astronomer is mad.
    ....
The soul of man was made to walk the skies;
Delightful outlet of her prison here!

---

translated by M. J. Crowe and published in Crowe, *Debate*, p. 140.
[1] *The Complete Poetical Works of Alexander Pope*, ed. by Henry W. Boynton (Cambridge, Mass.: Riverside Press, 1903), p. 138.
[2] *Complete Poetical Works of Alexander Pope*, p. 142.

There, disincumbered from her chains, the ties
Of toys terrestrial, she can rove at large. . . .[1]

James Thomson, "The Seasons" (1746), "Summer," lines 1703–5, 1714, 1717–19
Amid the radiant orbs,
That more than deck, that animate the sky,
The life-infusing suns of other worlds. . . .
The enlighten'd few . . . they in their powers exult,
That wondrous force of thought, which mounting spurns
This dusky spot, and measures all the sky.[2]

Gotthold Ephraim Lessing, "Die Planetenbewohner"
In charming fancies to take delight,
To populate the planets of the night,
Before on safe grounds you can infer
That on those planets wine is there:
That is called populating too soon.

Friend, you should first divine
Whether in new worlds there is wine,
As on our earth we know there be:
Then, believe me, any child will see
That those worlds will have their drinkers.[3]

Thomas Wright, *An Original Theory or New Hypothesis of the Universe* (1750)

In this great Celestial Creation, the Catastrophy of a World, such as ours, or even the total Dissolution of a System of Worlds, may possibly be no more to the great Author of Nature, than the most common Accident in Life with us, and in all Probability such final and general Doom-Days may be as

---

[1] *Young's Night Thoughts,* ed. George Gilfillan (Edinburgh, 1853), pp. 277, 278, and 285.
[2] James Thomson, *The Seasons* and *The Castle of Indolence,* ed. James Sambroook (Oxford: Clarendon, 1972), p. 84.
[3] From *Lessings Werke,* vol. 1, ed. Georg Witkowski (Leipzig, 1911), pp. 67–8 and as translated by M. J. Crowe and published in Crowe, *Extraterrestrial Life Debate,* p. 143.

frequent there, as even Birth-Days, or Mortality with us upon the Earth.

This Idea has something so chearful in it, that I own I can never look upon the Stars without wondering why the whole World does not become Astronomers....[1]

Immanuel Kant (1724–1804)

From Kant's *Allgemeine Naturgeschichte und Theorie des Himmels* (1755)

The teachers of the mechanical production of the structure of the world referred to, derive all the order which may be perceived in it from mere chance which made the atoms to meet in such a happy concourse that they constituted a well-ordered whole. Epicurus had the hardihood to maintain that the atoms diverged from their straight motion without a cause, in order that they might encounter one another. All these theorizers pushed this absurdity so far that they even assigned the origin of all animated creatures to this blind concourse, and actually derived reason from the irrational. In my system, on the contrary, I find matter bound to certain necessary laws. Out of its universal dissolution and dissipation I see a beautiful and orderly whole quite naturally developing itself. This does not take place by accident, or of chance; but it is perceived that natural qualities necessarily bring it about. And are we not thereby moved to ask, why matter must just have had laws which aim at order and conformity? Was it possible that many things, each of which has its own nature independent of the others, should determine each other of themselves just in such a way that a well-ordered whole should arise therefrom; and if they do this, is it not an undeniable proof of the community of their origin at first, which must have been a universal Supreme Intelligence, in which the natures of things were devised for common combined purposes?

Matter, which is the primitive constituent of all things, is therefore bound to certain laws, and when it is freely

---

[1] Wright, *Original Theory*, intro. by Michael Hoskin (New York: Science History, 1971), p. 72 of text.

abandoned to these laws it must necessarily bring forth beautiful combinations. It has no freedom to deviate from this perfect plan. Since it is thus subject to a supremely wise purpose, it must necessarily have been put into such harmonious relationships by a First Cause ruling over it; and *there is a God, just because nature even in chaos cannot proceed otherwise than regularly and according to order.*[1]

From Kant's *Kritik der Praktischen Vernunft* (1788)

Two things fill my mind with ever new and increasing wonder and awe, the more often and persistently I reflect upon them: *the starry heaven above me and the moral law within me.* I should not seek and conjecture these two as entities hidden in obscurity or in the boundlessness beyond my sight. I see them in front of me and unite them immediately with the consciousness of my own existence. The first originates in the place that I occupy in the external sense world and extends the nexus in which I stand to an immense size with world upon worlds and systems of systems, and extends the nexus even further into the limitless time of the periodic movement of these systems, of their beginning and duration. The second originates in my invisible self, my personality, and presents me in a world that has true infinity, only discernible by the faculty of understanding. I am united to this world (thereby also simultaneously to all those visible worlds) not, as I was in the external-sense worlds, in a merely accidental nexus but rather in a universal and necessary one. The first view of an innumerable multitude of worlds nullifies as it were my importance, as an *animal creature* that must again surrender to the planet (a mere point in the cosmos), the matter from which it arose after the animal for a short time (we know not how) has been endowed with life-force. The second, on the contrary, elevates infinitely my worth, as an *intelligence,* through my personality, in which the moral law reveals to me a life independent of animality and even of the whole sense-world, at least as far as can be gleaned by means of this law from the

---

[1] From Kant's *Universal Natural History and Theory of the Heavens,* as translated by W. Hastie in W. Hastie (ed.), *Kant's Cosmogony* (Glasgow, John Maclehose and Sons, 1900), pp. 25–6.

appropriate destiny of my existence, a destiny not confined to the conditions and limits of this life but rather continuing into infinity.[1]

John Adams, from his diary for 30 April 1756

Astronomers tell us ... that not only all the Planets and Satellites in our Solar System, but all the unnumbered Worlds that revolve round the fixt Starrs are inhabited.... If this is the Case all Mankind are no more in comparison of the whole rational Creation of God, than a point to the Orbit of Saturn. Perhaps all these different Ranks of Rational Beings have in a greater or less Degree, committed moral Wickedness. If so, I ask a Calvinist, whether he will subscribe to this Alternitive (sic), "either God almighty must assume the respective shapes of all these different Species and suffer the Penalties of their Crimes, in their Stead, or else all these Being[s] must be consigned to everlasting Perdition?"[2]

Friedrich Klopstock, from his "Psalm" (1789), based on the "Our Father"
Around suns circle moons
Earths around suns
Hosts of all suns travel
Around a great sun:
"Our Father, who art in heaven!"

On all these worlds, illuminating and illuminated,
Live species varying in vitality and bodily frames;
But all contemplate and take delight in God.
"Hallowed be Thy name!"[3]

---

[1] As translated for this volume by Joseph T. Ross from Kant, *Kritik der praktischen Vernunft*, 2 Theil. in *Kants gesammelte Schriften*, part 1, vol. 5 (Berlin: Georg Reimer, 1913), pp. 161–2.

[2] *Diary and Autobiography of John Adams*, ed. by L. H. Butterfield, vol. 1, Diary 1755-1770 (New York: Atheneum, 1964), p. 22. For a statement Adams made in the years after he had become U.S. President, see *Epilogue* item for 1825.

[3] From *Klopstocks Werke*, pt. 3, ed. R. Hamel, Deutsche National-Litteratur edition (Berlin, n.d.), pp. 178–9, and as translated by M. J. Crowe and published in Crowe, *Extraterrestrial Life Debate*, p. 147.

Jérôme Lalande
I have searched through the heavens, and nowhere have I found a trace of God.[1]

Thomas Paine, *The Age of Reason*, Part I (1794)
There may be many systems of religion that, so far from being morally bad, are in many respects morally good; but there can be but ONE that is true; and that one necessarily must, as it ever will, be in all things consistent with the ever-existing word of God that we behold in his works. But such is the strange construction of the Christian system of faith that every evidence the heavens afford to man either directly contradicts it or renders it absurd.[2]

Thomas Paine, *The Age of Reason*, Part I (1794)
From whence, then, could arise the solitary and strange conceit that the Almighty, who had millions of worlds equally dependent on his protection, should quit the care of all the rest and come to die in our world because, they say, one man and one woman had eaten an apple! And, on the other hand, are we to suppose that every world in the boundless creation had an Eve and an apple, a serpent, and a redeemer? In this case, the person who is irreverently called the Son of God, and sometimes God himself, would have nothing else to do than to travel from world to world, in an endless succession of death, with scarcely a momentary interval of life.[3]

Pierre Simon Laplace, *Exposition du Système du Monde* (1796)
Contemplated as one grand whole, astronomy is the most beautiful monument of the human mind; the noblest record of its intelligence. Seduced by the illusions of the senses, and of self-love, man considered himself, for a long time, as the centre of the motion of the celestial bodies, and his pride was justly punished by the vain terrors they inspired. The labour of many

---

[1] As quoted in Ludwig Büchner, *Force and Matter*, reprint of the 4th English ed. (New York: Peter Eckler, 1920), pp. 105–6.
[2] Thomas Paine, *Representative Selections*, ed. Harry Hayden Clark, rev. ed. (New York: Hill and Wang, 1962), p. 284.
[3] Paine, *Selections*, p. 283.

ages has at length withdrawn the veil which covered the system. Man appears, upon a small planet, almost imperceptible in the vast extent of the solar system, itself only an insensible point in the immensity of space. The sublime results to which this discovery has led, may console him for the limited place assigned him in the universe. Let us carefully preserve, and even augment the number of these sublime discoveries, which form the delight of thinking beings.[1]

Jean Paul, from the chapter of his *Siebenkäs* (1796) entitled "Speech of the Dead Christ" which is set in a graveyard in which bodiless dead wander; one of these, addressed as Christ, is asked: "Is there no God?" Christ answers "There is none" and adds: "I have traversed the worlds, I have risen to suns, with the milky ways, I have passed athwart the great waste spaces of the sky; there is no God. And I descended to where the very shadow cast by Being dies out and ends, and I gazed into the gulf beyond, and cried, 'Father, where art Thou?' But answer came there none.... And when I looked up to the boundless universe for the Divine eye, behold, it glared at me from a socket empty and bottomless. ... Shriek on, then, discords, shatter the shadows with your shrieking din, for HE IS NOT!"[2]

Nineteenth Century

Charles Lamb in his 2 January 1810 letter to Thomas Manning
Nothing puzzles me more than time and space; and yet nothing troubles me less....[3]

Percy Bysshe Shelley, note added to his *Queen Mab* (1813)
The plurality of worlds,—the indefinite immensity of the universe, is a most awful subject of contemplation. He who

---

[1] Pierre Simon Laplace, *System of the World*, trans. by John Pond, vol. 2 (London: Richard Phillips, 1809), pp. 373–4.
[2] Jean Paul, *Flower, Fruit, and Thorn Pieces; or, The Wedded Life, Death, and Marriage of Firmian Stanislaus Siebenkaes*, trans. Alexander Ewing (London, 1895), pp. 262–3.
[3] As quoted in *The Oxford Dictionary of Quotations*, 2nd ed. (London: Oxford University Press, 1955), p. 307.

rightly feels its mystery and grandeur is in no danger of seduction from the falsehoods of religious systems, or of deifying the principle of the universe. It is impossible to believe that the Spirit that pervades this infinite machine begat a son upon the body of a Jewish woman; or is angered at the consequences of that necessity, which is a synonym of itself. All this miserable tale of the Devil, and Eve, and an Intercessor, with the childish mummeries of the God of the Jews, is irreconcilable with the knowledge of the stars. The works of His fingers have born witness against Him.[1]

G. F. W. Hegel, *Enzyklopädie der philosophischen Wissenschaften* (1817):
It has been rumoured round the town that I have compared the stars to a rash on an organism . . . or to an ant-heap. . . . In fact, I do rate what is concrete higher than what is abstract, and an animality that develops into no more than a slime, higher than the starry host.[2]

Some passages from Thomas Chalmers, *Discourses on the Christian Revelation Viewed in Connection with the Modern Astronomy* (1817)
[After referring to the discovery of nebulae:] This carries us upward through another ascending step in the scale of magnificence, and there leaves us wildering in the uncertainty whether even here the wonderful progression is ended; and at all events, fixes the assured conclusion in our minds, that to an eye which could spread itself over the whole, the mansion which accommodates our species might be so very small as to lie wrapped in microscopical concealment; and in reference to the only Being who possesses this universal eye, well might we say, "What is man, that thou art mindful of him? and the son of man, that thou visitest him?"

. . . .

---

[1]Shelley, *The Complete Poetical Works*, ed. Neville Rogers, vol. 1 (Oxford: Clarendon, 1972), p. 296.
[2]From Hegel's *Enzyklopädie* as trans. and published as *Hegel's Philosophy of Nature*, trans. A. V. Miller (Oxford: Clarendon, 1970), p. 297.

And what is this world in the immensity which teems with them; and what are they who occupy it? The universe at large would suffer as little, in its splendor and variety, by the destruction of our planet, as the verdure and sublime magnitude of a forest would suffer by the fall of a single leaf.

....

Now, it is this littleness, and this insecurity, which make the protection of the Almighty so dear to us, and bring with such emphasis to every pious bosom the holy lessons of humility and gratitude. The God who sitteth above, and presides in high authority over all worlds, is mindful of man; and though at this moment his energy is felt in the remotest provinces of creation, we may feel the same security in his providence as if we were the objects of his undivided care.[1]

John Adams in an 1825 letter to Thomas Jefferson, advising him on why he should not hire European professors for the University of Virginia faculty:
They all believe that that great Principle which has produced this boundless universe, Newton's universe and Herschell's [sic] universe, came down to this little ball, to be spit upon by the Jews. And until this awful blasphemy is got rid of, there will never be any liberal science in the world.[2]

Ralph Waldo Emerson (1803–1882)

From Emerson's sermon "Astronomy," first preached in 1832

An important result of the study of astronomy has been to correct and exalt our views of God, and humble our view of ourselves.

....

I regard it as the irresistible effect of the Copernican astronomy to have made the theological *scheme of Redemption* absolutely incredible. The great geniuses who studied the mechanism of the heavens became unbelievers in the popular doctrine. . . .

In spite of the awful exhibition of wisdom and might

---

[1]Chalmers, *Christian Revelation* (New York: Religious Tract Society, 185?), pp. 42, 44, 46–7.
[2]*The Works of John Adams*, vol. 10 (Boston, 1856), p. 415.

disclosed to their eyes ... the incongruity between what [the astronomers] beheld and the gross creeds which were called religion and Christianity by their fellow countrymen so revolted them, the profound astronomers of France rejected the hope and consolation of man and in the face of that divine mechanism which they explored denied a cause and adopted the belief of an eternal Necessity, as if that very external necessity were anything else than God, an intelligent cause.

....

And finally, what is the effect upon the doctrine of the New Testament which these contemplations produce? It is not contradiction but correction. It is not denial but purification. It proves the sublime doctrine of One God, whose offspring we all are and whose care we all are. On the other hand, it throws into the shade all temporary, all indifferent, all local provisions. Here is neither tithe nor priest nor Jerusalem nor Mount Gerizim. Here is no mystic sacrifice, no atoning blood.[1]

From Emerson's *Journal* for 23 May 1832
[Given modern astronomy,] Who can be a Calvinist or who an Atheist[?]—[2]

From Emerson's "Nature" (1836)
But if a man would be alone, let him look at the stars.... One might think the atmosphere was made transparent with this design, to give man, in the heavenly bodies, the perpetual presence of the sublime. ... If the stars should appear one night in a thousand years, how would men believe and adore; and preserve for many generations the remembrance of the city of God which had been shown![3]

---

[1] From *Young Emerson Speaks: Unpublished Discourses on Many Subjects*, ed. Arthur Cushman McGiffert (Boston: Houghton Mifflin, 1938), pp. 172, 174–5, 177.
[2] *The Journals and Miscellaneous Notebooks of Ralph Waldo Emerson*, vol. 4, ed. Alfred R. Ferguson (Cambridge: Harvard University Press, 1964), p. 24.
[3] Ralph Waldo Emerson, *Selected Prose and Poetry*, ed. Reginald L. Cook (New York: Rinehart, 1959), pp. 4–5.

**Illustration by J. J. Grandville
for Taxile Delord's *Un autre monde* (1844)**

Daniel Webster, words dictated on the day before his death in 1852 and appearing on his tombstone:
Philosophical argument, especially that drawn from the vastness of the universe, in comparison with the apparent insignificance of this globe, has sometimes shaken my reason for the faith which is in me; but my heart has always assured

and reassured me that the gospel of Jesus Christ must be Divine Reality. The Sermon on the Mount cannot be a mere human production. This belief enters into the very depth of my conscience. The whole history of man proves it.[1]

[William Whewell], *Of the Plurality of Worlds: An Essay* (1853)
One school of moral discipline, one theatre of moral action, one arena of moral contests for the highest prizes, is a sufficient centre for innumerable hosts of stars and planets, globes of fire and earth, water and air, whether or not tenanted by corals and madrepores, fishes and creeping things. So great and majestic are those names of Right and Good, Duty and Virtue, that all mere material or animal existence is worthless in the comparison.[2]

Walt Whitman, from "Night on the Prairies" (1860)
I was thinking this globe enough, till there sprang out so noiseless around me myriads of other globes.
Now, while the great thoughts of space and eternity fill me, I will measure myself by them;
And now, touch'd with the lives of other globes, arrived as far along as those of the earth,
Or waiting to arrive, or pass'd on farther than those of the earth,
I henceforth no more ignore them than I ignore my own life,
Or the lives of the earth arrived as far as mine, or waiting to arrive.[3]

Attributed to Thomas Carlyle
Margaret Fuller: "I accept the universe." Carlyle: "Gad! She better!"[4]

---

[1] Bartlett, *Familiar Quotations*, p. 342.
[2] [William Whewell], *Of the Plurality of Worlds: An Essay* (London, 1853), p. 256.
[3] Walt Whitman, *Leaves of Grass: Comprehensive Reader's Edition*, ed. Harold W. Blodgett and Sculley Bradley (New York: New York University Press, 1965), p. 452.
[4] *The Oxford Dictionary of Quotations*, 2nd ed. (London: Oxford

Alfred Lord Tennyson (1809–1892)

While a child, Tennyson said to his brother, who suffered from shyness: "Fred, think of Herschel's great star-patches, and you will soon get over all that."[1]

From Tennyson's "Locksley Hall" (1842), lines 7–10
Many a night from yonder ivied casement, ere I went to rest,
Did I look on great Orion sloping slowly to the west.
Many a night I saw the Pleiads, rising thro' the mellow shade,
Glitter like a swarm of fireflies tangled in a silver braid.[2]

From Tennyson's "Ode on the Death of the Duke of Wellington" (1852), lines 258–65
And Victor he must ever be.
For tho' the Giant Ages heave the hill
And break the shore, and evermore
Make and break, and work their will,
Tho' world on world in myriad myriads roll
Round us, each with different powers,
And other forms of life than ours,
What know we greater than the soul?[3]

From his "Epilogue" (1885)
The fires that arch this dusky dot—
Yon myriad-worlded way—
The vast sun-clusters' gather'd blaze,
World-isles in lonely skies,
Whole heavens within themselves, amaze
Our brief humanities.[4]

University Press, 1955), p. 127.
[1] Hallam Lord Tennyson, *Alfred Lord Tennyson: A Memoir*, vol. 1 (New York, 1897), p. 20.
[2] *The Poetical Works of Tennyson*, ed. G. Robert Strange (Boston: Houghton Mifflin, 1974), p. 90.
[3] Tennyson, *Works*, p. 226.
[4] Tennyson, *Works*, p. 510.

From Tennyson's "Vastness" (1889)

I

Many a hearth upon our dark globe sighs after many a vanish'd face,
Many a planet by many a sun may roll with the dust of a vanish'd race.

II

Raving politics, never at rest—as this poor earth's pale history runs—
What is it all but a trouble of ants in the gleam of a million million of suns?[1]

Feodor Dostoyevsky (1821–1881)

From Dostoyevsky's *The Possessed* (1871), Part Two, Ch. 1, section V.

"Suppose you had lived in the moon," Stravrogin interrupted, not listening, but pursuing his own thoughts, "and suppose you had done all these nasty and ridiculous things.... You know from here for certain that they will laugh at you and hold you in scorn for a thousand years as long as the moon lasts. But now you are here, and looking at the moon from here. You don't care here for anything you've done there, and that the people there will hold you in scorn for a thousand years, do you?"[2]

From Dostoyevsky's *Diary of a Writer* (Entry under January, 1876). Referring to Goethe and to the suicide of Werther in Goethe's *Sorrows of the Young Werther,* Dostoyevsky wrote:

The self-destroyer Werther, when committing suicide, in the last lines left by him, expresses regret that he would nevermore behold "the beautiful constellation of the Great Bear" and he bids it farewell. Oh, how was the then still youthful Goethe revealed in this little trait! Why were these constellations so dear to young Werther?—Because, whenever contemplating them, he realized that he was by no means an

---

[1] Tennyson, *Works,* p. 533.
[2] Dostoyevsky, *The Possessed,* trans. Constance Garnett (New York: Random House, 1936), pp. 238–9.

atom and nonentity compared with them; that all these numberless mysterious, divine miracles were in no sense higher than his thought and consciousness; not higher than the ideal of beauty confined in his soul, and, therefore, they were equal to him and made him akin to the infinity of being.... And for the happiness to perceive this great thought which reveals to him who he is, he is indebted exclusively to *his human image*.

'Great Spirit, I thank Thee for the human image bestowed on me by Thee.'

Such must have been the lifelong prayer of the great Goethe. In our midst this image bestowed upon man is being smashed quite simply and with no German tricks, while no one would think of bidding farewell not only to the Great, but even to the Little Bear, and even if one should think of it, he would not do it: he would feel too much ashamed.[1]

From Dostoyevsky's "The Dream of a Strange Man: A Fantastic Story" (1877)
[Transported through space by some mysterious being, the central character records what happened, especially his visit to a planet identical in form to the earth:]
We were sweeping through dark and unknown spaces. I had long since ceased to perceive the constellations familiar to my eyes. I knew that there were stars in the heavenly expanse whose rays reached the earth in thousands and millions of years.... suddenly I saw our sun! I knew that this could not have been *our* sun which had generated *our* earth. ...

"And are such duplications really possible in the universe? ..."
[After landing on the planet, he discovers:] This was an earth not defiled by sin ... a paradise similar to that in which, according to the tradition of all mankind, lived our fallen forefathers....

[The inhabitants] pointed at stars and spoke to me something about them which I was unable to grasp, but I am quite sure that through some means they communicated, as it were, with these celestial bodies,—only not through thought but through some

---

[1] Dostoievsky, *Diary of a Writer*, trans. Boris Brasol (Haslemere, Surrey: Ianmead, 1984), p. 158.

live medium.

... up to the present I have been concealing the full truth, but now I am going to complete my story. The point is that I have ... debauched them all!

[After recounting the debased form their society took after he had led them to their fall, the "ridiculous man" adds:] I implored them to crucify me; I taught them how to make the cross. ... But they merely laughed at me and finally, they began to consider me crazy.[1]

From Dostoyevsky's *The Brothers Karamazov* (1879), Bk. VI, Ch. 3.

Father Zossima states:

"Much on earth is hidden from us, but to make up for that we have been given a precious mystic sense of our living bond with the other world, with the higher heavenly world and the roots of our thoughts and feelings are not here but in other worlds. That is why the philosophers say that we cannot apprehend the reality of things on earth.

"God took seeds from different worlds and sowed them on this earth, and his garden grew up and everything came up that could come up, but what grows lives and is alive only through the feeling of its contact with other mysterious worlds. If that feeling grows weak or is destroyed in you, the heavenly growth will die away in you. Then you will be indifferent to life and even grow to hate it. That's what I think."[2]

From Dostoyevsky's *The Brothers Karamazov*, end of Bk. VII, Ch. 4.

What was he [Alyosha] weeping over?

Oh! in his rapture he was weeping even over those stars, which were shining to him from the abyss of space, and "he was not ashamed of that ecstasy." There seemed to be threads from all those innumerable worlds of God, li[n]king his soul to them, and it was trembling all over "in contact with other worlds." He longed to forgive every one and for everything,

---

[1] Dostoievsky, *Diary*, pp. 681–3, 686, 688.
[2] Dostoyevsky, *The Brothers Karamazov*, trans. Constance Garnett (New York: Random House, 1950), pp. 384–5.

and to beg forgiveness. Oh, not for himself, but for all men, for all and for everything. "And others are praying for me too," echoed again in his soul. But with every instant he felt clearly, and as it were tangibly, that something firm and unshakable as that vault of heaven had entered into his soul. It was as though some idea has seized the sovereignty of his mind—and it was for all his life and for ever and ever. He had fallen on the earth a weak boy, and he rose up a resolute champion, and he knew and felt it suddenly at the very moment of his ecstasy. And never, never, all his life long, could Alyosha forget that minute.

"Some one visited my soul in that hour," he used to say afterwards, with implicit faith in his words.

Within three days he left the monastery in accordance with the words of his elder, who had bidden him "sojourn in the world."[1]

From Dostoyevsky's *Diary of a Writer* (Entry of August, 1880).
...whence can the ideal of civic organization in human society be derived?—Trace it historically and you will forthwith perceive whence it is derived. You will see that it is solely the product of moral self-betterment of individual entities; it has its inception there. Thus it has been from time immemorial, and thus it always will be. In the origin of every people, of every nationality, the moral idea invariably preceded the organization of the nationality itself, *since the former created the latter*. The moral idea always emanated from mystical ideas, from the conviction that man is eternal, that he is not a mere earthly animal, but that he is tied to other worlds and eternity.[2]

William Kingdon Clifford, from "The Influence upon Morality of a Decline in Religious Belief" in his *Lectures and Essays* (1879)
A little field-mouse, which busies itself in the hedge, and does not mind my company, is more to me than the longest ichthyosaurus that ever lived, even if he lived a thousand

---

[1]Dostoyevsky, *Brothers*, pp. 436–7.
[2]Dostoievsky, *Diary*, pp. 1000–1.

years. When we look at a starry sky, the spectacle whose awfulness Kant compared with that of the moral sense, does it help out our poetic emotion to reflect that these specks are really very very big, and very very hot, and very very far away? Their heat and their bigness oppress us; we should like them to be taken still farther away, the great blazing lumps. But when we think of the unseen planets that surround them, of the wonder of life, of reason, of love that may dwell therein, then indeed there is something sublime in the sight. Fitness and kinship: these are the truly great things for us, not force and massiveness and length of days.[1]

Thomas Hardy, from *Two on a Tower* (1882)
"Until a person has thought out the stars and their interspaces, he has hardly learnt that there are things much more terrible than monsters of shape, namely, monsters of magnitude without known shape. Such monsters are the voids and waste places of the sky. Look, for instance, at those pieces of darkness in the Milky Way," he went on, pointing with his finger to where the galaxy stretched across over their heads with the luminousness of a frosted web. "You see that dark opening in it near the Swan? There is a still more remarkable one south of the equator, called the Coal Sack.... In these our sight plunges quite beyond any twinkler we have yet visited. Those are deep wells for the human mind to let itself down into...."[2]

George Meredith, from "Meditations under Stars" (1888), line 1–2
What links are ours with orbs that are
     So resolutely far; ....[3]

---

[1] Clifford, *Lectures and Essays*, ed. Leslie Stephen and Frederick Possock (London, 1886), p. 390.
[2] Hardy, *Two on a Tower* (New York: St Martin's Press, 1977), p. 57.
[3] *The Poems of George Meredith*, ed. Phyllis B. Bartlett, vol. 1 (New Haven: Yale University Press, 1978), p. 452.

Vincent van Gogh, *The Starry Night* (1889)[1]

Twentieth Century

Alice Meynell, "Christ in the Universe" (ca. 1910)
    With this ambiguous earth
His dealings have been told us. These abide:
The signal of a maid, the human birth,
The lesson, and the young Man crucified.
    But not a star of all
The innumerable host of stars has heard
How he administered this terrestrial ball.
Our race have kept their Lord's entrusted Word.
    Of His earth-visiting feet

---

[1] For information on the astronomical background of van Gogh's *The Starry Night*, see Albert Boime, "Van Gogh's *Starry Night:* A History of Matter and a Matter of History," *Arts Magazine* (Dec., 1984), 86–103 and Charles A. Whitney, "The Skies of Vincent van Gogh," *Art History*, 9 (Sept., 1986), 351–62.

None knows the secret, cherished, perilous,
The terrible, shamefast, frightened, whispered, sweet,
Heart-shattering secret of His way with us.
    No planet knows that this
Our wayside planet, carrying land and wave,
Love and life multiplied, and pain and bliss,
Bears, as chief treasure, one forsaken grave.
    Nor, in our little day,
May His devices with the heavens be guessed,
His pilgrimage to thread the Milky Way,
Or His bestowals there be manifest.
    But, in the eternities,
Doubtful we shall compare together, hear
A million alien Gospels, in what guise
He trod the Pleiades, the Lyre, the Bear.
    O, be prepared, my soul!
To read the inconceivable, to scan
The million forms of God those stars unroll
When, in our turn, we show to them a Man.[1]

Harlow Shapley (1921)
Thus the significance of man and the earth in the sidereal scheme has dwindled with advancing knowledge of the physical world....[2]

James Branch Cabell, *The Silver Stallion* (1926), Bk iv, ch. 26
The optimist proclaims that we live in the best of all possible worlds; and the pessimist fears this is true.[3]

Myles Connolly from *Mr. Blue* (1928)
    "I think," he [Mr. Blue] whispered half to himself, "my heart would break with all this immensity if I did not know that God Himself once stood beneath it, a young man, as small

---

[1] *The Poems of Alice Meynell* (New York: Charles Scribner's Sons, 1923), p. 92.
[2] Harlow Shapley, "The Scale of the Universe," *Bulletin of the National Research Council*, 2 (1921), 171–93:172.
[3] James Branch Cabell, *The Silver Stallion* (New York: Robert M. McBride, 1926), p. 129.

as I. .... Did it ever occur to you that it was Christ Who humanized infinitude, so to speak? When God became man He made me and the rest of us pretty important people. He not only redeemed us. He saved us from the terrible burden of infinity."[1]

....

His [Mr. Blue's] eyes were glowing in the dark. He threw his hands up toward the stars: "My hands, my feet, my poor little brain, my eyes, my ears, all matter more than the whole sweep of these constellations!" he burst out. "God Himself, the God to Whom this whole universe-specked display is as nothing, God Himself had hands like mine, and eyes, and brain, and ears!...." He looked at me intently. "Without Christ we would be little more than bacteria breeding on a pebble in space, or glints of ideas in a whirling void of abstractions. Because of Him, I can stand here out under this cold immensity and know that my infinitesimal pulse-beats and acts and thoughts are of more importance than this whole show of a universe. Only for Him, I would be crushed beneath the weight of all these worlds. Only for Him, I would tumble dazed into the gaping chasms of space and time. Only for Him, I would be confounded before the awful fertility and intricacy of all of life. Only for Him, I would be the merest of animalcules crawling on the merest of motes in a frigid Infinity." He turned away from me, turned toward the spread of night behind the parapet. "But behold," he said, his voice rising with exultancy, "behold! God wept and laughed and dined and wined and suffered and died even as you and I. Blah!—for the immensity of space! Blah!—for those who would have me a microcosm in the meaningless tangle of an endless evolution! I'm no microcosm. I, too, am a Son of God!"[2]

Albert Schweitzer, from "Religion and Modern Civilization" (1934)

Thinking which keeps contact with reality must look up to the heavens, it must look over the earth, and dare to direct its gaze to the barred windows of a lunatic asylum. Look to the

---

[1] Myles Connolly. *Mr. Blue* (New York: Macmillan, 1928), p. 38.
[2] Connolly. *Mr. Blue*, pp. 39–41.

stars and understand how small our earth is in the universe. Look upon earth and know how minute man is upon it. In the history of the universe, man is on earth for but a second. . . .we must not place man in the center of the universe. And our gaze must be fixed on the barred windows of a lunatic asylum, in order that we may remember the terrible fact that the mental and spiritual are also liable to destruction. . . .

All thinking must renounce the attempt to explain the universe. We cannot understand what happens in the universe. What is glorious in it is united with what is full of horror. What is full of meaning is united with what is senseless. The spirit of the universe is at once creative and destructive—it creates while it destroys and destroys while it creates, and therefore it remains to us a riddle. And we must inevitably resign ourselves to this.[1]

Arthur Stanley Eddington, *New Pathways in Science* (1935)
As for man—it seems unfair to be always raking up against Nature her one little inadvertence. By a trifling hitch of machinery—not of any serious consequence in the development of the universe—some lumps of matter of the wrong size have occasionally been formed. These lack the purifying protection of intense heat or the equally efficacious absolute cold of space. Man is one of the gruesome results of this occasional failure of antiseptic precautions.[2]

Bertrand Russell, from "My Mental Development" (1944)
When I was young I hoped to find religious satisfaction in philosophy; even after I had abandoned Hegel, the eternal Platonic world gave me something non-human to admire. I thought of mathematics with reverence, and suffered when Wittgenstein led me to regard it as nothing but tautologies. I have always ardently desired to find some justification for the

---

[1] Albert Schweitzer, "Religion in Modern Civilization" as quoted from *Christian Century*, 51 (Nov. 28, 1934), p. 1520 in *Albert Schweitzer: An Anthology*, ed. by Charles R. Joy, rev. and enlarged ed. (Boston: Beacon Press, 1967), pp. 4–5.
[2] Eddington, *New Pathways in Science* (New York: Macmillan, 1935), pp. 309–10.

emotions inspired by certain things that seemed to stand outside human life and to deserve feelings of awe. I am thinking in part of very obvious things, such as the starry heavens and a stormy sea on a rocky coast; in part of the vastness of the scientific universe both in space and time, as compared to the life of mankind; in part of the edifice of impersonal truth, especially truth which, like that of mathematics, does not merely describe the world that happens to exist. Those who attempt to make a religion of humanism, which recognizes nothing greater than man, do not satisfy my emotions. And yet I am unable to believe that, in the world as known, there is anything that I can value outside human beings, and, to a much lesser extent, animals. Not the starry heavens, but their effects on human percipients, have excellence; to admire the universe for its size is slavish and absurd; impersonal non-human truth appears to be a delusion.[1]

C. S. Lewis, from *Mid-Twentieth Century*
The vast distances between solar systems may be a form of divine quarantine: they prevent the spiritual infection of a fallen species from spreading; they block it from playing the role of the serpent in the Garden of Eden.[2]

Robinson Jeffers, from "Star Swirls" (1954)
There is nothing like astronomy to pull the stuff out of man. His stupid dreams and red-rooster importance: let him count the star-swirls.[3]

Harlow Shapley (1958)
In the beginning was the word, and the word was hydrogen. . . .[4]

---

[1] Bertrand Russell, "My Mental Development" in *The Philosophy of Bertrand Russell*, ed. P. A. Schilpp (Evanston: Northwestern University Press, 1944), pp. 19–20.
[2] As quoted in James E. Gunn, *The Listeners* (New York: New American Library, 1972), p. 85.
[3] Robinson Jeffers, *The Beginning and the End* (New York: Random House, 1954), p. 18.
[4] As quoted in Gunn, *Listeners*, p. 39.

Loren Eiseley in his *The Immense Journey* (1959), reflecting on the fact that evolution takes place through chance variation:
[N]owhere in all space or on a thousand worlds will there be men to share our loneliness. There may be wisdom; there may be power; somewhere across space great instruments, handled by strange, manipulative organs, may stare vainly at our floating cloud wrack, their owners yearning as we yearn. Nevertheless, in the nature of life and in the principles of evolution we have had our answer. Of men elsewhere, and beyond, there will be none forever.[1]

Boris Pasternak, from "Night" (1963)
And there, with frightful listing
Through emptiness, away
Through unknown solar systems
Revolves the Milky Way.[2]

Woody Allen, from *Getting Even* (1966)
The universe is merely a fleeting idea in God's mind—a pretty uncomfortable thought, particularly if you've just made a down payment on a house.[3]

Astronomy and Astrophysics for the 1970's, vol. 1, "Report of the Astronomy Survey Committee," National Academy of Sciences, 1972
Through the vast reaches of space and time, part of the matter of the universe has evolved into living matter, of which a tiny part is in the form of brains capable of intelligent reasoning. As a result, the universe is now able to reflect upon itself. In this respect, at least, the whole evolutionary chain of events is endowed with meaning.[4]

---

[1] Loren Eiseley, *The Immense Journey* (New York: Random House, 1957), p. 162.
[2] Boris Pasternak, *Fifty Poems,* trans. Lydia Pasternak Slater (London: George Allen & Unwin, 1963), p. 76.
[3] Woody Allen, *Getting Even* (New York: Random House, 1978), p. 24.
[4] "Report of the Astronomy Survey Committee" in *Astronomy and Astrophysics for the 1970's,* vol. 1 (Washington, D.C., National Academy of Sciences, 1972), p. 18.

James Gunn, from *The Listeners* (1972)
"AS THE FIRST ASTRONAUT TO SET FOOT ON THE PLANET MARS AND RETURN, TELL US—IS THERE LIFE ON MARS?"
"WELL, THERE IS A LITTLE ON SATURDAY NIGHT, BUT THE REST OF THE WEEK IT'S PRETTY DULL."[1]

Arthur Koestler, *Bricks to Babel* (1980)
Astrophysics is a science which should be pursued in the following manner. Drink some vodka on a clear, cold night, wrap your feet in a warm blanket, sit down on your balcony and stare into the sky. Preferred localities are mountainous regions with a faint thunder of avalanches in the distance; preferred time, the hours of melancholy. If this prescription is not followed, the science in question will appear as a petrified forest of numbers and equations; but he who observes the prescription will experience a curious state of trance. The algebraic signs will change into violin clefs and out of the bizarre equations will emerge the symphony of the rise and decline of the universe....[2]

Stephen W. Hawking, *A Brief History of Time* (1988)
... if we do discover a complete theory [of the universe], it should in time be understandable in broad principle by everyone, not just a few scientists. Then we shall all, philosophers, scientists, and just ordinary people, be able to take part in the discussion of the question of why it is that we and the universe exist. If we find the answer to that, it would be the ultimate triumph of human reason—for then we would know the mind of God.[3]

---

[1] Gunn, *Listeners*, p. 48.
[2] Koestler, *Bricks to Babel* (New York: Random House, 1980), p. 75.
[3] Hawking, *A Brief History of Time: From the Big Bang to Black Holes* (Toronto: Bantam, 1988), p. 175.

Humanity and the Universe: Some Calculations

It is possible to enrich one's appreciation of what some of the distance determinations made by astronomers mean in regard to humanity's position in the universe by making the following calculations. Although they entail calculating with very large numbers, the calculations are in principle quite simple.

(1) Current research sets the age of the universe at about fifteen billion (1.5×10^{10}) years. Assuming this is correct, let us map this period onto a year long calendar so as to place it on a scale comprehensible to us. On such a scale, the universe would be 7.5×10^9 years old (half its age) on July 1st. Let us now assume that humanity first appeared in its present form one hundred thousand (10^5) years ago, that Ptolemy died 1,800 years ago, and that Newton's *Principia* was published 300 years ago. Calculate the date and time (correct to the nearest second) for each of these events on our year calendar. Hint: begin by calculating the number of seconds in a year.

Modern humans appeared:

Month:____Day:____Hour:____Minute:____Second:____

Death of Ptolemy:

Month:____Day:____Hour:____Minute:____Second:____

Newton's *Principia* published:

Month:____Day:____Hour:____Minute:____Second:____

(2) The most remote observable celestial objects are about ten billion light-years away; consequently, the diameter of the observable universe is about twenty billion (2×10^{10}) LY. Let us map the universe onto a map of the continental United States, assuming that its width is about 3000 miles. Given that our Milky Way galaxy is about one hundred thousand (10^5) LY wide, calculate the length, correct to the nearest foot, that would be needed to show it on our map, which is the size of the U.S.

Width of Milky Way in relation to the size of observable universe if the latter is set to the scale of a map which is the size of the United States: _____ miles _____ feet.

(3) The nearest star is 4.3 LY away. What distance (correct to the nearest inch) would be needed on our U. S. map representing the universe to show this distance?

_____ miles _____ feet _____ inches

(4) It is sometimes maintained that in a sufficiently large universe the laws of probability demand that anything conceivable will become a reality. Recent calculations by Frank W. Cousins[1] shed light on this claim. Cousins shows that were a monkey to type randomly at the rate of one character per second on a typical typewriter, it would happen to hammer out *Hamlet* on the average every $10^{460,000}$ seconds. Suppose a billion (10^9) galaxies comparable to the Milky Way exist, that each contains a billion stars, that each star is surrounded by ten planets, that on every planet a billion monkeys have been typing since the beginning of the universe (about 10^{18} seconds ago). Suppose a friend were to offer to bet you a billion dollars to your one dollar, the terms being that your friend wins if the monkeys fail to type *Hamlet*, whereas you win if they succeed. Under these conditions, would it be wise for you to take the bet? Select one:
a. A cinch bet; my friend might as well give me the money.
b. A good bet, well worth the risk.
c. About even.
d. Very little chance.
e. The bet would be absurdly bad; I might as well give her/him the buck.
Explain your reasoning.

---

[1] Frank W. Cousins, *The Solar System* (London: John Baker, 1972), p. 263.

Appendix

Laboratory Exercise on Nebulae and on the Milky Way

Note: This laboratory exercise has been designed to accompany chapter 3.

The ideal laboratory experience for a person to have while studying the early history of the observation of nebulae would consist in proceeding to eighteenth-century England to peer through the giant telescope of William Herschel and then to nineteenth-century Ireland for the even larger telescope of Lord Rosse. This being impossible, we can approximate this experience by means of the photographs given on the next two pages. The first provides a good idea of what Herschel was able to see; the second serves a similar function in regard to Lord Rosse. The photographs are quite recent, but because distant objects are pictured, they represent what earlier astronomers saw of nearer objects.

Part I
Nebulae and Sir William Herschel—First Photograph

Examine the first photograph. It is of the Corona Borealis region and approximates the view that William Herschel obtained of some of the nearer areas in the cosmos. Note that some of the features are of instrumental origin, for example, the spikes around the larger stars. Now proceed to the exercises below.

1. Remembering that stars appear as circles, can you decide which objects are most probably nebulae? What criteria should be used in deciding which are nebulae? Circle the five objects on the photograph that you are most certain are nebulae.

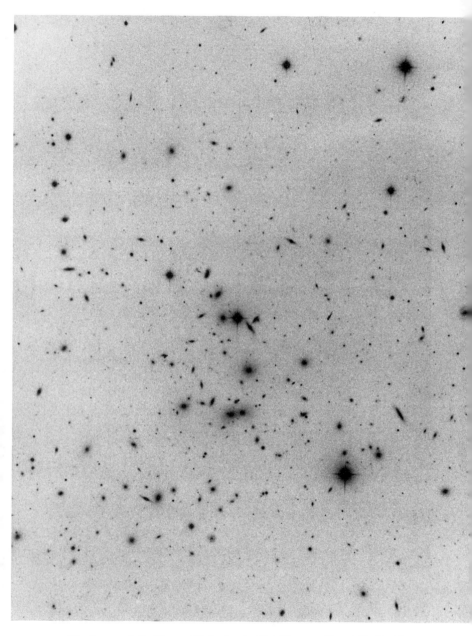

Photograph 1: Some Nebulae in the Corona Borealis Region

2. Attempt to classify the nebulae into groups. Sketch in the boxes below the different types you are able to discern, noting the distinguishing characteristics on the lines below the boxes. Also place the letters A, B, C ... next to the objects on the photograph that most clearly represent each type.

Type A	Type B	Type C	Type D

_____   _____   _____   _____

_____   _____   _____   _____

_____   _____   _____   _____

3. Can you draw any conclusions about which objects in the photograph are nearest; which most distant? If so, on what grounds? If not, why not?

_____

4. Do the nebulae in this photograph appear to be homogeneously distributed or do they cluster in certain areas? If you detect clustering, draw a line around the areas in which you see clustering.

_____

5. Can you formulate any additional questions to ask concerning this photograph?

_____

_____

Part II
Observation of the Andromeda and of the Orion Nebulae

Using a telescope or good binoculars, examine the Andromeda and Orion nebulae. Draw what you are able to see in the spaces below and also give brief written descriptions, indicating whether you were able to resolve either or both into individual stars.

Andromeda Nebula

Orion Nebula

_____ _____

_____ _____

_____ _____

Part III
Nebulae and Lord Rosse—Second Photograph

Examine the second photograph, which shows various nebulae in the Hercules region. It should give you some idea of the view that Lord Rosse was able to attain during the 1840s with the 72-inch-aperture reflecting telescope he erected on his estate in Ireland.
1. As before, circle the five most prominent nebulae.
2. Using your earlier classification of nebulae, mark the prime examples in this photograph of each of the types located in the first photograph by placing an A, B, and C ... next to nebulae of those types. If you are able to discern any new types in this photograph, draw them below, labeling them S, T, . . .

Briefly describe each of your new types.

Type S	Type T	Type	Type

3. Can you formulate any additional questions to ask concerning this photograph?

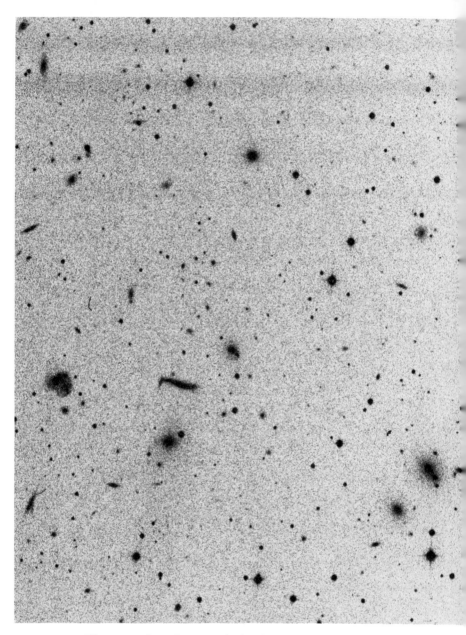

Photograph 2: Some Nebulae in the Hercules Region

Part IV
The Milky Way

This part requires inspection of the Lund Observatory map of the entire sky. The farthest left point on the map represents an object immediately adjacent to an object at the far right. The map is drawn so that the Milky Way lies along the map's equator. It is of sufficient quality that you should be able to locate some important objects on it. The map uses galactic coordinates to designate positions. The major axis of the ellipse (the horizontal line bisecting the map) is 0° latitude. Above the line and below it and more or less parallel to it are other latitude lines, running from 10° to 90° up and from –10° to –90° down. The longitude lines, which are more or less vertical and roughly perpendicular to the latitude lines, run from 0° to 360°, proceeding to the left. Note that 0° longitude is the fourth line to the left of center.

1. Locate the Orion Nebula on the map. It is known as M42 (the 42nd Messier object). It is located at 173° longitude, –16° latitude. You should be able to find a definite object on the map at this location. Mark its location on the map by placing arrows labeled M42 on the bottom and side of the map. After doing this, check with the instructor to be sure that you are proceeding correctly.
2. Locate the Andromeda Nebula (M31) at 89° longitude, –20° latitude. Mark it with M31 arrows.
3. Locate the Large Magellanic Cloud (LMC) at 247° longitude and –33° latitude. Mark it with LMC arrows.
4. Locate the Small Magellanic Cloud (SMC) at 269° longitude and –45° latitude. Mark it with SMC arrows.
5. Locate our North Celestial Pole (Polaris) at 90° longitude and 28° latitude. Mark it with NCP arrows. Then determine the location of our South Celestial Pole and mark it with SCP arrows.
6. Locate Sirius (the brightest star) at 195° longitude, –8° latitude. Mark it.

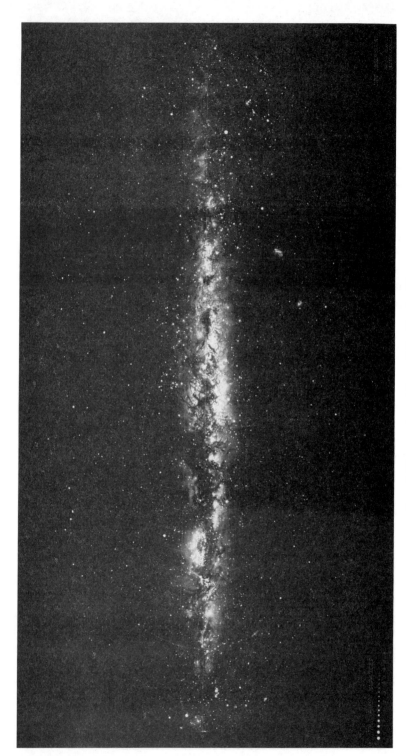

Lund Observatory Map of the Entire Sky

7. Suppose the Milky Way were a hamburger-shaped object composed of vast numbers of stars. From examining the map, can you infer where in this drawing you should position the earth? Hint: At what longitude in the map do you find the largest number of stars? Where fewest? How can you use your hamburger hypothesis to explain this?

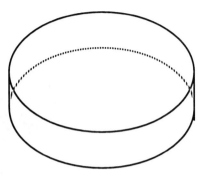

_____

_____

_____

8. The so-called **Coal Sack** centers on 271° longitude and −1° latitude. Suggest at least two hypotheses concerning why this region appears so dark.

_____

_____

9. Another dark region, the Great Rift, runs from about 340° to about 25° longitude along the equator. Can you suggest two hypotheses as to why this area is dark?

_____

_____

_____

10. Were you to locate the 25 brightest spiral nebulae (a type discovered by Lord Rosse), you would find that none lie on or near the plane of the Milky Way; they all lie above or below it, that is, they have galactic latitudes greater than 10° or less than −10° and generally rather high galactic latitudes.

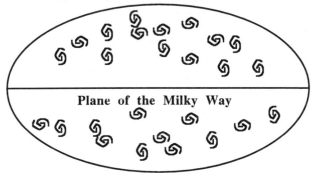

Diagram Showing Distribution of Spirals in the Milky Way

Suppose the spirals are island universes, i.e., other Milky Ways. If this is the case, how should they be distributed in relation to the Milky Way? Should they cluster in particular areas or be randomly distributed? What inference can you draw from the actual pattern of their distribution as to whether or not they are island universes?

Selected Bibliography of the History of Stellar Astronomy

Abbreviations Used for Frequently Cited Journals:
JHA = *Journal for the History of Astronomy*
PA = *Popular Astronomy*
RASCJ = *Royal Astronomical Society of Canada Journal*
RASQJ = *Royal Astronomical Society Quarterly Journal*
S&T = *Sky and Telescope*
VA = *Vistas in Astronomy*

Part I: Some General Sources: Bibliographical, Historical, and Philosophical

The best single source for bibliography on the history of astronomy is:

DeVorkin, David H., *The History of Astronomy and Astrophysics: A Selected, Annotated Bibliography* (New York: Garland, 1982).

Among the many general historical treatments of astronomy, the following deserve special notice:

Clerke, Agnes, *A Popular History of Astronomy during the Nineteenth Century*, 2nd ed. (Edinburgh: Adam and Charles Black, 1887).

Herrmann, Dieter B., *The History of Astronomy from Herschel to Hertzsprung*, trans. and revised by Kevin Krisciunas (Cambridge: Cambridge Univ. Press, 1984).

Pannekoek, A., *A History of Astronomy* (New York: Dover reprint of the 1961 original, 1989).

One of the few book length studies of the methodology of astronomy is:

Hetherington, Norriss S., *Science and Objectivity: Episodes in the History of Astronomy* (Ames: Iowa State Univ. Press, 1988). For reviews, see *JHA*, 20 (1989), 150-2 and *Isis*, 79 (1988), 704–5. Many of the episodes discussed are from the history of stellar astronomy.

A rich source of information on cosmology, including its history and philosophical aspects, is:

Hetherington, Norriss S., *Encyclopedia of Cosmology: Historical, Philosophical, and Scientific Foundations of Modern Cosmology* (New York: Garland, 1993).

Part II: General Surveys on the History of Stellar Astronomy

For a useful survey of the history of stellar astronomy, see:

Jaki, Stanley L., *The Milky Way: An Elusive Road for Science* (New

York: Science History, 1972). For reviews, see: *JHA*, 4 (1973), 200–1 and *History of Science*, 12 (1974), 299–306. The latter review compares Jaki's book with

Whitney, Charles A., *The Discovery of Our Galaxy* (New York: Alfred A. Knopf, 1971). Whitney's book is the less scholarly of the presentations. The next book, consisting primarily of essays previously published, contains much information on the general history of stellar astronomy:

Hoskin, M. A., *Stellar Astronomy: Historical Studies* (Bucks, England: Science History, 1982). This work contains excellent essays on Bradley, Wright, Lambert, Goodricke, W. Herschel, the Great Debate, and various other relevant topics. See also:

de Sitter, W., *Kosmos* (Cambridge, Mass: Harvard Univ. Press, 1932). A study of the history of research on the structure of the universe; especially good on Kapteyn.

Gore, J. Ellard, *The Visible Universe* (London: Crosby Lockwood and Son, 1893). Consists of a presentation of stellar astronomy from a historical point of view.

Hoffleit, Dorrit, "A History of Variable Star Astronomy to 1900 and Slightly Beyond," *Journal of the American Association of Variable Star Observers*, 15 (1986), 77–106.

Hoskin, M. A., "The Milky Way from Antiquity to Modern Times" in *The Milky Way Galaxy*, ed. by Hugo van Woerden et. al. (Dordrecht: Reidel, 1985), 11–24.

Lundmark, Knut, "On Metagalactic Distance Indicators," *VA*, 2 (1956), 1607–19.

Woolley, Richard, "The Stars and the Structure of the Galaxy," *RASQJ*, 11 (1970), 403–28. Covers the history from the seventeenth to the twentieth century.

Part III: From Galileo to the Death of William Herschel (1822)

For a **general survey** of sidereal astronomy in this period, see:

Hoskin, Michael, "Newton and the Beginnings of Stellar Astronomy" in G. V. Coyne, M. Heller, and J. Zycinski (eds.), *Newton and New Directions in Science* (Vatican City: Specola Vaticana, 1988), pp. 55–63.

Hoskin, M. A., "Sidereal Astronomy in Adolescence" in *Avant, Avec, Après Copernic* (Paris: Blanchard, 1975), 315–42.

For analyses of Wright, Kant, Lambert, and Herschel from a somewhat different perspective, see:

Crowe, Michael J., *The Extraterrestrial Life Debate 1750–1900: The Idea of a Plurality of Worlds from Kant to Lowell* (Cambridge, England: Cambridge Univ. Press, 1986), pp. 41–70.

Bibliography 407

On the early history of stellar astronomy, see:

Hoskin, M. A., "Newton, Providence, and the Universe of Stars," *JHA*, 8 (1977), 77–101.

Hoskin, M. A., "Novae and Variables from Tycho to Bullialdus," *Sudhoffs Archiv*, 61 (1977), 195–204; also in M. Hoskin, *Stellar Astronomy*, pp. 22–8.

Williams, Mari, "Was There Such a Thing as Stellar Astronomy in the Eighteenth Century?" *History of Science*, 21 (1983), 369–85; for commentary on this paper, see:

Hoskin, M. A., "Stellar Astronomy in the Eighteenth Century: A Comment," *History of Science*, 21 (1983), 385–8.

On the early history of the study of the most prominent **nebulae**, the best sources are:

Ashworth, William, "John Bevis and His *Uranographia* (ca 1750)," *Proceedings of the American Philosophical Society*, 125 (1981), 52–73.

Gingerich, O., "Abbe Lacaille's List of Clusters and Nebulae," *S&T*, 19 (1960), 207–8.

Gingerich, O., "The Mysterious Nebulae, 1610–1924," *RASCJ*, 81 (1987), 113–27.

Gingerich, O., "Messier and His Catalogue," *S&T*, 12 (1953), 255–8.

Gingerich, O., "The Missing Messier Objects," *S&T*, 20 (1960), 196–9.

Harrison, Thomas G., "The Orion Nebula: Where in History Is It?," *RASQJ*, 25 (1984), 65–79. Espouses the view that the Orion nebula has changed.

Hogg, Helen Sawyer, "Catalogues of Nebulous Objects in the Eighteenth Century," *RASCJ*, 41 (1947), 265–73.

Hogg, Helen Sawyer, "Halley's List of Nebulous Objects," *RASCJ*, 41 (1947), 60–71.

Hogg, Helen Sawyer, "Derham's Catalogue of Nebulous Objects from Hevelius' Prodomus," *RASCJ*, 41 (1947), 333–8.

Hogg, Helen Sawyer, "Halley's List of Nebulous Objects," *RASCJ*, 41 (1947), 60–71.

Jones, Kenneth Glyn, *Messier's Nebulae and Star Clusters*, 2nd ed. (Cambridge: Cambridge Univ. Press, 1991), which also contains some useful bibliography.

Jones, Kenneth Glyn, *The Search for the Nebulae* (New York: Science History, 1975), which consists of a collection of articles first published as "The Search for the Nebulae" in *Journal of the British Astronomical Association*, 78 and 79, (1968).

Mallas, John H., and Evered Kreimer, *The Messier Album* (Cambridge: Cambridge Univ. Press, 1978). Both this and the first

Jones book contain photographs of Messier's nebulae. This volume reprints Messier's catalogue, whereas Jones's volume is richer in historical information.

Serio, G. F., L. Indorato, and P. Nastasi, "G. B. Hodierna's Observations of Nebulae and His Cosmology," *JHA*, 16 (1985), 1–36. Hodierna was a seventeenth-century Italian astronomer.

Warner, Deborah J., *The Sky Explored: Celestial Cartography 1500–1800* (New York: Alan Liss, 1979).

For **Wright**, the best sources are the editions by M. A. Hoskin of Wright's *Clavis Coelestis*, *Original Theory*, and *Second Thoughts*. These three books were reviewed in *Isis*, 63 (1972), 235–41. See also:

Hoskin, M., "The English Background to the Cosmology of Wright and Herschel" in W. Yourgrau and A. D. Breck (eds.), *Cosmology, History and Theology* (New York: Plenum, 1977), 219–31.

Hoskin, M. A., "The Cosmology of Thomas Wright of Durham," *JHA*, 1 (1970), 44–52; reprinted in Hoskin's *Stellar Astronomy*.

Hughes, Edward, "The Early Journal of Thomas Wright of Durham," *Annals of Science*, 7 (1951), 1–24.

Robinson, J. Hedley, "Thomas Wright, Herschel and the Galaxy," *British Astronomical Association Journal*, 100 (1990), 93–4.

Schaffer, Simon, "The Phoenix of Nature: Fire and Evolutionary Cosmology in Wright and Kant," *JHA*, 9 (1978), 180–200.

For **Kant**, see the new translation of his main relevant work:

Kant, Immanuel, *Universal Natural History and Theory of the Heavens*, trans. by S. L. Jaki (Edinburgh: Scottish Academic Press, 1981; paperback ed., 1992); includes extensive commentary as well as translation of the parts left out in the earlier translation by Hastie. See also the prefaces or introductions (by W. Ley, G. J. Whitrow, and M. Munitz) to the three reprintings of Hastie's incomplete translation of Kant's *General Natural History and Theory of the Heavens*. See also:

Hetherington, Norriss S., "Sources of Kant's Model of the Stellar System," *Journal of the History of Ideas*, 34 (1973), 461–2.

Jones, K. G., "The Observational Basis for Kant's Cosmogony: A Critical Analysis," *JHA*, 2 (1971), 29–34.

Palmquist, Stephen, "Kant's Cosmology Revisited," *Studies in History and Philosophy of Science*, 18 (1987), 255–69

Whitrow, G. J., "Kant and the Extragalactic Nebulae," *RASQJ*, 8 (1967), 48–56.

On **Lambert**, see:

Lambert, Johann H., *Cosmological Letters on the Arrangement of the World-Edifice*, trans. by Stanley L. Jaki (New York: Science History, 1976). See also:

Hoskin, M. A., "The Cosmology of J. H. Lambert" in Hoskin, *Stellar Astronomy*, pp. 117–23.

Hoskin, M. A., "Lambert and Herschel," *JHA*, 9 (1978), 140–2.

Hoskin, M. A., "Newton and Lambert," *Colloque International et interdisciplinaire Jean-Henri Lambert* (Paris: Editions Ophrys, 1979), 363–70.

Hoskin, M. A., "Newton and Lambert," *VA*, 22 (1978), 483–4.

Jaki, Stanley L., "Lambert and the Watershed of Cosmology," *Scientia*, 113 (1978), 75–95.

On **Michell**, see:

Gower, B., "Astronomy and Probability: Forbes versus Michell on the Distribution of the Stars," *Annals of Science*, 39 (1982), 145–60.

Hardin, C. L., "The Scientific Work of John Michell," *Annals of Science*, 22 (1966), 27–47.

McCormmach, Russell, "John Michell and Henry Cavendish: Weighing the Stars," *British Journal for the History of Science*, 4 (1968), 126–55.

On **Goodricke**, see:

Hoskin, M. A., "Goodricke, Pigott and the Quest for Variable Stars" in Hoskin, *Stellar Astronomy*, pp. 37–55; see also *JHA*, 10 (1979), 23–41.

Materials on **William Herschel** are very extensive; the following are among the most important items:

Bennett, J. A., "The Telescopes of William Herschel," *JHA*, 7 (1976), 75–108.

Hendry, John, "Mayer, Herschel and Prévost on the Solar Motion," *Annals of Science*, 39 (1982), 61–75.

Hoskin, M. A., "Herschel's Determination of the Solar Apex," *JHA*, 11 (1980), 153–63; also in M. Hoskin, *Stellar Astronomy*, pp. 56–66.

Hoskin, Michael A., "William Herschel and the Construction of the Heavens," *Proceedings of the American Philosophical Society*, 133 (Sept., 1989), 427–33.

Hoskin, M. A., *William Herschel and the Construction of the Heavens* (London: Oldbourne, 1963), which contains many of Herschel's most important papers. For a review, see *History of Science*, 3 (1964), 91–101.

Hoskin, M. A., "William Herschel and the Making of Modern Astronomy," *Scientific American, 254* (Feb., 1986), 106–12.

Hoskin, M. A., *William Herschel: Pioneer of Sidereal Astronomy* (New York: Sheed and Ward, 1959).

Hoskin, M. A., "William Herschel's Early Investigations of Nebulae: A Reassessment," *JHA, 10* (1979), 165–76; also in Michael Hoskin, *Stellar Astronomy*, pp. 125–36.

Jones, R. V., "Through Music to the Stars: William Herschel, 1738–1822," *Notes and Records of the Royal Society, 33* (1978), 37–56.

Lovell, Bernard, "Herschel's Work on the Structure of the Universe," *Notes and Records of the Royal Society, 33* (1978), 57–75.

Lubbock, Constance, *The Herschel Chronicle* (Cambridge, Cambridge Univ. Press, 1933). The most detailed biography of William Herschel.

Millman, Peter M., "The Herschel Dynasty: Part I: William Herschel," *RASCJ , 74* (1980), 134–46.

Schaffer, Simon, "'The Great Laboratories of the Universe': William Herschel on Matter Theory and Planetary Life," *JHA, 11* (1980), 81–111.

Schaffer, Simon, "Herschel in Bedlam: Natural History and Stellar Astronomy," *British Journal for the History of Science, 13* (1980), 211–39.

Subsection on the Seven Correctional Factors

In general, the most useful source is:

Robert Grant's *History of Physical Astronomy* (London, 1852).

For scientific details, see:

Smart, W. M., *Text-Book on Spherical Astronomy*, 5th ed. (Cambridge: Cambridge Univ. Press, 1965).

Precession

This is discussed in many histories of ancient astronomy, e.g.:

Dreyer, J. L. E., *A History of Astronomy from Thales to Kepler* (New York: Dover, 1953).

Proper Motion

See the discussion in Angus Armitage, *Edmond Halley* (London: Nelson, 1966).

Aberration and Nutation

Bradley, James, *Miscellaneous Works and Correspondence*, ed. with a supplement by S. P. Rigaud (New York, Johnson Reprint, 1972 of the 1832–33 original).

"James Bradley, 1693–1762; Bicentenary Contributions," *RASQJ*, 4 (1963), 38–61. Consists of articles by W. H. McCrea, D. E. Blackwell, R. Woolley, H. R. Calvert, P. S. Laurie, and D. H. Waters.

Hoskin, M. A., "Hooke, Bradley and the Aberration of Light" in Hoskin, *Stellar Astronomy*, pp. 29–36.

Sarton, George, "Discovery of the Aberration of Light," *Isis*, 16 (1931), 233–265. Includes the classic paper.

Sarton, George, "Discovery of the Main Nutation of the Earth's Axis," *Isis*, 17 (1932), 333–83. Includes Bradley's paper.

Stewart, Albert B., "The Discovery of Stellar Aberration," *Scientific American*, 210 (March, 1964), 100–8.

Turner, H. H., "Bradley's Discoveries of the Aberration of Light and of the Nutation of the Earth's Axis" in Turner's *Astronomical Discovery* (Berkeley: Univ. of California Press reprinting of the 1904 original, 1963), pp. 86–120.

Williams, Mari, "James Bradley and the Eighteenth Century 'Gap' in Attempts to Measure Annual Stellar Parallax," *Notes and Records of the Royal Society*, 37 (1982), 83–100.

Personal Equation

Duncombe, Raynor, "Personal Equation in Astronomy," *PA*, 53 (1945), 2–13; 63–76; 110–121 with extensive bibliography.

Schaffer, Simon, "Astronomers Mark Time: Discipline and the Personal Equation," *Science in Context*, 2 (1988), 115–45.

Parallax

Fernie, J. D., "The Historical Search for Stellar Parallax," *RASCJ*, 69 (August, 1975), 153–61; 222–39.

Hetherington, N., "The First Measurements of Stellar Parallax," *Annals of Science*, 28 (1972), 319–25.

Hoffleit, Doris, "The Quest for Stellar Parallax," *PA*, 57 (June 1949), 259–73.

Hoskin, M. A., "Stellar Distances: Galileo's Method and Its Subsequent History," *Indian Journal for the History of Science*, 1 (1966), 22–9.

Joy, Alfred H., "A Century's Progress in Determining Stellar Distances," *Astronomical Society of the Pacific Leaflet* 173 (July, 1943).

Pedersen, K. M., "Römer, Flamsteed, and the Search for a Stellar Parallax," *VA*, 20 (1976), 165–9.

Serio, Giorgia Foderà, "Giuseppe Piazzi and the Discovery of the Proper Motion of 61 Cygni," *JHA*, 21 (1990), 275–82.

Strand, K. Aa., "Determination of Stellar Distances," *Science, 144* (June, 1964), 1299–1309.
Struve, Otto, "The First Determination of Stellar Parallax," *S&T, 16* (1956), 9–12,69–72.
Struve, Otto, "The First Stellar Parallax Determinations" in *Men and Moments in the History of Science*, ed. by H. M. Evans (Seattle: Univ. of Washington Press, 1959), pp. 177–206. The fullest treatment of this topic.
Williams, Mari E. W., "Flamsteed's Alleged Measurement of Annual Parallax for the Pole Star," *JHA, 10* (1979), 102–16.
Williams, Mari E. W., *Attempts to Measure Annual Stellar Parallax: Hooke to Bessel* (a 1981 Imperial College, London, doctoral dissertation).

Part IV: From 1822 (Death of William Herschel) to 1900
In general: no detailed survey of nineteenth-century stellar astronomy has ever been written; some of the useful general sources are:
Clerke, Agnes, *The System of the Stars* (London: Longmans, Green and Co., 1890); 2nd ed. (London: Adam and Charles Black, 1905).
Gore, J. E., *The Visible Universe* (London: Crosby Lockwood and Son, 1893).
Hoskin, M. A., "The Nebulae from Herschel to Huggins" in Hoskin's *Stellar Astronomy*, pp. 137–53.
North, J. D., *The Measure of the Universe* (New York: Dover reprint of the 1965 original); see Ch. I: "Nineteenth-Century Astronomy: The Nebulae."
Proctor, R. A., *Old and New Astronomy*, completed by A. C. Ranyard (London: Longmans, Green and Co., 1895).
Rohlfs, Kristen, "Galactic Astronomy in Continental Europe in the Nineteenth Century in the Time Succeeding Herschel," *VA, 32* (1989), 215-23.
Smith, R. W., "Studies of the Milky Way 1850–1930: Some Highlights" in *The Milky Way Galaxy*, ed. by Hugo van Woerden, R. J. Allen, and W. B. Burton (Dordrecht: Reidel, 1985), pp. 43–58.

On **John Herschel**, the only biography is:
Buttmann, G., *The Shadow of the Telescope* (New York: Charles Scribner's Sons, 1970). See also:
Cannon, Walter F., "John Herschel and the Idea of Science," *Journal of the History of Ideas*, 22 (1961), 215–39. A fine study of the place of John Herschel in Victorian science.

Crowe, M. J., "Introduction" to Crowe (ed.), *The Letters and Papers of Sir John Herschel: A Guide to the Manuscripts and Microfilm* (in the series: *Collections from the Royal Society*), (Bethesda, Maryland: University Publications of America, 1991), pp. v–xxxviii.

Evans, David S., "The Astronomical Work of Sir John Herschel at the Cape" in "Sir John Herschel at the Cape, 1834–1838," a special issue of *The South African Public Library Quarterly Bulletin*, 13:3 (Dec., 1957), 44–58.

Evans, David S., "Dashing and Dutiful: Herschel and Maclear Made a Strange If Effective Team in Their Astronomical Work at the Cape of Good Hope," *Science*, 127 (1958), 935–48.

Evans, David S., et al., *Herschel at the Cape* (Austin: Univ. of Texas Press, 1969).

Hoskin, M. A., "The Cosmology of Sir John Herschel," *JHA*, 17 (1987), 1–34.

Millman, Peter M., "The Herschel Dynasty: Part II: John Herschel," *RASCJ*, 74 (1980), 203–15.

Proctor, Richard A., "Sir John Herschel" and "Sir John Herschel as a Theorist in Astronomy" in Proctor's *Essays in Astronomy* (London: Longmans, Green and Co., 1872), pp. 1–7; 8–23.

Schweber, S. S., "John F. W. Herschel: A Prefatory Essay," in S. S. Schweber (ed), *Aspects of the Life and Thought of Sir John Frederick Herschel*, vol. I (New York: Arno, 1981), 1–158.

Warner, Brian, "Sir John Herschel's Description of His 20–feet Reflector," *VA*, 23 (1979), 75–107.

Warner, Brian and Nancy, *Maclear & Herschel: Letters & Diaries at the Cape of Good Hope 1834–1838* (Cape Town: Balkema, 1984).

On **William Parsons, Lord Rosse**, the best single source is:

Moore, Patrick, *Astronomy of Birr Castle* (London: Mitchell Beazley, 1971). See also:

Ashbrook, Joseph, "Spiral Structures in Galaxies," *S&T*, 35 (1968), 366–7.

Bennett, J. A., "Lord Rosse and the Giant Reflector" in *Science in Ireland, 1800–1930: Tradition and Reform (Proceedings of an International Symposium Held at Trinity College Dublin, March 1986*, ed. by J. R. Nudde, N. D. McMillan, D. L. Weaire, S. M. P. McKenna Lawlor (Dublin: Trinity College Dublin, 1988), pp. 105-113.

Bennett, J. A., and M. A. Hoskin, "The Rosse Papers and Instruments," *JHA*, 12 (1981), 216–29.

Dewhirst, David, and M. A. Hoskin, "The Rosse Spirals," *JHA*, 22 (1991), 257–66.

Hetherington, Norriss S., "The Earl of Rosse's Experiments on Reflecting Telescopes," *British Astronomical Association Journal*, 87 (1977), 472–7.

Hoskin, M. A., "Apparatus and Ideas in Mid–Nineteenth Century Cosmology" in *VA*, 9 (1967), 79–85.

Hoskin, M. A., "The First Drawing of a Spiral Nebula," *JHA*, 13 (1982), 97–101.

Hoskin, M. A., "Rosse, Robinson, and the Resolution of the Nebulae, *JHA*, 21 (1990), 331-44.

Ryden, Barbara, "The Earl and His Leviathan," *Griffith Observer*, 53 (Nov., 1989), 2–14.

On **Olbers and his paradox**, see especially:

Jaki, Stanley L., *The Paradox of Olbers' Paradox* (New York: Herder and Herder, 1969) and the review of it in *History of Science, 10:* 128–32. See also:

Harrison, Edward, *Darkness at Night: A Riddle of the Universe* (Cambridge: Harvard Univ. Press, 1987).

Harrison, H. R., "Why the Sky Is Dark at Night," *Physics Today*, 27 (Feb., 1974), 30–6.

Hoskin, M. A., "Dark Skies and Fixed Stars," *British Astronomical Association Journal*, 83 (1973), 254–62.

Hoskin, M. A., "Stukeley's Cosmology and the Newtonian Origin of Olbers's Paradox," *JHA*, 16 (1985), 77–112.

Hoskin, Michael, "Cosmology and Theology: Newton and the Paradoxes of an Infinite Universe of Stars," in Sergio Rossi (ed.), *Science and Imagination in 18th Century British Culture* (Milan: Unicopli, 1987), pp. 237–40.

Jaki, Stanley L., "New Light on Olbers' Dependence on Chéseaux," *JHA*, 1 (1970), 53–5.

Jaki, Stanley L., *Olbers Studies: With Three Unpublished Manuscripts by Olbers* (Tucson: Pachart, 1991).

Struve, Otto, "Some Thoughts on Olbers' Paradox," *S&T*, 25 (1963), 140–2.

On **Wilhelm Struve**, see especially:

Batten, Alan H., *Resolute and Undertaking Characters: The Lives of Wilhelm and Otto Struve* (Dordrecht: Reidel, 1988).

Szanser, Adam J., "F. G. W. Struve (1793–1864). Astronomer at the Pulkovo Observatory," *Annals of Science*, 28 (1972), 327–46.

On **parallax**, see the section of this bibliography on the seven correctional factors.

On **the design of reflecting telescopes**, see:
Chapin, S. L., "'In a Mirror Brightly': French Attempts to Build Reflecting Telescopes Using Platinum," *JHA*, 3 (1972), 87–104. On the method developed by Léon Foucault and C. A. von Steinheil for preparing telescope mirrors by electroplating glass.
Hogg, Arthur, "The Last of the Specula," *Astronomical Society of the Pacific Leaflet 364* (Oct., 1959).

On **astronomical photography**, see:
Bell, Trudy E., "History of Astrophotography," *Astronomy*, 4 (1976), 66–79.
Hoffleit, D., *Some Firsts in Astronomical Photography* (Cambridge, Mass., 1950).
Lankford, John, "The Impact of Photography on Astronomy" in *Astrophysics and Twentieth-Century Astronomy*, Part A, ed. by Owen Gingerich (Cambridge: Cambridge Univ. Press, 1984), pp. 16–39.
Norman, D., "The Development of Astronomical Photography," *Osiris*, 5 (1938), 560–94.
Russel, H. C., "Progress of Astronomical Photography," *PA*, 2 (1894), 101–5, 170–6; (1895), 310–316, 457–63.
Tenn, Joseph S., "The Rise and Fall of Astrophotography," *Griffith Observer*, 53:8 (Aug., 1989), 1–9.
Vaucouleurs, Gérard de, *Astronomical Photography*, trans R. Wright (London: Faber and Faber, 1961).
Weaver, Harold F., "The Development of Astronomical Photography," *PA*, 54 (1946), 211–30; 287–99; 339–51; 389–404; 451–64; 504–26.

On **astronomical spectroscopy**, see:
Dingle, Herbert, "A Hundred Years of Spectroscopy," *British Journal for the History of Science*, 1 (1963), 199–216.
Hearnshaw, J. B., *The Analysis of Starlight: One Hundred and Fifty Years of Astronomical Spectroscopy* (Cambridge: Cambridge Univ. Press, 1986).
Huggins, William, "The New Astronomy: A Personal Retrospect," *Nineteenth Century*, 41 (1897), 907–29, which consists of the recollections of one of the great pioneers.
Leitner, A., "Joseph Fraunhofer," *American Journal of Physics*, 43 (Jan., 1975), 59–68.

McCarthy, M. F., "Fr. Secchi and Stellar Spectra," *PA*, *58* (1950), 153–69.
McGucken, William, *Nineteenth-Century Spectroscopy* (Baltimore: Johns Hopkins Univ. Press, 1969).
Meadows, A. J., "The New Astronomy" in *Astrophysics and Twentieth-Century Astronomy*, Part A, ed. by Owen Gingerich (Cambridge: Cambridge Univ. Press, 1984), pp. 59–72.
Meadows, A. J., "The Origins of Astrophysics" in *Astrophysics and Twentieth-Century Astronomy*, Part A, ed. by Owen Gingerich (Cambridge: Cambridge Univ. Press, 1984), pp. 3–15.
Menzel, D. H., "The History of Astronomical Spectroscopy," *Annals of the New York Academy of Sciences*, *198* (1972), 225–44. Coverage extends into the twentieth century.
See, T. J. J., and Hector MacPherson, "Tribute to the Memory of William Huggins," *PA*, *18* (1910), 287–98; 398–401.
von Rohr, Moritz, "Fraunhofer's Work and Its Present Day Significance," *Transactions of the Optical Society*, *27* (1926), 277–94.
Woolf, Harry, "Astrophysics in the Early Nineteenth Century," *Actes XIe Congrès international d'histoire de sciences*, *3* (1965; pub. 1968), 127–35.
Woolf, Harry, "The Beginning of Astronomical Spectroscopy," *Mélanges Alexandre Koyré*, vol. I (Paris, 1964), pp. 619–34. Covers the period up to Fraunhofer.

Miscellaneous

Ashbrook, Joseph, "Spiral Structure in Galaxies," *S&T*, *36* (1968), 366–7.
Hearnshaw, J. B., "Origins of the Stellar Magnitude Scale," *S&T*, *83* (1992), 494–9.
Hoffleit, Dorrit, "A History of Variable Star Astronomy to 1900 and Slightly Beyond," *Journal of the American Association of Variable Star Observers*, *15:2* (1986), 77–105.
Holden, Edward S., *Monograph of the Nebula of Orion* in *Astronomical and Meteorological Observations Made ... at the United States Naval Observatory* (Washington: Government Printing Office, 1878), Appendix: pp. 1–230. A rich source for history of studies of the Orion nebula.
Jones, Derek, "Norman Pogson and the Definition of Stellar Magnitude," *Astronomical Society of the Pacific Leaflet 469* (July, 1968).
Hussey, William J., "Notes on the Progress of Double Star Astronomy," *Publications of the Astronomical Society of the Pacific*, *12* (1900), 91–103.

Jones, Kenneth Glyn, "S Andromedae, 1885: An Analysis of Contemporary Reports and a Reconstruction," *JHA*, 7 (1976), 27–40.

Mayer, Annemarie, "On the History of the Stellar Magnitude Scale," *Journal of the American Association of Variable Star Observers*, 15:2 (1986), 283–5.

Nha, Il-Seong, "Calibration of the Stellar Magnitude Scale Prior to Pogson," *Historia Scientariarum*, 35 (1988), 39–44.

Osterbrock, Donald E., *James E. Keeler: Pioneer American Astrophysicist* (Cambridge: Cambridge Univ. Press, 1984).

Struve, Otto, "Milestones in Double Star Astronomy," *S&T*, 24 (July, 1962), 17–19.

Vaucouleurs, Gérard de, "Discovering M31's Spiral Shape," *S&T*, 73 (1987), 595–8.

Williams, Mari, "Beyond the Planets: Early Nineteenth-Century Studies of Double Stars," *British Journal for the History of Science*, 17 (1981), 295–309.

Part V: From 1900 to the Expanding Universe

In general, see:

Berendzen, R., R. Hart, and D. Seeley, *Man Discovers the Galaxies* (New York: Science History, 1975). An excellent study of the 1900–1930 period.

Fernie, J. D., "The Historical Quest for the Nature of the Spiral Nebulae," *Astronomical Society of the Pacific Publications*, 82 (1970), 1189–1230.

Kienle, Hans, "Historical Development of Our Ideas Concerning the Structure of the Galaxy" in *Structure and Evolution of the Galaxy* (Proceedings of the NATO Advanced Study Institute Held in Athens, Sept. 8–19, 1969), (Dordrecht: Reidel, 1971), pp. 1–16.

Lang, Kenneth R., and Owen Gingerich (eds.), *A Source Book in Astronomy and Astrophysics, 1900–1975* (Cambridge: Harvard Univ. Press, 1979). Contains many major papers in twentieth-century stellar astronomy along with commentaries.

Parker, Barry, "Discovery of the Expanding Universe," *S&T*, 72 (Sept, 1986), 227–30.

Shapley, Harlow (ed.), *Source Book in Astronomy 1900–1950* (Cambridge: Harvard Univ. Press, 1960).

Smith, Robert W., *The Expanding Universe: Astronomy's 'Great Debate'* (Cambridge: Cambridge Univ. Press, 1982). This is the best overall survey of the developments in the great debate. Contains a very useful bibliography.

On **Kapteyn, statistical astronomy,** and **galactic rotation,** see:

Oort, J. H., "Development of Our Insight into the Structure of Our Galaxy between 1920 and 1940," *Annals of the New York Academy of Science,* 193 (1972), 255–66.

Paul, E. Robert, "The Death of a Research Programmme: Kapteyn and the Dutch Astronomical Community," *JHA,* 12 (June, 1981), 77–94.

Paul, E. Robert, "Kapteyn and Statistical Astronomy" in *The Milky Way Galaxy,* ed. by Hugo van Woerden, R. J. Allen, and W. B. Burton (Dordrecht: Reidel, 1985), 25–42.

Paul, Erich Robert, *The Milky Way Galaxy and Statistical Cosmology 1890–1924* (Cambridge: Cambridge Univ. Press, 1993).

Thoren, V., G. Gow, and K. Honeycutt, "An Early View of Galactic Rotation," *Centaurus,* 18 (1974), 301–14.

On **Leavitt** and the **period-luminosity relationship,** see:

Bailey, S., "Henrietta Swan Leavitt," *PA,* 30 (1922), 197–9.

Fernie, J. D., "The Period-Luminosity Relation: A Historical Review," *Astronomical Society of the Pacific Publications,* 81 (1969), 707–31.

Mitchel, Helen Buss, "Henrietta Swan Leavitt and Cepheid Variables," *Physics Teacher,* 14 (March, 1976), 162–7.

On the **Hertzsprung-Russell diagram,** see:

DeVorkin, David H., "The Origins of the Hertzsprung-Russell Diagram" in Philip, A. G. Davis, and David H. DeVorkin (eds.), *In Memory of Henry Norris Russell* (Albany: Dudley Observatory, 1977), pp. 61–77.

DeVorkin, David H., "Stellar Evolution and the Origin of the Hertzsprung-Russell Diagram" in *Astrophysics and Twentieth-Century Astronomy,* Part A, ed. by Owen Gingerich (Cambridge: Cambridge Univ. Press, 1984), pp. 90–108.

DeVorkin, David H., "Steps toward the Hertzsprung-Russell Diagram," *Physics Today,* 31 (1978), 32–9.

Kenat, Ralph, and David H. DeVorkin, "Quantum Physics and the Stars (III): Henry Norris Russell and the Search for a Rational Theory of Stellar Spectra," *JHA,* 21 (1990), 157–86.

Nielsen, Axel V., "The History of the Hertzsprung-Russell Diagram," *Centaurus,* 9 (1963), 219–53.

Sitterly, Bancroft W., "Changing Interpretations of the Hertzsprung-Russell Diagram: A Historical Note," *VA,* 12 (1970), 357–66.

Strand, K. Aa., "Hertzsprung's Contributions to the HR Diagram" in Philip, A. G. Davis and David H. DeVorkin (eds.), *In Memory of Henry Norris Russell* (Albany: Dudley Observatory, 1977), 55–9.

Waterfield, R. L., "The Story of the Hertzsprung-Russell Diagram," *British Astronomical Association Journal*, 67 (1956), 1–24.

On van Maanen and rotations of spiral nebulae, see:

Berendzen, R., and R. Hart, "Adriaan van Maanen's Influence on the Island Universe Theory," *JHA*, 4 (1973), 46–56; 73–98.

Brashear, Ronald W., and Norriss Hetherington, "The Hubble–van Maanen Conflict over Internal Motions in Spiral Nebulae: Yet More Information on an Already Old Topic," *VA*, 35 (1991), 415–23.

Hart, Richard Cullen, *Adriaan van Maanen's Influence on the Island Universe Theory* (a 1973 doctoral dissertation at Boston University).

Hetherington, Norriss S., "Additional Shapley–van Maanen Correspondence," *JHA*, 7 (1976), 73–4.

Hetherington, Norriss S., "Adriaan van Maanen and Internal Motions of Spiral Nebulae: A Historical Review," *RASQJ*, 13 (1972), 25–39.

Hetherington, Norriss S., "Adriaan van Maanen on the Significance of Internal Motions in Spiral Nebulae," *JHA*, 5 (1974), 52–3.

Hetherington, Norriss S., "Edwin Hubble on Adriaan van Maanen's Internal Motions in Spiral Nebulae," *Isis*, 65 (1974), 390–3.

Hetherington, Norriss S., "Edwin Hubble's Examination of Internal Motions in Spiral Nebulae," *RASQJ*, 15 (1974), 392–418.

Hetherington, Norriss S., "New Source Material on Shapley, van Maanen, and Lundmark" *JHA*, 7 (1976), 73–4.

Hetherington, Norriss S., "The Simultaneous 'Discovery' of Internal Motions in Spiral Nebulae," *JHA*, 6 (1975), 115–25.

Hetherington, Norriss S., "Walter S. Adams and the Imposed Settlement between Edwin Hubble and Adriaan van Maanen," *JHA*, 23 (1992), 53–6.

On Curtis, Shapley, and the Great Debate, see:

Aitken, R. O., "Heber Doust Curtis 1872–1942," *Biographical Memoirs of the National Academy of Sciences*, 22 (1943), 275–94.

Bok, Bart, "Harlow Shapley and the Discovery of Our Galaxy's Center" in J. Neyman (ed.), *The Heritage of Copernicus* (Cambridge: Massachusetts Institute of Technology Press, 1974), pp. 26–62.

Gingerich, Owen, and Barbara Welther, "Harlow Shapley and the Cepheids," *S&T*, 70 (1985), 540–2.

Gingerich, Owen, "Shapley's Impact" in *The Harlow-Shapley Symposium on Globular Cluster Systems in Galaxies*, ed. by J. Grindlay and A. G. Davis Philip (Dordrecht: Kluwer, 1988), pp. 23–30; see also pp. 31–3.

Hetherington, Norriss R., "The Shapley-Curtis Debate," *Astronomical Society of the Pacific Leaflet 490* (1970).

Hogg, Helen Sawyer, "Harlow Shapley and Globular Clusters," *Astronomical Society of the Pacific Publications*, 77 (1965), 336–46.

Hogg, Helen Sawyer, "Shapley's Era" in *The Harlow-Shapley Symposium on Globular Cluster Systems in Galaxies*, ed. by J. Grindlay and A. G. Davis Philip (Dordrecht: Kluwer, 1988), pp. 11–22.

Hoskin, M. A., "The Great Debate: What Really Happened," *JIIA*, 7 (1976), 169–82.

Hoskin, Michael, "Shapley's Debate" in *The Harlow-Shapley Symposium on Globular Cluster Systems in Galaxies*, ed. by J. Grindlay and A. G. Davis Philip (Dordrecht: Kluwer, 1988), pp. 3–9.

McCarthy, Martin F., "Harlow Shapley and Red Giant Stars" in *The Harlow-Shapley Symposium on Globular Cluster Systems in Galaxies*, ed. by J. Grindlay and A. G. Davis Philip (Dordrecht: Kluwer, 1988), pp. 481–2.

Peterson, Charles J., "Harlow Shapley and the University of Missouri" in *The Harlow-Shapley Symposium on Globular Cluster Systems in Galaxies*, ed. by J. Grindlay and A. G. Davis Philip (Dordrecht: Kluwer, 1988), pp. 479–80.

Shapley, Harlow, *Through Rugged Ways to the Stars* (New York: Charles Scribner's Sons, 1969). This is Shapley's autobiography. It should be read in conjunction with R. Berendzen's review in *JHA*, 1 (1970), 85–7.

Smith, Robert, "The Great Debate Revisited," *S&T*, 65 (1983), 28–9.

Struve, Otto, "A Historic Debate about the Universe," *S&T*, 19 (May, 1960), 398–401.

Struve, Otto, and Velta Zebergs, *Astronomy in the Twentieth Century* (New York: Macmillan, 1962), esp. chs. 19 and 20.

Welther, Barbara L., "Harlow Shapley: A View from the Harvard Archives" in *The Harlow-Shapley Symposium on Globular Cluster Systems in Galaxies*, ed. by J. Grindlay and A. G. Davis Philip (Dordrecht: Kluwer, 1988), pp. 477-8.

On **Hubble**, see:

Berendzen, Richard, and M. A. Hoskin, "Hubble's Announcement of Cepheids in Spiral Nebulae," *Astronomical Society of the Pacific Leaflet 504* (1971).

Goldsmith, Donald, "Edwin Hubble and the Universe outside Our Galaxy" in J. Neyman (ed.), *The Heritage of Copernicus* (Cambridge: Massachusetts Institute of Technology Press, 1974), 65–94.

Hart, R., and R. Berendzen, "Hubble's Classification on Non-Galactic Nebulae 1922–1926," *JHA*, 2 (1971), 109–19.

Hart, R., and R. Berendzen, "Hubble, Lundmark and the Classification on Non-Galactic Nebulae," *JHA*, 2 (1971), 200.

Hetherington, Norriss S., "Edwin Hubble and a Relativistic Expanding Model of the Universe," *Astronomical Society of the Pacific Leaflet 509* (1971).

Hetherington, Norriss S. (ed.), *The Edwin Hubble Papers: Previously Unpublished Papers on the Extragalactic Nature of the Spiral Nebulae* (Tucson: Pachart, 1990).

Hetherington, Norriss S., "Edwin Hubble's Cosmology" in *Edwin Hubble Centennial Symposium, University of California, 1989. Evolution of the Universe of Galaxies; Edwin Hubble Centennial Symposium*, ed. by Richard C. Kron (San Francisco, Calif.: Astronomical Society of the Pacific, 1990), 22–4.

Hetherington, Norriss S., "Edwin Hubble's Examination of Internal Motions in Spiral Nebulae," *RASQJ*, 15 (1974), 392–418.

Hetherington, Norriss S., "Hubble's Cosmology," *American Scientist*, 78 (1990), 142–51.

Hetherington, Norriss S., "Philosophical Values and Observation in Edwin Hubble's Choice of a Model of the Universe," *Historical Studies in the Physical Sciences*, 13 (1982), 41–67.

Hetherington, Norriss S., *The Development and Early Application of the Velocity-Distance Relationship* (a 1970 doctoral dissertation at Indiana University).

Hoskin, M. A., "Edwin Hubble and the Existence of External Galaxies," *XIIe Congrès international d'histoire de sciences*, vol. 5 (Paris, 1971), 49–53.

Jones, Brian, "The Legacy of Edwin Hubble," *Astronomy*, 17 (Dec., 1989), 38–44.

Maddison, Ron, "Edwin Hubble and the Idea of an Expanding Universe," *Astronomy Now*, 3 (Aug., 1989), 43–7.

Mayall, N. U., "Edwin Hubble: Observational Cosmologist," *S&T*, 13 (Jan., 1954), 78–80, 85.

Mayall, N. U., "Edwin Powell Hubble," *Biographical Memoirs of the*

National Academy of Sciences, 41 (1970), 173–214.
Sandage, Allan, "Edwin Hubble 1889–53," *RASCJ, 83* (1989), 351–62.
Smith, Robert W., "The Origins of the Velocity-Distance Relation," *JHA, 10* (1979), 133–65.
Smith, Robert, "Edwin P. Hubble and the Transformation of Cosmology," *Physics Today, 43* (April, 1990), 52–8.
van den Bergh, Sidney, "Golden Anniversary of Hubble's Classification," *S&T, 52* (1976), 410–14.

On **interstellar absorption and matter**, see:
Noonan, Thomas W., "Galactic Absorption," *Astronomical Society of the Pacific Leaflet 506* (1971).
Seeley, D., and R. Berendzen, "The Development of Research in Interstellar Absorption, c. 1900–1930," *JHA, 3* (1972), 52–64; 75–86.
Verschuur, G.L., "Barnard's 'Dark' Dilemma," *Astronomy, 17* (Feb., 1989), 30–8.
Verschuur, G.L., *Interstellar Matters: Essays on Curiosity and Astronomical Discovery* (New York: Springer–Verlag, 1989).

Miscellaneous
Arny, Thomas, "The Star Makers: A History of the Theories of Stellar Structure and Evolution," *VA, 33* (1990), 211–33. Covers the period from 1848–1966.
Baade, Walter, *Evolution of Stars and Galaxies* (Cambridge: Harvard Univ. Press, 1963). Uses a historical approach and includes both his own work and that of his predecessors.
Bondi, Hermann, "Fact and Inference in Theory and in Observation," *VA, 1* (1955), 155–62. Discusses whether observations or theories have more frequently proved erroneous in the historical development of astronomy.
Gingerich, Owen, "The Discovery of the Milky Way's Spiral Arms," *S&T, 68* (1984), 10–12.
Gingerich, Owen, "The Discovery of the Spiral Arms of the Milky Way" in *The Milky Way Galaxy*, ed. by Hugo van Woerden, R. J. Allen, and W. B. Burton (Dordrecht: Reidel, 1985), pp. 59–70.
Gordon, Kurtiss J., "History of the Understanding of a Spiral Galaxy: Messier 33," *RASQJ, 10* (1969), 293–307.
Hetherington, Norriss S., "Just How Objective Is Science?" *Nature, 306* (Dec. 22/29, 1983), 727–730. Discusses this issue largely in the context of the history of stellar astronomy; builds on Bondi's paper cited above.
Hetherington, Norriss S., "The Measurement of Radial Velocities of Spiral Nebulae," *Isis, 62* (1971), 309–13.

Hoskin, M. A., "Ritchey, Curtis and the Discovery of Novae in Spiral Nebulae," *JHA*, 7 (1976), 47–53.

Lundmark, Knut, "On Metagalactic Distance Indicators," *VA*, 2 (1956), 1607–19.

Struve, Otto, "Stellar Radial Velocities and Their Observation," *S&T*, 22 (1961), 132–5. A historical survey.

Teerikorpi, Pekka, "Lundmark's 1922 Unpublished Nebulae Classification," *JHA*, 20 (1989), 165–70.

Vaucouleurs, Gérard de, "Who Discovered the Local Supercluster of Galaxies?" *Observatory, 109* (1989), 237–8.

Wright, Helen, *Explorer of the Universe: A Biography of George Ellery Hale* (New York: Dutton, 1966).

Wright, Helen, J. N. Warnow, and Charles Weiner (eds), *The Legacy of George Ellery Hale* (Cambridge: Massachusetts Institute of Technology Press, 1972).

Index

Note: I am indebted to John P. Bransfield for having prepared this index.

Abbe, Cleveland 222
Aberration
    chromatic 7, 13, 14
    spherical 7, 14
Aberration of light 24–26
Adams, John 372, 376
Adams, Walter 290, 293, 294, 305, 315
Addison, Joseph 366–367
Airy, George 14, 186
Alexander, Stephen 174, 208–209
Allen, Woody 391
Arc measure 12
Aristotle 36, 361
Baade, Walter 346, 349–352
Bacon, Roger 193
Ballot. *See* Buys-Ballot, C. H.
Barnard, E. E. 243
Berendzen, Richard 354
Bessel, Friedrich Wilhelm 19, 27, 28, 151, 154–158
Bliss, Nathaniel 14
Blue stars 272, 286
Bohlin, Karl 237
Bond, William Cranch 151
    life 172–173
    Orion resolved 173
Bondi, Hermann 5, 149
Boss, Benjamin 252, 254, 293, 310
Bradley, James 14, 25–26, 28, 31, 58, 152, 155
Brahe, Tycho 13, 19, 23
Brewster, David 180
Brinkley, John 153
Bruno, Giordano 34

Bunsen, Wilhelm 178, 181
    Cesium and Rubidium, discovery of 181
    spectroscope, use of 181
Buys-Ballot, C. H. 193
Cabell, James Branch 387
Calendrelli, D. 153
Campbell, W. W. 254, 259, 356
Cannon, Annie Jump 183
Caryle, Thomas 379
Cassini, Jean-Dominique 13, 19
Celestial equator 15
Celoria, Giovanni 196, 210, 213
Cepheid variables 227, 228, 237, 271, 281, 284, 291–295
Chalmers, Thomas 375
Chamberlin, T. C. 236, 264
Charles II, King 14
Charlier, C. V. 274, 275, 281, 287
Chéseaux, J.-P. L. de 42
Cicero 362
Clark, Alvan 184
Cleomedes 18
Clerke, Agnes Mary 196, 201, 235
Clifford, William Kingdon 384
Cluster variables 284–285
Coal Sack 196, 203, 208, 347, 348
Colbert, Jean-Baptiste 13
Color index 284, 286, 334
Colure, equinoctial 23
Comte, Auguste 146–149, 180
Connolly, Myles 387–388
Copernicus, Nicholas 20, 24, 25, 273
Crimean Observatory 11, 12
Crommelin, A. C. D. 237, 241
Curtis, Heber D. 233, 240, 242, 260, 263, 266, 269, 271–272, 274, 281, 282, 290, 291, 327
    galaxies, five main types of 301
    life 271
    obscuring matter 348
    portrait 271

Darwin, Charles 159, 161, 199
Declination 17
Democritus 36, 53, 202
Derham, William 42, 70
Descartes, René 19
Diffraction grating 180
Digges, Thomas 33
Disk theory 196, 204
Dollond, John 14
Donati, Giovanni Battista 182
Donne, John 363
Doppler, Christian 190, 193
Doppler principle 182, 190–194
Dostoyevsky, Feodor 381–384
Double stars
    gravitational doubles 113
    optical doubles 113
Draper, Henry 183
Draper, John 180
Dreyer, J. L. E. 195
Duncan, John C. 329
Easton, Cornelius 209, 210, 212, 219, 238
Ecliptic 16
    obliquity of 22
Ecliptic system of celestial coordinates 17
Eddington, Arthur Stanley 76, 239, 252, 255, 256, 264–268, 274, 275, 287, 294, 300, 389
Eiseley, Loren 391
Emerson, Ralph Waldo 376
Epicurus 49, 53, 70
Equator system of celestial coordinates 17
Equinox
    autumnal 16
    vernal 16
Euler, Leonhard 34
Ferguson, James 72
Fizeau, Hippolyte 193
Flamsteed, John 14, 21, 152
Focal length 6, 11

Fontenelle, Bernard de 366
Francis of Assisi 363
Fraunhofer, Josef 151, 178, 180
Galilei, Galileo 6, 8–13, 37–38, 202
Globular clusters 281, 285, 302
Gould, B. A. 213
Grandville, J. J. 378–379
Greenwich Observatory 14
Gregory, James 32
Guinand, Pierre 151
Hall, Chester Moor 14
Halley, Edmond 14, 20–24, 39–42, 77
Hardy, Thomas 385
Harper, W. 266
Harriot, Thomas 6, 13
Hawking, Stephen 392
Hegel, G. F. W. 375
Heliometer 155
Helmholtz, Hermann von 188
Henry Draper Catalogue 183
Herschel, Caroline Lucretia 72, 74, 75, 160
    portrait 72
Herschel, John 75, 92, 146, 154, 159, 166, 168, 175, 176, 179, 195, 203, 207, 216–217, 218–219, 358
Herschel, Mary (Pitt) 75
Herschel, William 10, 11–12, 14, 179, 186, 203, 204–206, 207, 217, 218, 266, 358, 395
    impact on later astronomers 146, 152, 153, 159, 164, 166, 176–177
    late-life theories 136–145
    life 71–75
    methodology 93
    nebulae, types of 94
    nebulous stars 124, 125
    planetary nebulae 92, 108, 110, 135
    portrait 71
    strata 80–91
    telescopes 73, 74, 75
    Uranus, discovery of 73–74
Hertzsprung, Ejnar 238, 242, 248, 250, 252, 274, 275, 294

Hertzsprung-Russell diagrams 239, 350
Hetherington, Norriss 5
Hevelius, Johannes 13
Hipparchus of Nicea 19, 22
Holden, Edward S. 204
Hooke, Robert 25, 152
Hubble Constant 336
Hubble, Edwin Powell 1–5, 350
    classification of nebulae 335
    critique of van Maanen's work 345–347
    Hubble's law 336
    impact Hubble's 1925 paper on Curtis, Shapley, and Russell 329
    life 328–329
    methodology 1–5, 356–359
    portrait 328
    Silliman Lectures 355
    telescope, importance of 356
    use of Leavitt's and of Shapley's work 329
Huggins, William 178, 181, 182, 183–194, 257, 358
    life 183
    measuring radial motions, method of 190–194
Humason, Milton 344
Hutton, James 77
Huygens, Christiaan 13, 31
Janssen, Jules 183
Jantar Mantar Observatory 14
Jean Paul 374
Jeans, James 240, 267, 345
Jeffers, Robinson 390
Jefferson, Thomas 376
Joy, James H. 293
Jupiter 13
Kant, Immanuel 4, 45–70, 79, 358, 370–372
Kapteyn, Jacobus 237, 239, 240, 248, 252, 254, 259, 275, 279, 287–288, 292, 294, 300, 304, 316, 347
Keats, John 73
Keeler, J. E. 195, 209, 236, 257, 260, 263
Kepler, Johannes 8, 13, 19
Kinnebrook, David 26–27

Kirchhoff, Gustav Robert 178, 181, 184
    Cesium and Rubidium, discovery of 181
    spectroscope, use of 181
Klinkerfues, W. 193
Klopstock, Friedrich 372
Kobold, H. 267
Koestler, Arthur 392
Kostinsky, S. 240, 267
Lalande, Jérôme 373
Lamb, Charles 374
Lambert, Johann 46–47
Laplace, Pierre Simon 76, 77, 175, 373
Latitude, celestial 17
Leavitt, Henrietta Swan 222, 226–232, 237, 309
Lenses 6
    achromatic 14
Leucippus 53
Lewis, C. S. 390
Light-gathering power 10
Light-year, defined 158
Lipperhey, Hans 8
Locke, John 72
Lockyer, J. Norman 183
Longitude, celestial 17
Louis XIV, King 13
Lowell, Percival 236
Lucretius 70, 362
Luminosity 226, 285
Lundmark, Knut 243, 281, 341, 354
M13 317
M31 35, 36, 38, 40, 42, 99, 107, 173, 176, 189, 195, 215, 216, 235, 236, 237, 243, 257, 265, 319, 323, 324, 325, 326, 329, 330, 331, 337, 345, 349, 350, 351, 358, 398
M33 297, 329, 330, 331, 345
M42 23, 35, 36, 40, 78, 92, 107, 126, 129, 167, 172, 173, 174, 176, 186–189, 217, 224, 398, 401
M101 240, 242, 255, 262, 267, 297
Magellanic Clouds 222, 223, 228, 235, 237, 266, 401
Magnification 11

Magnitude
    absolute 226, 283, 298
    apparent 226, 283, 298
    photographic 286, 331
    photovisual 284, 286
    stellar 233
Manning, Thomas 374
Marius, Simon 38
Maskelyne, Nevil 14, 26–27
Maunder, E. W. 202
Maupertuis, Pierre de 42, 59, 67
Maury, Antonia 183
Maxwell, James Clerk 193
Mayall, N. K. 328
Mechain, P. 82, 91
Meredith, George 385
Messier, Charles 42, 77, 78, 82, 91, 195
Methodology of astronomy 1–5
Meynell, Alice 386
Michell, John 112, 113, 116
Mill, John Stuart 161
Miller, William 184–185
Moon 13
Moore, J. 266
Moulton, F. R. 236, 264
Mount Wilson Observatory 5, 11, 12
Newcomb, Simon 196, 211, 212, 274, 275, 295, 300
Newton, Isaac 4, 9, 19, 20, 21, 179
Nichol, J. P. 168
Nicholson, Seth 264, 346
Nutation 26
Olbers's paradox 149–150, 199
Olbers, H. W. M. 149, 154
Paine, Thomas 373
Palomar Observatory 11, 12
Pappus Alexandrinus 24
Parallax, stellar 24–26, 152–158
Paris Observatory 13
Parsec, defined 158

Pascal, Blaise 363, 365
Pasternak, Boris 391
Pease, F. 266, 267
Perrine, C. 260, 263
Photography 178
Piazzi, Giuseppe 153
Picard, Jean 31
Pickering, E. C. 226
Planetismal theory 236
Plassmann, Joseph 195, 210, 236
Plato 4, 361
Plummer, H. 287
Plutarch 362
Pogson, Norman 234
Pond, John 14, 153–154
Pope, Alexander 368
Population I Stars 350
Population II Stars 350
Positivism 147
Precession of equinoxes 19–20, 22, 26
Proctor, Richard A. 195, 208, 215, 219, 220
Proper motion 20–24, 77, 155, 293, 294, 299
Ptolemy, Claudius 18, 19, 20, 22, 23, 36, 273, 296
Quarra, Thabit Ibn 20
Radial motions 190, 274
Radial velocity 223, 239, 266, 274, 299, 337
Ranyard, A. C. 212
Reduction of observation 28
Refraction, atmospheric 18–19
Resolving power 12
Reynolds, J. H. 243, 297
Right ascension 17
Ritchey, G. W. 240, 260, 263, 264
Ritter, J. W. 179
Roberts, Isaac 235
Robinson, T. R. 167
Roemer, Olaus 13, 152, 193

Rosse, Lord 186, 395
    impact on later astronomers 174
    life 166–168
    spiral nebulae, the discovery of 168–172
Rowland, Henry 183
Roy, A. 293
Russell, Bertrand 3, 389
Russell, H. C. 203, 212
Russell, Henry Norris 238, 254, 270, 271, 275, 287, 294, 314, 330
S Andromedae 235, 243, 324
Sandage, Allan 1, 355
Sarton, George 2, 356
Saturn 13
Scheiner, Christopher 8, 13
Scheiner, Julius 236
Schouten, W. 240, 306, 315
Schultz, H. 267
Schweitzer, Albert 388
Seares, F. H. 292, 297
Secchi, Angelo 183
Sestini, B. 193
Shakespeare, William 363
Shapley, Harlow 5, 233, 240, 242, 269–270, 271–272, 300, 303, 308, 310–314, 315, 317–319, 350, 387, 390
    commentary on his argument 283–297
    commentary on his theory 297
    Great Debate 273
    Great Debate, reflections on 352, 355
    life 269–270
    portrait 269
    reaction to Hubble's discoveries 352–355
    reliance on van Maanen's work 346, 347, 354
    transparency of space 348
Shelley, Percy Bysshe 374–375
Singh, Maharaja Jai 14
Sirius 23, 401
Sitter, Willem de 337
Slipher, Vesto 238, 239, 259, 265, 266, 267, 336, 358

Solstice
  summer 16
  winter 16
South, James 160
Spectrograph 183
Spectroscopic method 299
Spectroscopy 178
Spectrum-luminosity diagram 238, 254, 271
Spencer, Herbert 175–176, 186, 197, 220
Spiral nebulae 168
Stars
  distance of 25
  motions of 15–28
Stebbins, Joel 329
Strömberg, G. 292, 341
Struve, Friedrich Georg Wilhelm 151, 156, 158, 176, 206
Struve, Otto 186, 336
Telescope
  Cassegrainian 9
  Galilean 8
  Herschelian 9
  Keplerian 8, 13
  Newtonian 9, 13
Telescopes 6–14
Tennyson, Alfred Lord 380–381
Terminology in astronomy 5
Thomson, James 369
Trumpler, Robert J. 348
Tycho's Nova 324
van Gogh, Vincent 386
van Maanen, Adriaan 5, 240, 256, 264, 266, 267, 293, 297, 345
Venus 13
von Auwers, G. F. J. 310
von Humboldt, Alexander 175
von Seeliger, Hugo 210, 237, 267
Wallace, Alfred Russel 199–201
Whewell, William 164, 175, 176, 216, 379
Whitman, Walt 379
Whitrow, G. J. 336

Williams, M. E. W. 146
Wilson, R. E. 266
Wirtz, C. 267
Wolf, Max 197, 217, 219, 237, 258, 265, 300
Wollaston, W. H. 179
Wright, Thomas 4, 43–45, 58, 59, 61, 69, 204, 358, 369–370
Young, Edward 34, 368–369
Young, R. 266
Zebergs, Velta 336

A CATALOG OF SELECTED
DOVER BOOKS
IN ALL FIELDS OF INTEREST

A CATALOG OF SELECTED DOVER BOOKS IN ALL FIELDS OF INTEREST

CONCERNING THE SPIRITUAL IN ART, Wassily Kandinsky. Pioneering work by father of abstract art. Thoughts on color theory, nature of art. Analysis of earlier masters. 12 illustrations. 80pp. of text. 5⅜ x 8½. 23411-8

ANIMALS: 1,419 Copyright-Free Illustrations of Mammals, Birds, Fish, Insects, etc., Jim Harter (ed.). Clear wood engravings present, in extremely lifelike poses, over 1,000 species of animals. One of the most extensive pictorial sourcebooks of its kind. Captions. Index. 284pp. 9 x 12. 23766-4

CELTIC ART: The Methods of Construction, George Bain. Simple geometric techniques for making Celtic interlacements, spirals, Kells-type initials, animals, humans, etc. Over 500 illustrations. 160pp. 9 x 12. (Available in U.S. only.) 22923-8

AN ATLAS OF ANATOMY FOR ARTISTS, Fritz Schider. Most thorough reference work on art anatomy in the world. Hundreds of illustrations, including selections from works by Vesalius, Leonardo, Goya, Ingres, Michelangelo, others. 593 illustrations. 192pp. 7⅛ x 10¼. 20241-0

CELTIC HAND STROKE-BY-STROKE (Irish Half-Uncial from "The Book of Kells"): An Arthur Baker Calligraphy Manual, Arthur Baker. Complete guide to creating each letter of the alphabet in distinctive Celtic manner. Covers hand position, strokes, pens, inks, paper, more. Illustrated. 48pp. 8¼ x 11. 24336-2

EASY ORIGAMI, John Montroll. Charming collection of 32 projects (hat, cup, pelican, piano, swan, many more) specially designed for the novice origami hobbyist. Clearly illustrated easy-to-follow instructions insure that even beginning papercrafters will achieve successful results. 48pp. 8¼ x 11. 27298-2

THE COMPLETE BOOK OF BIRDHOUSE CONSTRUCTION FOR WOODWORKERS, Scott D. Campbell. Detailed instructions, illustrations, tables. Also data on bird habitat and instinct patterns. Bibliography. 3 tables. 63 illustrations in 15 figures. 48pp. 5¼ x 8½. 24407-5

BLOOMINGDALE'S ILLUSTRATED 1886 CATALOG: Fashions, Dry Goods and Housewares, Bloomingdale Brothers. Famed merchants' extremely rare catalog depicting about 1,700 products: clothing, housewares, firearms, dry goods, jewelry, more. Invaluable for dating, identifying vintage items. Also, copyright-free graphics for artists, designers. Co-published with Henry Ford Museum & Greenfield Village. 160pp. 8¼ x 11. 25780-0

HISTORIC COSTUME IN PICTURES, Braun & Schneider. Over 1,450 costumed figures in clearly detailed engravings–from dawn of civilization to end of 19th century. Captions. Many folk costumes. 256pp. 8⅜ x 11¾. 23150-X

CATALOG OF DOVER BOOKS

STICKLEY CRAFTSMAN FURNITURE CATALOGS, Gustav Stickley and L. & J. G. Stickley. Beautiful, functional furniture in two authentic catalogs from 1910. 594 illustrations, including 277 photos, show settles, rockers, armchairs, reclining chairs, bookcases, desks, tables. 183pp. 6½ x 9¼. 23838-5

AMERICAN LOCOMOTIVES IN HISTORIC PHOTOGRAPHS: 1858 to 1949, Ron Ziel (ed.). A rare collection of 126 meticulously detailed official photographs, called "builder portraits," of American locomotives that majestically chronicle the rise of steam locomotive power in America. Introduction. Detailed captions. xi+ 129pp. 9 x 12. 27393-8

AMERICA'S LIGHTHOUSES: An Illustrated History, Francis Ross Holland, Jr. Delightfully written, profusely illustrated fact-filled survey of over 200 American lighthouses since 1716. History, anecdotes, technological advances, more. 240pp. 8 x 10¾. 25576-X

TOWARDS A NEW ARCHITECTURE, Le Corbusier. Pioneering manifesto by founder of "International School." Technical and aesthetic theories, views of industry, economics, relation of form to function, "mass-production split" and much more. Profusely illustrated. 320pp. 6⅛ x 9¼. (Available in U.S. only.) 25023-7

HOW THE OTHER HALF LIVES, Jacob Riis. Famous journalistic record, exposing poverty and degradation of New York slums around 1900, by major social reformer. 100 striking and influential photographs. 233pp. 10 x 7⅞. 22012-5

FRUIT KEY AND TWIG KEY TO TREES AND SHRUBS, William M. Harlow. One of the handiest and most widely used identification aids. Fruit key covers 120 deciduous and evergreen species; twig key 160 deciduous species. Easily used. Over 300 photographs. 126pp. 5⅜ x 8½. 20511-8

COMMON BIRD SONGS, Dr. Donald J. Borror. Songs of 60 most common U.S. birds: robins, sparrows, cardinals, bluejays, finches, more–arranged in order of increasing complexity. Up to 9 variations of songs of each species.
Cassette and manual 99911-4

ORCHIDS AS HOUSE PLANTS, Rebecca Tyson Northen. Grow cattleyas and many other kinds of orchids–in a window, in a case, or under artificial light. 63 illustrations. 148pp. 5⅜ x 8½. 23261-1

MONSTER MAZES, Dave Phillips. Masterful mazes at four levels of difficulty. Avoid deadly perils and evil creatures to find magical treasures. Solutions for all 32 exciting illustrated puzzles. 48pp. 8¼ x 11. 26005-4

MOZART'S DON GIOVANNI (DOVER OPERA LIBRETTO SERIES), Wolfgang Amadeus Mozart. Introduced and translated by Ellen H. Bleiler. Standard Italian libretto, with complete English translation. Convenient and thoroughly portable–an ideal companion for reading along with a recording or the performance itself. Introduction. List of characters. Plot summary. 121pp. 5¼ x 8½. 24944-1

TECHNICAL MANUAL AND DICTIONARY OF CLASSICAL BALLET, Gail Grant. Defines, explains, comments on steps, movements, poses and concepts. 15-page pictorial section. Basic book for student, viewer. 127pp. 5⅜ x 8½. 21843-0

THE CLARINET AND CLARINET PLAYING, David Pino. Lively, comprehensive work features suggestions about technique, musicianship, and musical interpretation, as well as guidelines for teaching, making your own reeds, and preparing for public performance. Includes an intriguing look at clarinet history. "A godsend," *The Clarinet,* Journal of the International Clarinet Society. Appendixes. 7 illus. 320pp. 5⅜ x 8½. 40270-3

HOLLYWOOD GLAMOR PORTRAITS, John Kobal (ed.). 145 photos from 1926-49. Harlow, Gable, Bogart, Bacall; 94 stars in all. Full background on photographers, technical aspects. 160pp. 8⅜ x 11¼. 23352-9

THE ANNOTATED CASEY AT THE BAT: A Collection of Ballads about the Mighty Casey/Third, Revised Edition, Martin Gardner (ed.). Amusing sequels and parodies of one of America's best-loved poems: Casey's Revenge, Why Casey Whiffed, Casey's Sister at the Bat, others. 256pp. 5⅜ x 8½. 28598-7

THE RAVEN AND OTHER FAVORITE POEMS, Edgar Allan Poe. Over 40 of the author's most memorable poems: "The Bells," "Ulalume," "Israfel," "To Helen," "The Conqueror Worm," "Eldorado," "Annabel Lee," many more. Alphabetic lists of titles and first lines. 64pp. 5 3/16 x 8¼. 26685-0

PERSONAL MEMOIRS OF U. S. GRANT, Ulysses Simpson Grant. Intelligent, deeply moving firsthand account of Civil War campaigns, considered by many the finest military memoirs ever written. Includes letters, historic photographs, maps and more. 528pp. 6⅛ x 9¼. 28587-1

ANCIENT EGYPTIAN MATERIALS AND INDUSTRIES, A. Lucas and J. Harris. Fascinating, comprehensive, thoroughly documented text describes this ancient civilization's vast resources and the processes that incorporated them in daily life, including the use of animal products, building materials, cosmetics, perfumes and incense, fibers, glazed ware, glass and its manufacture, materials used in the mummification process, and much more. 544pp. 6⅛ x 9¼. (Available in U.S. only.) 40446-3

RUSSIAN STORIES/RUSSKIE RASSKAZY: A Dual-Language Book, edited by Gleb Struve. Twelve tales by such masters as Chekhov, Tolstoy, Dostoevsky, Pushkin, others. Excellent word-for-word English translations on facing pages, plus teaching and study aids, Russian/English vocabulary, biographical/critical introductions, more. 416pp. 5⅜ x 8½. 26244-8

PHILADELPHIA THEN AND NOW: 60 Sites Photographed in the Past and Present, Kenneth Finkel and Susan Oyama. Rare photographs of City Hall, Logan Square, Independence Hall, Betsy Ross House, other landmarks juxtaposed with contemporary views. Captures changing face of historic city. Introduction. Captions. 128pp. 8¼ x 11. 25790-8

AIA ARCHITECTURAL GUIDE TO NASSAU AND SUFFOLK COUNTIES, LONG ISLAND, The American Institute of Architects, Long Island Chapter, and the Society for the Preservation of Long Island Antiquities. Comprehensive, well-researched and generously illustrated volume brings to life over three centuries of Long Island's great architectural heritage. More than 240 photographs with authoritative, extensively detailed captions. 176pp. 8¼ x 11. 26946-9

NORTH AMERICAN INDIAN LIFE: Customs and Traditions of 23 Tribes, Elsie Clews Parsons (ed.). 27 fictionalized essays by noted anthropologists examine religion, customs, government, additional facets of life among the Winnebago, Crow, Zuni, Eskimo, other tribes. 480pp. 6⅛ x 9¼. 27377-6

CATALOG OF DOVER BOOKS

FRANK LLOYD WRIGHT'S DANA HOUSE, Donald Hoffmann. Pictorial essay of residential masterpiece with over 160 interior and exterior photos, plans, elevations, sketches and studies. 128pp. 9¼ x 10¾. 29120-0

THE MALE AND FEMALE FIGURE IN MOTION: 60 Classic Photographic Sequences, Eadweard Muybridge. 60 true-action photographs of men and women walking, running, climbing, bending, turning, etc., reproduced from rare 19th-century masterpiece. vi + 121pp. 9 x 12. 24745-7

1001 QUESTIONS ANSWERED ABOUT THE SEASHORE, N. J. Berrill and Jacquelyn Berrill. Queries answered about dolphins, sea snails, sponges, starfish, fishes, shore birds, many others. Covers appearance, breeding, growth, feeding, much more. 305pp. 5¼ x 8¼. 23366-9

ATTRACTING BIRDS TO YOUR YARD, William J. Weber. Easy-to-follow guide offers advice on how to attract the greatest diversity of birds: birdhouses, feeders, water and waterers, much more. 96pp. 5³/₁₆ x 8¼. 28927-3

MEDICINAL AND OTHER USES OF NORTH AMERICAN PLANTS: A Historical Survey with Special Reference to the Eastern Indian Tribes, Charlotte Erichsen-Brown. Chronological historical citations document 500 years of usage of plants, trees, shrubs native to eastern Canada, northeastern U.S. Also complete identifying information. 343 illustrations. 544pp. 6½ x 9¼. 25951-X

STORYBOOK MAZES, Dave Phillips. 23 stories and mazes on two-page spreads: Wizard of Oz, Treasure Island, Robin Hood, etc. Solutions. 64pp. 8¼ x 11. 23628-5

AMERICAN NEGRO SONGS: 230 Folk Songs and Spirituals, Religious and Secular, John W. Work. This authoritative study traces the African influences of songs sung and played by black Americans at work, in church, and as entertainment. The author discusses the lyric significance of such songs as "Swing Low, Sweet Chariot," "John Henry," and others and offers the words and music for 230 songs. Bibliography. Index of Song Titles. 272pp. 6½ x 9¼. 40271-1

MOVIE-STAR PORTRAITS OF THE FORTIES, John Kobal (ed.). 163 glamor, studio photos of 106 stars of the 1940s: Rita Hayworth, Ava Gardner, Marlon Brando, Clark Gable, many more. 176pp. 8⅜ x 11¼. 23546-7

BENCHLEY LOST AND FOUND, Robert Benchley. Finest humor from early 30s, about pet peeves, child psychologists, post office and others. Mostly unavailable elsewhere. 73 illustrations by Peter Arno and others. 183pp. 5⅜ x 8½. 22410-4

YEKL and THE IMPORTED BRIDEGROOM AND OTHER STORIES OF YIDDISH NEW YORK, Abraham Cahan. Film Hester Street based on *Yekl* (1896). Novel, other stories among first about Jewish immigrants on N.Y.'s East Side. 240pp. 5⅜ x 8½. 22427-9

SELECTED POEMS, Walt Whitman. Generous sampling from *Leaves of Grass*. Twenty-four poems include "I Hear America Singing," "Song of the Open Road," "I Sing the Body Electric," "When Lilacs Last in the Dooryard Bloom'd," "O Captain! My Captain!"–all reprinted from an authoritative edition. Lists of titles and first lines. 128pp. 5³/₁₆ x 8¼. 26878-0

CATALOG OF DOVER BOOKS

THE BEST TALES OF HOFFMANN, E. T. A. Hoffmann. 10 of Hoffmann's most important stories: "Nutcracker and the King of Mice," "The Golden Flowerpot," etc. 458pp. 5⅜ x 8½. 21793-0

FROM FETISH TO GOD IN ANCIENT EGYPT, E. A. Wallis Budge. Rich detailed survey of Egyptian conception of "God" and gods, magic, cult of animals, Osiris, more. Also, superb English translations of hymns and legends. 240 illustrations. 545pp. 5⅜ x 8½. 25803-3

FRENCH STORIES/CONTES FRANÇAIS: A Dual-Language Book, Wallace Fowlie. Ten stories by French masters, Voltaire to Camus: "Micromegas" by Voltaire; "The Atheist's Mass" by Balzac; "Minuet" by de Maupassant; "The Guest" by Camus, six more. Excellent English translations on facing pages. Also French-English vocabulary list, exercises, more. 352pp. 5⅜ x 8½. 26443-2

CHICAGO AT THE TURN OF THE CENTURY IN PHOTOGRAPHS: 122 Historic Views from the Collections of the Chicago Historical Society, Larry A. Viskochil. Rare large-format prints offer detailed views of City Hall, State Street, the Loop, Hull House, Union Station, many other landmarks, circa 1904-1913. Introduction. Captions. Maps. 144pp. 9⅜ x 12¼. 24656-6

OLD BROOKLYN IN EARLY PHOTOGRAPHS, 1865-1929, William Lee Younger. Luna Park, Gravesend race track, construction of Grand Army Plaza, moving of Hotel Brighton, etc. 157 previously unpublished photographs. 165pp. 8⅞ x 11¾. 23587-4

THE MYTHS OF THE NORTH AMERICAN INDIANS, Lewis Spence. Rich anthology of the myths and legends of the Algonquins, Iroquois, Pawnees and Sioux, prefaced by an extensive historical and ethnological commentary. 36 illustrations. 480pp. 5⅜ x 8½. 25967-6

AN ENCYCLOPEDIA OF BATTLES: Accounts of Over 1,560 Battles from 1479 B.C. to the Present, David Eggenberger. Essential details of every major battle in recorded history from the first battle of Megiddo in 1479 B.C. to Grenada in 1984. List of Battle Maps. New Appendix covering the years 1967-1984. Index. 99 illustrations. 544pp. 6½ x 9¼. 24913-1

SAILING ALONE AROUND THE WORLD, Captain Joshua Slocum. First man to sail around the world, alone, in small boat. One of great feats of seamanship told in delightful manner. 67 illustrations. 294pp. 5⅜ x 8½. 20326-3

ANARCHISM AND OTHER ESSAYS, Emma Goldman. Powerful, penetrating, prophetic essays on direct action, role of minorities, prison reform, puritan hypocrisy, violence, etc. 271pp. 5⅜ x 8½. 22484-8

MYTHS OF THE HINDUS AND BUDDHISTS, Ananda K. Coomaraswamy and Sister Nivedita. Great stories of the epics; deeds of Krishna, Shiva, taken from puranas, Vedas, folk tales; etc. 32 illustrations. 400pp. 5⅜ x 8½. 21759-0

THE TRAUMA OF BIRTH, Otto Rank. Rank's controversial thesis that anxiety neurosis is caused by profound psychological trauma which occurs at birth. 256pp. 5⅜ x 8½. 27974-X

A THEOLOGICO-POLITICAL TREATISE, Benedict Spinoza. Also contains unfinished Political Treatise. Great classic on religious liberty, theory of government on common consent. R. Elwes translation. Total of 421pp. 5⅜ x 8½. 20249-6

CATALOG OF DOVER BOOKS

MY BONDAGE AND MY FREEDOM, Frederick Douglass. Born a slave, Douglass became outspoken force in antislavery movement. The best of Douglass' autobiographies. Graphic description of slave life. 464pp. 5⅜ x 8½. 22457-0

FOLLOWING THE EQUATOR: A Journey Around the World, Mark Twain. Fascinating humorous account of 1897 voyage to Hawaii, Australia, India, New Zealand, etc. Ironic, bemused reports on peoples, customs, climate, flora and fauna, politics, much more. 197 illustrations. 720pp. 5⅜ x 8½. 26113-1

THE PEOPLE CALLED SHAKERS, Edward D. Andrews. Definitive study of Shakers: origins, beliefs, practices, dances, social organization, furniture and crafts, etc. 33 illustrations. 351pp. 5⅜ x 8½. 21081-2

THE MYTHS OF GREECE AND ROME, H. A. Guerber. A classic of mythology, generously illustrated, long prized for its simple, graphic, accurate retelling of the principal myths of Greece and Rome, and for its commentary on their origins and significance. With 64 illustrations by Michelangelo, Raphael, Titian, Rubens, Canova, Bernini and others. 480pp. 5⅜ x 8½. 27584-1

PSYCHOLOGY OF MUSIC, Carl E. Seashore. Classic work discusses music as a medium from psychological viewpoint. Clear treatment of physical acoustics, auditory apparatus, sound perception, development of musical skills, nature of musical feeling, host of other topics. 88 figures. 408pp. 5⅜ x 8½. 21851-1

THE PHILOSOPHY OF HISTORY, Georg W. Hegel. Great classic of Western thought develops concept that history is not chance but rational process, the evolution of freedom. 457pp. 5⅜ x 8½. 20112-0

THE BOOK OF TEA, Kakuzo Okakura. Minor classic of the Orient: entertaining, charming explanation, interpretation of traditional Japanese culture in terms of tea ceremony. 94pp. 5⅜ x 8½. 20070-1

LIFE IN ANCIENT EGYPT, Adolf Erman. Fullest, most thorough, detailed older account with much not in more recent books, domestic life, religion, magic, medicine, commerce, much more. Many illustrations reproduce tomb paintings, carvings, hieroglyphs, etc. 597pp. 5⅜ x 8½. 22632-8

SUNDIALS, Their Theory and Construction, Albert Waugh. Far and away the best, most thorough coverage of ideas, mathematics concerned, types, construction, adjusting anywhere. Simple, nontechnical treatment allows even children to build several of these dials. Over 100 illustrations. 230pp. 5⅜ x 8½. 22947-5

THEORETICAL HYDRODYNAMICS, L. M. Milne-Thomson. Classic exposition of the mathematical theory of fluid motion, applicable to both hydrodynamics and aerodynamics. Over 600 exercises. 768pp. 6⅛ x 9¼. 68970-0

SONGS OF EXPERIENCE: Facsimile Reproduction with 26 Plates in Full Color, William Blake. 26 full-color plates from a rare 1826 edition. Includes "The Tyger," "London," "Holy Thursday," and other poems. Printed text of poems. 48pp. 5¼ x 7. 24636-1

OLD-TIME VIGNETTES IN FULL COLOR, Carol Belanger Grafton (ed.). Over 390 charming, often sentimental illustrations, selected from archives of Victorian graphics–pretty women posing, children playing, food, flowers, kittens and puppies, smiling cherubs, birds and butterflies, much more. All copyright-free. 48pp. 9¼ x 12¼. 27269-9

PERSPECTIVE FOR ARTISTS, Rex Vicat Cole. Depth, perspective of sky and sea, shadows, much more, not usually covered. 391 diagrams, 81 reproductions of drawings and paintings. 279pp. 5⅜ x 8½. 22487-2

DRAWING THE LIVING FIGURE, Joseph Sheppard. Innovative approach to artistic anatomy focuses on specifics of surface anatomy, rather than muscles and bones. Over 170 drawings of live models in front, back and side views, and in widely varying poses. Accompanying diagrams. 177 illustrations. Introduction. Index. 144pp. 8⅜ x11¼. 26723-7

GOTHIC AND OLD ENGLISH ALPHABETS: 100 Complete Fonts, Dan X. Solo. Add power, elegance to posters, signs, other graphics with 100 stunning copyright-free alphabets: Blackstone, Dolbey, Germania, 97 more–including many lower-case, numerals, punctuation marks. 104pp. 8⅛ x 11. 24695-7

HOW TO DO BEADWORK, Mary White. Fundamental book on craft from simple projects to five-bead chains and woven works. 106 illustrations. 142pp. 5⅜ x 8. 20697-1

THE BOOK OF WOOD CARVING, Charles Marshall Sayers. Finest book for beginners discusses fundamentals and offers 34 designs. "Absolutely first rate . . . well thought out and well executed."–E. J. Tangerman. 118pp. 7¾ x 10⅝. 23654-4

ILLUSTRATED CATALOG OF CIVIL WAR MILITARY GOODS: Union Army Weapons, Insignia, Uniform Accessories, and Other Equipment, Schuyler, Hartley, and Graham. Rare, profusely illustrated 1846 catalog includes Union Army uniform and dress regulations, arms and ammunition, coats, insignia, flags, swords, rifles, etc. 226 illustrations. 160pp. 9 x 12. 24939-5

WOMEN'S FASHIONS OF THE EARLY 1900s: An Unabridged Republication of "New York Fashions, 1909," National Cloak & Suit Co. Rare catalog of mail-order fashions documents women's and children's clothing styles shortly after the turn of the century. Captions offer full descriptions, prices. Invaluable resource for fashion, costume historians. Approximately 725 illustrations. 128pp. 8⅜ x 11¼. 27276-1

THE 1912 AND 1915 GUSTAV STICKLEY FURNITURE CATALOGS, Gustav Stickley. With over 200 detailed illustrations and descriptions, these two catalogs are essential reading and reference materials and identification guides for Stickley furniture. Captions cite materials, dimensions and prices. 112pp. 6½ x 9¼. 26676-1

EARLY AMERICAN LOCOMOTIVES, John H. White, Jr. Finest locomotive engravings from early 19th century: historical (1804–74), main-line (after 1870), special, foreign, etc. 147 plates. 142pp. 11⅜ x 8¼. 22772-3

THE TALL SHIPS OF TODAY IN PHOTOGRAPHS, Frank O. Braynard. Lavishly illustrated tribute to nearly 100 majestic contemporary sailing vessels: Amerigo Vespucci, Clearwater, Constitution, Eagle, Mayflower, Sea Cloud, Victory, many more. Authoritative captions provide statistics, background on each ship. 190 black-and-white photographs and illustrations. Introduction. 128pp. 8⅞ x 11¾. 27163-3

CATALOG OF DOVER BOOKS

LITTLE BOOK OF EARLY AMERICAN CRAFTS AND TRADES, Peter Stockham (ed.). 1807 children's book explains crafts and trades: baker, hatter, cooper, potter, and many others. 23 copperplate illustrations. 140pp. 4⅝ x 6.   23336-7

VICTORIAN FASHIONS AND COSTUMES FROM HARPER'S BAZAR, 1867–1898, Stella Blum (ed.). Day costumes, evening wear, sports clothes, shoes, hats, other accessories in over 1,000 detailed engravings. 320pp. 9⅜ x 12¼.   22990-4

GUSTAV STICKLEY, THE CRAFTSMAN, Mary Ann Smith. Superb study surveys broad scope of Stickley's achievement, especially in architecture. Design philosophy, rise and fall of the Craftsman empire, descriptions and floor plans for many Craftsman houses, more. 86 black-and-white halftones. 31 line illustrations. Introduction 208pp. 6½ x 9¼.   27210-9

THE LONG ISLAND RAIL ROAD IN EARLY PHOTOGRAPHS, Ron Ziel. Over 220 rare photos, informative text document origin (1844) and development of rail service on Long Island. Vintage views of early trains, locomotives, stations, passengers, crews, much more. Captions. 8⅞ x 11¾.   26301-0

VOYAGE OF THE LIBERDADE, Joshua Slocum. Great 19th-century mariner's thrilling, first-hand account of the wreck of his ship off South America, the 35-foot boat he built from the wreckage, and its remarkable voyage home. 128pp. 5⅜ x 8½.   40022-0

TEN BOOKS ON ARCHITECTURE, Vitruvius. The most important book ever written on architecture. Early Roman aesthetics, technology, classical orders, site selection, all other aspects. Morgan translation. 331pp. 5⅜ x 8½.   20645-9

THE HUMAN FIGURE IN MOTION, Eadweard Muybridge. More than 4,500 stopped-action photos, in action series, showing undraped men, women, children jumping, lying down, throwing, sitting, wrestling, carrying, etc. 390pp. 7⅞ x 10⅝.   20204-6 Clothbd.

TREES OF THE EASTERN AND CENTRAL UNITED STATES AND CANADA, William M. Harlow. Best one-volume guide to 140 trees. Full descriptions, woodlore, range, etc. Over 600 illustrations. Handy size. 288pp. 4½ x 6⅜.   20395-6

SONGS OF WESTERN BIRDS, Dr. Donald J. Borror. Complete song and call repertoire of 60 western species, including flycatchers, juncoes, cactus wrens, many more–includes fully illustrated booklet.   Cassette and manual 99913-0

GROWING AND USING HERBS AND SPICES, Milo Miloradovich. Versatile handbook provides all the information needed for cultivation and use of all the herbs and spices available in North America. 4 illustrations. Index. Glossary. 236pp. 5⅜ x 8½.   25058-X

BIG BOOK OF MAZES AND LABYRINTHS, Walter Shepherd. 50 mazes and labyrinths in all–classical, solid, ripple, and more–in one great volume. Perfect inexpensive puzzler for clever youngsters. Full solutions. 112pp. 8⅛ x 11.   22951-3

CATALOG OF DOVER BOOKS

PIANO TUNING, J. Cree Fischer. Clearest, best book for beginner, amateur. Simple repairs, raising dropped notes, tuning by easy method of flattened fifths. No previous skills needed. 4 illustrations. 201pp. 5⅜ x 8½. 23267-0

HINTS TO SINGERS, Lillian Nordica. Selecting the right teacher, developing confidence, overcoming stage fright, and many other important skills receive thoughtful discussion in this indispensible guide, written by a world-famous diva of four decades' experience. 96pp. 5⅜ x 8½. 40094-8

THE COMPLETE NONSENSE OF EDWARD LEAR, Edward Lear. All nonsense limericks, zany alphabets, Owl and Pussycat, songs, nonsense botany, etc., illustrated by Lear. Total of 320pp. 5⅜ x 8½. (Available in U.S. only.) 20167-8

VICTORIAN PARLOUR POETRY: An Annotated Anthology, Michael R. Turner. 117 gems by Longfellow, Tennyson, Browning, many lesser-known poets. "The Village Blacksmith," "Curfew Must Not Ring Tonight," "Only a Baby Small," dozens more, often difficult to find elsewhere. Index of poets, titles, first lines. xxiii + 325pp. 5⅜ x 8¼. 27044-0

DUBLINERS, James Joyce. Fifteen stories offer vivid, tightly focused observations of the lives of Dublin's poorer classes. At least one, "The Dead," is considered a masterpiece. Reprinted complete and unabridged from standard edition. 160pp. 5$\frac{3}{16}$ x 8¼. 26870-5

GREAT WEIRD TALES: 14 Stories by Lovecraft, Blackwood, Machen and Others, S. T. Joshi (ed.). 14 spellbinding tales, including "The Sin Eater," by Fiona McLeod, "The Eye Above the Mantel," by Frank Belknap Long, as well as renowned works by R. H. Barlow, Lord Dunsany, Arthur Machen, W. C. Morrow and eight other masters of the genre. 256pp. 5⅜ x 8½. (Available in U.S. only.) 40436-6

THE BOOK OF THE SACRED MAGIC OF ABRAMELIN THE MAGE, translated by S. MacGregor Mathers. Medieval manuscript of ceremonial magic. Basic document in Aleister Crowley, Golden Dawn groups. 268pp. 5⅜ x 8½. 23211-5

NEW RUSSIAN-ENGLISH AND ENGLISH-RUSSIAN DICTIONARY, M. A. O'Brien. This is a remarkably handy Russian dictionary, containing a surprising amount of information, including over 70,000 entries. 366pp. 4½ x 6⅛. 20208-9

HISTORIC HOMES OF THE AMERICAN PRESIDENTS, Second, Revised Edition, Irvin Haas. A traveler's guide to American Presidential homes, most open to the public, depicting and describing homes occupied by every American President from George Washington to George Bush. With visiting hours, admission charges, travel routes. 175 photographs. Index. 160pp. 8¼ x 11. 26751-2

NEW YORK IN THE FORTIES, Andreas Feininger. 162 brilliant photographs by the well-known photographer, formerly with *Life* magazine. Commuters, shoppers, Times Square at night, much else from city at its peak. Captions by John von Hartz. 181pp. 9¼ x 10¾. 23585-8

INDIAN SIGN LANGUAGE, William Tomkins. Over 525 signs developed by Sioux and other tribes. Written instructions and diagrams. Also 290 pictographs. 111pp. 6⅛ x 9¼. 22029-X

CATALOG OF DOVER BOOKS

ANATOMY: A Complete Guide for Artists, Joseph Sheppard. A master of figure drawing shows artists how to render human anatomy convincingly. Over 460 illustrations. 224pp. 8⅜ x 11¼. 27279-6

MEDIEVAL CALLIGRAPHY: Its History and Technique, Marc Drogin. Spirited history, comprehensive instruction manual covers 13 styles (ca. 4th century through 15th). Excellent photographs; directions for duplicating medieval techniques with modern tools. 224pp. 8⅜ x 11¼. 26142-5

DRIED FLOWERS: How to Prepare Them, Sarah Whitlock and Martha Rankin. Complete instructions on how to use silica gel, meal and borax, perlite aggregate, sand and borax, glycerine and water to create attractive permanent flower arrangements. 12 illustrations. 32pp. 5⅜ x 8½. 21802-3

EASY-TO-MAKE BIRD FEEDERS FOR WOODWORKERS, Scott D. Campbell. Detailed, simple-to-use guide for designing, constructing, caring for and using feeders. Text, illustrations for 12 classic and contemporary designs. 96pp. 5⅜ x 8½. 25847-5

SCOTTISH WONDER TALES FROM MYTH AND LEGEND, Donald A. Mackenzie. 16 lively tales tell of giants rumbling down mountainsides, of a magic wand that turns stone pillars into warriors, of gods and goddesses, evil hags, powerful forces and more. 240pp. 5⅜ x 8½. 29677-6

THE HISTORY OF UNDERCLOTHES, C. Willett Cunnington and Phyllis Cunnington. Fascinating, well-documented survey covering six centuries of English undergarments, enhanced with over 100 illustrations: 12th-century laced-up bodice, footed long drawers (1795), 19th-century bustles, 19th-century corsets for men, Victorian "bust improvers," much more. 272pp. 5⅜ x 8¼. 27124-2

ARTS AND CRAFTS FURNITURE: The Complete Brooks Catalog of 1912, Brooks Manufacturing Co. Photos and detailed descriptions of more than 150 now very collectible furniture designs from the Arts and Crafts movement depict davenports, settees, buffets, desks, tables, chairs, bedsteads, dressers and more, all built of solid, quarter-sawed oak. Invaluable for students and enthusiasts of antiques, Americana and the decorative arts. 80pp. 6½ x 9¼. 27471-3

WILBUR AND ORVILLE: A Biography of the Wright Brothers, Fred Howard. Definitive, crisply written study tells the full story of the brothers' lives and work. A vividly written biography, unparalleled in scope and color, that also captures the spirit of an extraordinary era. 560pp. 6⅛ x 9¼. 40297-5

THE ARTS OF THE SAILOR: Knotting, Splicing and Ropework, Hervey Garrett Smith. Indispensable shipboard reference covers tools, basic knots and useful hitches; handsewing and canvas work, more. Over 100 illustrations. Delightful reading for sea lovers. 256pp. 5⅜ x 8½. 26440-8

FRANK LLOYD WRIGHT'S FALLINGWATER: The House and Its History, Second, Revised Edition, Donald Hoffmann. A total revision–both in text and illustrations–of the standard document on Fallingwater, the boldest, most personal architectural statement of Wright's mature years, updated with valuable new material from the recently opened Frank Lloyd Wright Archives. "Fascinating"–*The New York Times*. 116 illustrations. 128pp. 9¼ x 10¾. 27430-6

PHOTOGRAPHIC SKETCHBOOK OF THE CIVIL WAR, Alexander Gardner. 100 photos taken on field during the Civil War. Famous shots of Manassas Harper's Ferry, Lincoln, Richmond, slave pens, etc. 244pp. 10⅝ x 8¼. 22731-6

FIVE ACRES AND INDEPENDENCE, Maurice G. Kains. Great back-to-the-land classic explains basics of self-sufficient farming. The one book to get. 95 illustrations. 397pp. 5⅜ x 8½. 20974-1

SONGS OF EASTERN BIRDS, Dr. Donald J. Borror. Songs and calls of 60 species most common to eastern U.S.: warblers, woodpeckers, flycatchers, thrushes, larks, many more in high-quality recording. Cassette and manual 99912-2

A MODERN HERBAL, Margaret Grieve. Much the fullest, most exact, most useful compilation of herbal material. Gigantic alphabetical encyclopedia, from aconite to zedoary, gives botanical information, medical properties, folklore, economic uses, much else. Indispensable to serious reader. 161 illustrations. 888pp. 6½ x 9¼. 2-vol. set. (Available in U.S. only.) Vol. I: 22798-7
Vol. II: 22799-5

HIDDEN TREASURE MAZE BOOK, Dave Phillips. Solve 34 challenging mazes accompanied by heroic tales of adventure. Evil dragons, people-eating plants, blood-thirsty giants, many more dangerous adversaries lurk at every twist and turn. 34 mazes, stories, solutions. 48pp. 8¼ x 11. 24566-7

LETTERS OF W. A. MOZART, Wolfgang A. Mozart. Remarkable letters show bawdy wit, humor, imagination, musical insights, contemporary musical world; includes some letters from Leopold Mozart. 276pp. 5⅜ x 8½. 22859-2

BASIC PRINCIPLES OF CLASSICAL BALLET, Agrippina Vaganova. Great Russian theoretician, teacher explains methods for teaching classical ballet. 118 illustrations. 175pp. 5⅜ x 8½. 22036-2

THE JUMPING FROG, Mark Twain. Revenge edition. The original story of The Celebrated Jumping Frog of Calaveras County, a hapless French translation, and Twain's hilarious "retranslation" from the French. 12 illustrations. 66pp. 5⅜ x 8½. 22686-7

BEST REMEMBERED POEMS, Martin Gardner (ed.). The 126 poems in this superb collection of 19th- and 20th-century British and American verse range from Shelley's "To a Skylark" to the impassioned "Renascence" of Edna St. Vincent Millay and to Edward Lear's whimsical "The Owl and the Pussycat." 224pp. 5⅜ x 8½. 27165-X

COMPLETE SONNETS, William Shakespeare. Over 150 exquisite poems deal with love, friendship, the tyranny of time, beauty's evanescence, death and other themes in language of remarkable power, precision and beauty. Glossary of archaic terms. 80pp. $5\frac{3}{16}$ x 8¼. 26686-9

THE BATTLES THAT CHANGED HISTORY, Fletcher Pratt. Eminent historian profiles 16 crucial conflicts, ancient to modern, that changed the course of civilization. 352pp. 5⅜ x 8½. 41129-X

CATALOG OF DOVER BOOKS

THE WIT AND HUMOR OF OSCAR WILDE, Alvin Redman (ed.). More than 1,000 ripostes, paradoxes, wisecracks: Work is the curse of the drinking classes; I can resist everything except temptation; etc. 258pp. 5⅜ x 8½. 20602-5

SHAKESPEARE LEXICON AND QUOTATION DICTIONARY, Alexander Schmidt. Full definitions, locations, shades of meaning in every word in plays and poems. More than 50,000 exact quotations. 1,485pp. 6½ x 9¼. 2-vol. set.
Vol. 1: 22726-X
Vol. 2: 22727-8

SELECTED POEMS, Emily Dickinson. Over 100 best-known, best-loved poems by one of America's foremost poets, reprinted from authoritative early editions. No comparable edition at this price. Index of first lines. 64pp. 5³⁄₁₆ x 8¼. 26466-1

THE INSIDIOUS DR. FU-MANCHU, Sax Rohmer. The first of the popular mystery series introduces a pair of English detectives to their archnemesis, the diabolical Dr. Fu-Manchu. Flavorful atmosphere, fast-paced action, and colorful characters enliven this classic of the genre. 208pp. 5³⁄₁₆ x 8¼. 29898-1

THE MALLEUS MALEFICARUM OF KRAMER AND SPRENGER, translated by Montague Summers. Full text of most important witchhunter's "bible," used by both Catholics and Protestants. 278pp. 6⅝ x 10. 22802-9

SPANISH STORIES/CUENTOS ESPAÑOLES: A Dual-Language Book, Angel Flores (ed.). Unique format offers 13 great stories in Spanish by Cervantes, Borges, others. Faithful English translations on facing pages. 352pp. 5⅜ x 8½. 25399-6

GARDEN CITY, LONG ISLAND, IN EARLY PHOTOGRAPHS, 1869–1919, Mildred H. Smith. Handsome treasury of 118 vintage pictures, accompanied by carefully researched captions, document the Garden City Hotel fire (1899), the Vanderbilt Cup Race (1908), the first airmail flight departing from the Nassau Boulevard Aerodrome (1911), and much more. 96pp. 8⅞ x 11¾. 40669-5

OLD QUEENS, N.Y., IN EARLY PHOTOGRAPHS, Vincent F. Seyfried and William Asadorian. Over 160 rare photographs of Maspeth, Jamaica, Jackson Heights, and other areas. Vintage views of DeWitt Clinton mansion, 1939 World's Fair and more. Captions. 192pp. 8⅞ x 11. 26358-4

CAPTURED BY THE INDIANS: 15 Firsthand Accounts, 1750-1870, Frederick Drimmer. Astounding true historical accounts of grisly torture, bloody conflicts, relentless pursuits, miraculous escapes and more, by people who lived to tell the tale. 384pp. 5⅜ x 8½. 24901-8

THE WORLD'S GREAT SPEECHES (Fourth Enlarged Edition), Lewis Copeland, Lawrence W. Lamm, and Stephen J. McKenna. Nearly 300 speeches provide public speakers with a wealth of updated quotes and inspiration–from Pericles' funeral oration and William Jennings Bryan's "Cross of Gold Speech" to Malcolm X's powerful words on the Black Revolution and Earl of Spenser's tribute to his sister, Diana, Princess of Wales. 944pp. 5⅜ x 8⅜. 40903-1

THE BOOK OF THE SWORD, Sir Richard F. Burton. Great Victorian scholar/adventurer's eloquent, erudite history of the "queen of weapons"–from prehistory to early Roman Empire. Evolution and development of early swords, variations (sabre, broadsword, cutlass, scimitar, etc.), much more. 336pp. 6⅛ x 9¼. 25434-8

CATALOG OF DOVER BOOKS

AUTOBIOGRAPHY: The Story of My Experiments with Truth, Mohandas K. Gandhi. Boyhood, legal studies, purification, the growth of the Satyagraha (nonviolent protest) movement. Critical, inspiring work of the man responsible for the freedom of India. 480pp. 5⅜ x 8½. (Available in U.S. only.) 24593-4

CELTIC MYTHS AND LEGENDS, T. W. Rolleston. Masterful retelling of Irish and Welsh stories and tales. Cuchulain, King Arthur, Deirdre, the Grail, many more. First paperback edition. 58 full-page illustrations. 512pp. 5⅜ x 8½. 26507-2

THE PRINCIPLES OF PSYCHOLOGY, William James. Famous long course complete, unabridged. Stream of thought, time perception, memory, experimental methods; great work decades ahead of its time. 94 figures. 1,391pp. 5⅜ x 8½. 2-vol. set.
Vol. I: 20381-6 Vol. II: 20382-4

THE WORLD AS WILL AND REPRESENTATION, Arthur Schopenhauer. Definitive English translation of Schopenhauer's life work, correcting more than 1,000 errors, omissions in earlier translations. Translated by E. F. J. Payne. Total of 1,269pp. 5⅜ x 8½. 2-vol. set. Vol. 1: 21761-2 Vol. 2: 21762-0

MAGIC AND MYSTERY IN TIBET, Madame Alexandra David-Neel. Experiences among lamas, magicians, sages, sorcerers, Bonpa wizards. A true psychic discovery. 32 illustrations. 321pp. 5⅜ x 8½. (Available in U.S. only.) 22682-4

THE EGYPTIAN BOOK OF THE DEAD, E. A. Wallis Budge. Complete reproduction of Ani's papyrus, finest ever found. Full hieroglyphic text, interlinear transliteration, word-for-word translation, smooth translation. 533pp. 6½ x 9¼. 21866-X

MATHEMATICS FOR THE NONMATHEMATICIAN, Morris Kline. Detailed, college-level treatment of mathematics in cultural and historical context, with numerous exercises. Recommended Reading Lists. Tables. Numerous figures. 641pp. 5⅜ x 8½. 24823-2

PROBABILISTIC METHODS IN THE THEORY OF STRUCTURES, Isaac Elishakoff. Well-written introduction covers the elements of the theory of probability from two or more random variables, the reliability of such multivariable structures, the theory of random function, Monte Carlo methods of treating problems incapable of exact solution, and more. Examples. 502pp. 5⅜ x 8½. 40691-1

THE RIME OF THE ANCIENT MARINER, Gustave Doré, S. T. Coleridge. Doré's finest work; 34 plates capture moods, subtleties of poem. Flawless full-size reproductions printed on facing pages with authoritative text of poem. "Beautiful. Simply beautiful."–*Publisher's Weekly.* 77pp. 9¼ x 12. 22305-1

NORTH AMERICAN INDIAN DESIGNS FOR ARTISTS AND CRAFTSPEOPLE, Eva Wilson. Over 360 authentic copyright-free designs adapted from Navajo blankets, Hopi pottery, Sioux buffalo hides, more. Geometrics, symbolic figures, plant and animal motifs, etc. 128pp. 8⅜ x 11. (Not for sale in the United Kingdom.) 25341-4

SCULPTURE: Principles and Practice, Louis Slobodkin. Step-by-step approach to clay, plaster, metals, stone; classical and modern. 253 drawings, photos. 255pp. 8⅛ x 11. 22960-2

THE INFLUENCE OF SEA POWER UPON HISTORY, 1660–1783, A. T. Mahan. Influential classic of naval history and tactics still used as text in war colleges. First paperback edition. 4 maps. 24 battle plans. 640pp. 5⅜ x 8½. 25509-3

CATALOG OF DOVER BOOKS

THE STORY OF THE TITANIC AS TOLD BY ITS SURVIVORS, Jack Winocour (ed.). What it was really like. Panic, despair, shocking inefficiency, and a little heroism. More thrilling than any fictional account. 26 illustrations. 320pp. 5⅜ x 8½. 20610-6

FAIRY AND FOLK TALES OF THE IRISH PEASANTRY, William Butler Yeats (ed.). Treasury of 64 tales from the twilight world of Celtic myth and legend: "The Soul Cages," "The Kildare Pooka," "King O'Toole and his Goose," many more. Introduction and Notes by W. B. Yeats. 352pp. 5⅜ x 8½. 26941-8

BUDDHIST MAHAYANA TEXTS, E. B. Cowell and others (eds.). Superb, accurate translations of basic documents in Mahayana Buddhism, highly important in history of religions. The Buddha-karita of Asvaghosha, Larger Sukhavativyuha, more. 448pp. 5⅜ x 8½. 25552-2

ONE TWO THREE . . . INFINITY: Facts and Speculations of Science, George Gamow. Great physicist's fascinating, readable overview of contemporary science: number theory, relativity, fourth dimension, entropy, genes, atomic structure, much more. 128 illustrations. Index. 352pp. 5⅜ x 8½. 25664-2

EXPERIMENTATION AND MEASUREMENT, W. J. Youden. Introductory manual explains laws of measurement in simple terms and offers tips for achieving accuracy and minimizing errors. Mathematics of measurement, use of instruments, experimenting with machines. 1994 edition. Foreword. Preface. Introduction. Epilogue. Selected Readings. Glossary. Index. Tables and figures. 128pp. 5⅜ x 8½. 40451-X

DALÍ ON MODERN ART: The Cuckolds of Antiquated Modern Art, Salvador Dalí. Influential painter skewers modern art and its practitioners. Outrageous evaluations of Picasso, Cézanne, Turner, more. 15 renderings of paintings discussed. 44 calligraphic decorations by Dalí. 96pp. 5⅜ x 8½. (Available in U.S. only.) 29220-7

ANTIQUE PLAYING CARDS: A Pictorial History, Henry René D'Allemagne. Over 900 elaborate, decorative images from rare playing cards (14th–20th centuries): Bacchus, death, dancing dogs, hunting scenes, royal coats of arms, players cheating, much more. 96pp. 9¼ x 12¼. 29265-7

MAKING FURNITURE MASTERPIECES: 30 Projects with Measured Drawings, Franklin H. Gottshall. Step-by-step instructions, illustrations for constructing handsome, useful pieces, among them a Sheraton desk, Chippendale chair, Spanish desk, Queen Anne table and a William and Mary dressing mirror. 224pp. 8⅛ x 11¼. 29338-6

THE FOSSIL BOOK: A Record of Prehistoric Life, Patricia V. Rich et al. Profusely illustrated definitive guide covers everything from single-celled organisms and dinosaurs to birds and mammals and the interplay between climate and man. Over 1,500 illustrations. 760pp. 7½ x 10⅛. 29371-8

Paperbound unless otherwise indicated. Available at your book dealer, online at **www.doverpublications.com**, or by writing to Dept. GI, Dover Publications, Inc., 31 East 2nd Street, Mineola, NY 11501. For current price information or for free catalogues (please indicate field of interest), write to Dover Publications or log on to **www.doverpublications.com** and see every Dover book in print. Dover publishes more than 500 books each year on science, elementary and advanced mathematics, biology, music, art, literary history, social sciences, and other areas.